The Challenge of

Interracial Unionism

The Challenge

DANIEL LETWIN

The University of
North Carolina Press

Chapel Hill and London

of

Interracial Unionism

Alabama Coal Miners, 1878–1921

The paper in this book meets the guidelines for
permanence and durability of the Committee on
Production Guidelines for Book Longevity of the
Council on Library Resources.

This book was set in Monotype Bulmer
by Keystone Typesetting, Inc.

Book design by April Leidig-Higgins.

Library of Congress Cataloging-in-Publication Data
Letwin, Daniel.   The challenge of interracial
unionism: Alabama coal miners, 1878–1921 / by
Daniel Letwin.   p. cm.   Includes bibliographical
references (p.   ) and index.
ISBN 0-8078-2377-5 (cloth : alk. paper)
ISBN 0-8078-4678-3 (pbk. : alk. paper)
1. Trade-unions—Coal miners—Alabama—History.
2. Afro-American coal miners—Alabama
—History. 3. Trade-unions—Afro-American
membership—Alabama—History. I. Title.
HD6515.M615L478   1998   97-9365
331.88′122334′09761—dc21   CIP

Portions of this book originally appeared as
"Interracial Unionism, Gender, and 'Social Equality'
in the Alabama Coalfields, 1878–1908," *Journal of
Southern History* 61, no. 3 (August 1995): 519–54.

02 01 00 99 98   5 4 3 2 1

To my parents,
Leon and Alita,
and to the memory of
my grandmother Bessie

Contents

Maps

Preface

THIS BOOK EXPLORES a question that has long concerned social critics, and now a growing number of American historians as well: How have black and white workers negotiated the dual identities of race and class, in settings where both were charged with meaning? In the introduction I discuss the narrative and themes of the book. Here I want to convey how I came to study the coal miners of Alabama, and to acknowledge those who have helped me along the way.

My interest in race and labor has deep roots. From my family I absorbed the trade union ethics of solidarity and social justice, values my father and mother had learned from their parents, Bessie and Lazar, Manya and Zell. I first become aware of a broader world amid the civil rights struggles of the 1960s. Like many others of my generation, I began to ponder what gave race such meaning and urgency in America. Central to my emerging worldview were visions of a revitalized labor movement with the commitment and power to overcome racism—a barrier that American unionism has at times transcended, but more often succumbed to or actively fortified.

If these concerns germinated in my family, so did my sense that social issues were complex and not always susceptible to dogmatic formulations. My political sensibilities were shaped by ongoing, lively discussion with my parents, Alita and Leon, and my brothers, Michael and David. My parents and my grandmother Bessie—the individuals to whom this book is dedicated—encouraged intellectual curiosity and critical thinking, as qualities that both tempered and reinforced one's social commitment.

My interest in issues of race and class deepened in college, where I encountered the dynamic scholarship then flowering in both labor and African American history. I arrived at graduate school undecided between these two fields. Eventually I came to see the dilemma as artificial. Blacks in the United States, after all, have in large part been workers (whether slaves, sharecroppers, or wage earners), and American working-class history has been profoundly affected by the presence of African Americans and by ideologies of race. For my dissertation topic I chose the Alabama coal fields during the rise of Jim Crow because it seemed a promising context in which to explore this relationship.

My work on the dissertation was supported by an Albert J. Beveridge Grant from the American Historical Association, a John F. Enders Research Assistance Grant from the graduate school of Yale University, and two travel grants from the Yale history department. Most of the writing took place during two years I spent teaching at Holy Cross College. Although mine was not a long-term position, the history department welcomed me as a colleague and sought in a variety of ways to facilitate my writing and my quest for employment. I want particularly to acknowledge the kind support of Robert Brandfon and William Green.

Several individuals made visits to Alabama far more pleasurable and productive than they might otherwise have been. Alan Heldman Jr. offered me a place to stay in Birmingham. Jeff Norrell, an accomplished historian of race and labor in Alabama, showed me around the district and helped me obtain access to useful manuscript collections. I conducted most of my archival research at the Birmingham Public Library Archives, whose director, Marvin Whiting, was unfailingly gracious and knowledgeable. I would also like to thank the staffs of the Alabama Department of Archives and History in Montgomery and of the William Stanley Hoole Special Collections Library at the University of Alabama at Tuscaloosa.

As the dissertation took shape, the following people—Yale faculty and graduate students, labor and black historians, and sundry friends—offered valuable criticism of work in progress: Patricia Cooper, Emilia Viotti Da Costa, Jeffrey Gould, James Grossman, Reeve Huston, Gerald Jaynes, Regina Kunzel, Barbara Laslett, John Laslett, Jerry Lembke, August Meier, Nick Salvatore, Karin Sawislak, Joe Trotter, and Jeffrey Wasserstrom. A spirited organizing drive and strike in 1984 by Yale's clerical and technical workers—and the support extended them from across the Yale and New Haven communities—was an instructive reminder of labor's continuing potential.

Several people played especially important roles during my graduate career. At Yale, I quickly joined the legions who have admired and learned from the knowledge, intellectual rigor, and moral vision of David Montgomery. As my advisor, he was generous with his time, raising illuminating questions and proposing fresh ways of thinking about the evidence. Paul Worthman, whose pioneering scholarship on interracial unionism in Birmingham remains among the finest work on race and labor in the New South, generously gave me access to his research notes, which steered me toward sources I would not otherwise have found so quickly. Through a daunting succession of drafts both during and since the dissertation stage, Eric Arnesen offered consistently sharp and searching criticism, leavened with amusing asides and genuine enthusiasm. (He also furnished me with a steady stream of pertinent documents discovered in his own prodigious research on black and white workers.) I have long valued our continuing dialogue on race and labor in American history. Karin Shapiro's probing analysis and extensive knowledge of New South labor and politics were of enormous importance; equally so, her unflagging personal loyalty and encouragement.

In the years since completing the dissertation and coming to Penn State I researched and wrote the last two chapters and substantially revised the rest. From Carolyn Lawes I received deep support, infused with warmth, humor, and invaluable perspective. Along the way, she subjected several drafts to keen editorial and scholarly critiques. Commentary from a number of historians—Alan Derickson, Lori Ginzberg, Alex Lichtenstein, Joe McCartin, Bruce Nelson, Kim Phillips, Judith Stein, Tom Sugrue, Nan Woodruff, and Robert Zieger—gave me much to think about as I developed the project. Anonymous readers for the University of North Carolina Press and the *Journal of Southern History* contributed thoughtful suggestions—not least, concerning where an often unwieldy manuscript might appropriately be condensed. Chapter 5 benefited from a vigorous discussion at the Penn State Labor History Workshop. Grants of research funding and course relief from the College of Liberal Arts at Penn State accelerated my progress. Gary Gallagher helped in many ways, not only in his capacity as department head, but also in first bringing this study to the attention of the University of North Carolina Press. In our department's faculty rock and jazz band, Irreconcilable Differences, I found a cherished respite from the rhythms of academic life.

I am grateful to Lewis Bateman, executive editor of the University of North Carolina Press, for his continuing interest as the study evolved from a dissertation into a book, and to Katherine Malin for her sharp eye and her patience in editing the manuscript for the Press. I also thank the *Journal of Southern History* for permission to reprint portions of my article "Interracial Unionism, Gender, and 'Social Equality' in the Alabama Coalfields, 1878–1908," *Journal of Southern History* 61, no. 3 (August 1995): 519–54.

My parents have provided unstinting support—intellectual, material, and moral—throughout, and the latest instance deserves special mention. Despite their crowded schedules, they agreed without hesitation to review the final draft and proceeded to offer extensive editorial and thematic advice. I want especially to recognize the painstaking effort by my father, who many years ago gave me my first humbling lessons in the murky business of writing. His sustained assault on all the vague and superfluous prose that had somehow survived thus far strengthened the manuscript immeasurably.

Eva Maczuga also enhanced the manuscript at the final stages, pressing for greater clarity and suggesting subtle improvements at many points. Still more, she enriched my life. With love and appreciation, I thank her.

The Challenge of

Interracial Unionism

Introduction

O NE SUMMER DAY IN 1878, Willis J. Thomas, an African American coal miner and an organizer for the Greenback-Labor Party in the Birmingham district, stopped in the town of Oxmoor to post signs announcing a public meeting where he was scheduled to speak. The notice addressed itself to both black and white voters. As he nailed a copy to the door of the Eureka Iron Company storehouse, Thomas was accosted by a group of men calling themselves "Democrats," who warned him that he was courting trouble and that he had best make himself scarce. Unperturbed, Thomas reached into his vest pocket and produced a piece of paper, perhaps documenting his association with the Greenback-Labor Party or his legal right to post the announcements. Nonplussed, the men handed it back, and left.

"Our country is going to the dogs!" one of the Democrats muttered afterwards to an acquaintance, unaware that he was speaking to a Greenbacker. "[T]o think that a Negro has that much authority in a good Democratic state, is enough to make a white man commit suicide." Such a development, he feared, threatened to undo the work of the Redeemers who had so recently "delivered" the South from Reconstruction: "[T]hree years ago if a Negro dared to say anything about politics, or public speaking, or sitting on a jury, or sticking up a notice, he would be driven out of the county, or shot, or hung in the woods." Especially alarming was the high esteem in which Thomas was held among whites. "Some white men that heard nigger Thomas," he fretted, "say he is the best speaker in Jefferson county, white or black."[1]

THE SCENE THAT so scandalized this Democrat was not an isolated one in the Alabama coal fields. Thomas was but the most prominent of a number of African Americans active in building the Greenback-Labor Party around the newly developed Birmingham mineral district. Nor did the rapid fading of Greenbackism spell the end of interracial organizing. Black and white miners joined forces on an even larger scale in the 1880s under the banner of the Knights of Labor. In the early 1890s the United Mine Workers of Alabama displaced the Knights as a bastion of labor interracialism in the New South. To be sure, the interracial sensibilities of these movements should not be exaggerated. Each went only so far in confronting the dominant assumptions of white supremacy. None wholly eliminated racial strains or hierarchies within their ranks or overcame a persisting aversion among many miners to organizing across the color line. Still, through their egalitarian rhetoric and their very existence as racially mixed enterprises, these associations stood as a conspicuous and, to many, unnerving exception to the rising tide of Jim Crow.

Recent years have brought growing attention to interracial organization across the South in the late nineteenth and early twentieth centuries. Historians have uncovered union collaboration among black and white workers in such disparate settings as the timberlands of Louisiana, the docks of New Orleans, the urban trades of Richmond or Birmingham, and the coal fields of southern and central Appalachia.[2] Much as C. Vann Woodward exploded the myth that segregation was somehow encoded in Southern history,[3] so this scholarship has shaken loose the classic depiction of black and white workers as locked in a cycle of exclusion, conflict, and mutual alienation. Rather, it has shown their relationship to have ranged along a spectrum from hostility to solidarity. Even if episodes of cooperation were less common and lasting than those of antagonism, they restore the question mark to the interaction of class and race among American workers.

Varied and complex, at times openly defiant of the prevailing norms and at times scrupulously bent to the contours of Jim Crow, labor interracialism has in recent years inspired spirited scholarly debate. Historians have locked horns over the extent, resiliency, motivations, impact, and overall significance of interracial unionism in the New South. Some, most prominently Herbert Hill, have discounted the phenomenon, deeming it too infrequent, too fleeting, too suspect in purpose, too mired in the assumptions of white supremacy to warrant serious consideration. According to such critics, those who grant interracial unionism credibility errantly subordinate the race consciousness of American workers to a romanticized class consciousness; to the contrary, they argue, the divisive pressures of race have overwhelmed the unifying capacities of class throughout American history.[4] In fact, current scholarship in the field remains far more nuanced than the Hill critique suggests. Few historians baldly afford class primacy over race in the consciousness of workers, black or white. For recent scholars, the corrosive effects of racism upon the relations of black and white workers do not negate the significance of interracial organization where it materialized, but instead render its exploration all the more compelling.[5]

The labor movement of the Birmingham mining district—spearheaded in turn by the Greenbackers, the Knights of Labor, and the United Mine Workers—was a part of this interracial tradition, a phenomenon that historians still are only beginning to reconstruct. This study explores how far organized labor in the Alabama coal fields was able, or inclined, to step beyond the codes of white supremacy that pervaded American society in the age of Samuel Gompers and Booker T. Washington, of imperial expansion and segregation. To the extent these movements did—or did not—depart from Jim Crow values, what were their motivations, and how did these vary by race? What was the degree of cooperation and mutual respect between black and white miners? What explains the ebb and flow in the fortunes of unions that crossed the color line?

Alabama's mineral district offers a rich setting for the study of race, class, and labor in the age of segregation. Birmingham was established in the southern tip of Appalachia in 1871, at the initiative of industrialists eager to develop the area's reserves of coal and iron. By the turn of the century the "Magic City" had become the industrial centerpiece of the New South. From the beginning, a diverse stream of people—African Americans primarily from the Black Belt and whites from northern Alabama and beyond—converged on the district to work in its various trades. The coal fields that enveloped Birmingham stood out especially for their racially mixed labor force. At the outset of large-scale coal production in the late 1870s, whites made up slightly less than three-fifths of the miners; at the end of the century, blacks made up a modest majority, and on the eve of World War I they represented approximately three-fifths.

The first two chapters provide the industrial and social background for miners' organization. The first section of chapter 1 traces the emergence of Alabama's coal and iron industry and the creation of a socially diverse labor force at the mines. The development of the coal and iron industry was a sporadic phenomenon, forever falling short of the breathless promotion that surrounded it. Although encouraged by the richness of the mineral deposits of central and northern Alabama, it remained perpetually hampered by weak regional demand, backward technology and infrastructure, an ambivalent planter class, insufficient capital, competition from the North, and unseasoned and reluctant labor. The halting character of Birmingham's growth would greatly affect the dynamics of race and labor in the coal fields. The remainder of chapter 1 explores how blacks and whites arrived at the mining district and how they came to terms with the industrial surroundings they encountered there.

Chapter 2 examines how the miners sorted through the varied meanings of race and class in their daily lives. Class conflict arose often, both in the communities and at the workplace, as the miners' drive to achieve an independent livelihood ran up against the operators' pursuit of stable, controllable labor, high productivity, and low costs. Day-to-day tensions flared on numerous fronts: the company store, wages, the cost and quality of housing, the weighing and screening of coal, the subcontracting of unskilled labor, physical conditions underground, power relations at the mines, the miners' high rates of mobility and transiency, and the operators' use of state and county convicts at the mines.

But if chronic battles with the operators encouraged a broad sense of common interest among the miners, their self-identities were also filtered through the divisive realities of race. The miners' world was, after all, not simply a coal district; they also inhabited the New South. Race relations in the community contrasted significantly with those underground. Above ground, black and white miners went separate ways. Such separation was less pervasive at the mines, where blacks and whites often worked in close contact, although to degrees and on terms that varied from setting to setting. These divergent patterns of race in different parts of the miners' lives affected relations with the operators throughout this period.

There is much about the interior world of the mining communities that we can only begin to know. The surviving evidence offers only glimpses of how people came to work and live in the coal fields. It throws little light on the ways miners and their families, black and white, experienced their home life, churches, fraternal lodges, saloons, and other such institutions. The subtle, informal kinds of class and racial resistance that have enlivened the study of American workers, slave and free[6]—recently cast by Robin Kelley as the "hidden transcripts" of labor conflict[7]—are largely inaccessible to students of the Birmingham mineral district. And yet, as Eric Arnesen reminds us, there remains an extensive "overt transcript" of union struggle in the New South that also warrants careful investigation.[8] It is in this arena that available documents allow for a textured study of race and class in the Alabama coal fields. Indeed, most of what we can know about the early miners of the district owes to sources generated by the organizing efforts of the Greenback-Labor Party, the Knights of Labor, and the United Mine Workers.

The core of this study recounts the series of labor drives that black and white miners pursued from the advent of large-scale coal production through the turbulent era of World War I. Chapter 3 examines the rise and decline of the Greenback-Labor Party (1878–82) and the Knights of Labor (1879–90). Overlapping in time, membership, and outlooks, these movements illuminate the miners' early reactions to life on the region's industrial frontier. The issues they raised in political campaigns and strikes—rates and regularity of pay, convict labor, mining conditions, and the company store—would become staples of organization in the coal fields for decades to come. Miners responded to the Greenbackers' and Knights' impassioned appeals for broad popular alliances, their denunciations of Gilded Age inequalities of wealth and power, and their challenges to the Redeemer elites of the New South. Neither movement, however, succeeded in building a lasting presence in the Birmingham district; each was finally overwhelmed by the superior resources of the operators, internal divisions over strategy, the smothering effects of regional boosterism, and their group's decline nationally.

Both the Greenbackers and the Knights called on black and white miners to subordinate racial differences to their common interests. In this endeavor, each organization enjoyed substantial success. Still, in key ways—the divergent social worlds of black and white miners, the ambivalent feelings between them, and regional hostility to interracialism of any sort—the "race question" haunted these movements. The ambiguous meanings of race would permeate the next, and most enduring, vehicle for organization—the United Mine Workers (UMW). The remainder of this narrative investigates three periods of UMW mobilization in the Birmingham district, each the focus of a chapter, from the union's founding in 1890 through the aftermath of World War I.

From the outset, the Alabama UMW rallied white and black miners around a number of work-related issues, most adopted from the campaigns of the Green-backers and the Knights. Its early efforts culminated in 1894 with a four-month, districtwide strike. The miners were routed by a hostile governor, a depressed economy, and an abundance of strikebreakers. Nonetheless, the union retained a modest presence through the dismal mid-1890s, providing the foundation for two subsequent waves of organization over the next quarter century—the first from 1898 to 1908, the second from 1917 to 1921. Each campaign materialized during flush periods for American labor; each declined in the aftermath of bitter strike defeats, in rhythm with the demise of unions nationally.

During each renewal of organization, the Alabama UMW challenged the hardening norms of segregation. African Americans figured prominently in its ranks and leadership. The labor press regularly denounced the operators' efforts to divide the miners along racial lines, and extolled the loyalties of black union members. But solidarity between black and white miners remained powerfully constrained by the Jim Crow order. These limits were reflected in the biracial (that is, either all-white or all-black) structure of union locals. Whites occupied a disproportionate number of leadership positions, even where blacks predominated among the members. Demeaning portrayals of African Americans infected the miners' press, even alongside tributes to black unionists. Conflicting racial currents thus swirled about the labor movement of the mining district: a fear of internal division and an impulse to submerge racial questions beneath issues the union considered paramount; a desire to retain credibility and legitimacy in the eye of a broadly racist public; and a persistent strain of white supremacist thought within the union itself.

Interracial unionism in the New South was a complex and often fluid phenomenon. It eludes the ahistorical dichotomies often employed to assess its meanings. This study of the Alabama miners suggests that the ways in which white and black workers dealt with each other were far more diverse, and often far more ambiguous, than we would conclude from scholarly inquiries into the relative importance of "race" and "class" consciousness, as if the one must displace the other. Related to the race-class dichotomy is a tendency to conceptualize relations between black and white workers as either harmonious or antagonistic, glossing over the gradations between those two poles. Finally, the motivations behind labor interracialism often have been arranged around an implied dichotomy between the visionary and the pragmatic. But, as the story of the Alabama miners vividly illustrates, determining where genuine fellowship ends and self-interest begins can be treacherously difficult and, ultimately, beside the point.[9]

Only after discarding such dichotomous assumptions can we comprehend what prompted and permitted black and white miners to join forces in labor campaigns, even while racial separation blanketed nearly all other areas of their lives. The answer proposed here lies in three currents, indissolubly mixed, of miners' consciousness. First, of course, there was an awareness that racial division weakened their hand against the operators—and that the latter knew it. But hard-headed calculation cannot wholly explain the interracial impulse. The words and actions of Alabama's black and white miners betray a broad sense of shared identity, rooted in common class experience, that spilled beyond the purely strategic. Finally, whether by instinct or design, the miners did much to inoculate their labor movement against race-baiting by adhering to key elements of the white supremacy tradition. In the Alabama coal fields, one more circumstance—the gender-specific composition of the labor force—contributed significantly, if quietly, to the viability of interracial unionism in a Jim Crow environment. In an occupation pursued only by men, the spectacle of blacks and whites engaged in collective activity was less viscerally threatening to segregationist sentiment than in settings where interracial association crossed the gender as well as the color line.

These factors provided the motivation and the breathing room for the collaboration of black and white miners. They did not, however, guarantee success. Alternating periods of ascent and defeat turned not so much on a waxing and waning of interracial cooperation—that impulse remained remarkably persistent, albeit within limits, throughout this period—as on an evolving constellation of external circumstances. The outcomes of the miners' organizing efforts were conditioned, at times perhaps decisively, by the racial atmosphere of the district and region. Yet, to a large extent the fortunes of unionism in the Alabama coal fields rose and fell in response to realities faced by workers everywhere, such as the state of the economy, the stance of the federal government, and the receptiveness of employers or the wider public to the labor movement. This study explores the relationship between the miners' interracialism, at once bold and adaptable, and the ever shifting circumstances it encountered.

1 The Rise of the Birmingham District

I N THE SPRING OF 1871 the Elyton Land Company of Jefferson County, Alabama, announced a public sale of lots from its recently acquired property in Jones Valley. The land was to be the site of a new "workshop town" at the projected intersection of the South & North and the Alabama & Chattanooga railroads. On the appointed day, droves of people descended on the verdant, sparsely settled farming district—some on horseback or muleback, others in wagons or carriages, still others on foot—to inspect the future metropolis. Nothing more than an inelegant expanse of corn stubble and mud, its streets were marked off by wooden pegs and subdivided into numbered lots. The sales that began that day continued briskly through the summer and fall. By the time a city charter was issued at the end of the year, more than 100 houses had been erected and more than 800 settlers—a colorful cast of land speculators and gamblers—had arrived. Birmingham, the self-styled "Magic City," was born.[1]

While most cities emerge along some prominent body of water, Birmingham had no such asset. It was the area's rich, untapped mineral deposits that attracted the railroads, land speculators, and early settlers to postbellum Jones Valley. Five miles wide and fifteen miles long, the valley was situated amid a remarkable concentration of coal, iron, and limestone; in some cases, both iron ore and coal could be mined from the same pit. "[B]eautifully diversified by fields and forests, springs and running streams," as the Birmingham *Iron Age* described it, the valley was one of several that formed a stretch of Appalachia's southern foothills. The area made up a central strip of Alabama's mineral belt, 10,000 square miles of rugged hill country embracing twenty-eight counties across the northern portion of the state. Along the southeast edge of Jones Valley loomed Red Mountain, a 100-mile-long ridge of hills rich in hematite iron. Enveloping the valley were three coal fields. By far the largest and most prized was the heavily forested Warrior field, extending 70 miles long and 65 miles wide along the northwest border of Jones Valley. The other two coal fields, each far smaller than the Warrior, were the Cahaba, which ran along the southeast face of Red Mountain, and the Coosa, farther to the south and east. "The vastness of this body of coal," wrote Reverend Benjamin F. Riley, an early promoter, "suggests that it will one day constitute one of the greatest industrial centers of the Union."[2]

The founding of Birmingham marked the first full-scale effort to develop the mineral district into an industrial center. In late 1870 a small group of influential Alabamians, chiefly financiers and railroad investors, purchased over 4,000 acres in Jones Valley. Shortly afterward they formed the Elyton Land Company, a realty business capitalized at $200,000, through which they planned to launch a new city. On January 26, 1871, the incorporators met in Montgomery at the bank of Josiah Morris, one of the key financial forces behind the company. After some discussion, they chose to name their nascent iron center after the formidable iron center of England. The company's bylaws duly noted: "The city to be built by the Elyton Land Company, near Elyton, in the county of Jefferson, State of Alabama, shall be called Birmingham."[3]

During its first two years the new town attracted many enterprising settlers and inspired much hyperbole. "The fact is plain," one observer declared, "Alabama is to be the manufacturing center of the habitable globe." Behind such breathless commentary lay a less glamorous reality. "Young Birmingham," according to an early biographer of the district, was "a staring, bold, mean, little

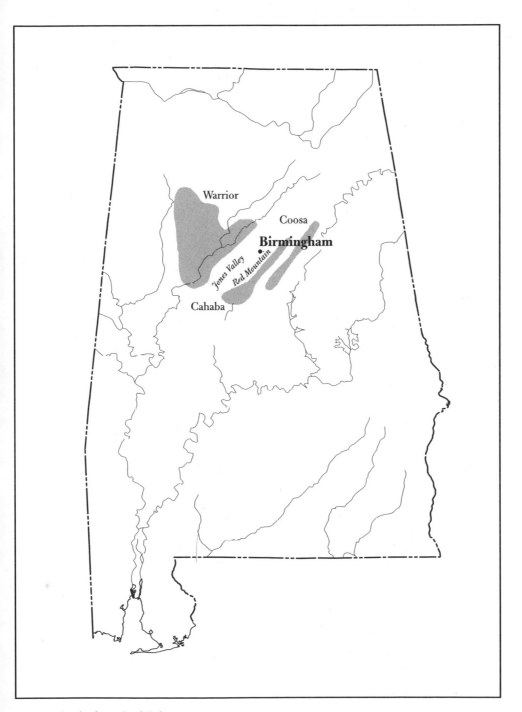

Map 1. Birmingham Coal Belt

town." During an 1875 tour of the South, Edward King found the Birmingham district "in many respects a wild country, sparsely populated, and rough in appearance." A day's trip along the South & North Railroad revealed "hardly any town of considerable size; in the forest clearings there were assemblages of rough board houses, and brawny men and scrawny women looked from the doors; now and then we passed a coal-shoot, and now long piles of iron ore." Nor did the expected boom materialize in these early years. Despite the completion of the South & North link in 1872 and the beginnings of coal and iron production, capital was slow in arriving. And whatever growth occurred was derailed in 1873 by two devastating setbacks—the economic panic that swept the nation and a local cholera epidemic. The district's development would not resume until the end of the decade.[4]

Heady, sobering, promising, uneven—Birmingham's sporadic progress over the 1870s set the pattern that would persist for decades. In the age of industrialization, the sheer abundance of coal, iron, and limestone was enough to fuel a continuing faith in the district's future as the "Pittsburgh of the South." Yet natural resources are only one of the factors that shape an area's industrial fortune. A host of other factors—the availability of capital, the supply of labor, access to markets, levels of knowledge and technology, the role of the state, competition from other regions, the disposition of different economic sectors and classes, the state of the economy, the resourcefulness of industrialists— proved unpredictable and problematic. Hence, Birmingham's perpetual cycles of achievement, anticipation, and disappointment.

No doubt, the district's growth during the late nineteenth century was impressive. In 1870, the Alabama mineral region produced a modest 13,000 tons of coal; in 1880, 323,000 tons; in 1889, 3.6 million tons; in 1900, 8.4 million tons.[5] The population of Jefferson County likewise rose dramatically with each decade, from 12,345 in 1870 to 23,272 in 1880, 88,501 in 1890, 140,420 in 1900.[6] By the end of the century, as it launched into steel production, Birmingham had emerged as the industrial trailblazer of the New South. Yet this was a mixed distinction in the "colonial economy" of late-nineteenth-century Dixie. The coal and iron industry of Alabama bore the marks of the region's chronic underdevelopment: a weak regional consumer market, unseasoned labor, scant technological and infrastructural resources, and a perennial lack of capital.[7] The mineral district's material potential and actual expansion kept alive visions of an industrial and civic maturity that would never fully arrive.

In vital ways, the ebb and flow of Birmingham's development affected the social experiences of those who converged to labor in the coal fields surrounding the "Magic City." It shaped the terms on which blacks and whites arrived there, the conditions of work and community that they encountered, and the ways they came to deal with each other and with the operators who hired them. This chapter explores the rise of Alabama's mineral district, and the heterogeneous labor force that emerged along with it.

THE EXISTENCE OF coal and iron deposits through much of northern and central Alabama had not escaped the attention of earlier inhabitants. White pioneers had begun utilizing the region's minerals as far back as the late eighteenth century, when the northern hill country was still the domain of the Cherokee, Chickasaw, Creek, and Choctaw nations.[8] The spread of white agricultural communities and slaveholding plantations across antebellum Alabama generated myriad uses for iron products, including shoes for horses, mules, and oxen; weapons; and a variety of farm implements and domestic utensils. The blacksmith shops that cropped up to serve these demands provided a market for small-scale mining. As early as the 1820s, coal from Jefferson and Walker counties was skimmed from the surfaces of outcroppings or lifted from shallow riverbeds onto boats. Two dozen blast furnaces and foundries appeared around northern and central Alabama in the decades before the Civil War. Though generally primitive and of modest size, they planted the seeds for future development.[9]

These enterprises inspired scattered visions of mineral development to rival the emerging coal and iron centers of the Pennsylvania-Ohio region. But various circumstances—limited knowledge of Alabama geology, the absence of a viable railroad system, high impurities in Alabama coal, and weak local markets—conspired to retard such progress during the antebellum period. There were also persisting divisions among Alabamians over the appropriate course of industrial development, although the nature of those divisions has been the subject of considerable historical debate.[10] Whatever the forces that promoted and constrained it, Alabama's antebellum manufacturing base revolved around light industry serving plantation society, characterized by small concerns such as grist mills, saw mills, cotton mills, and boot and shoe manufactures.[11]

During the 1850s, industrial boosters redoubled their efforts. Growing awareness of the mineral potential of central and northern Alabama, spurred by a visit from the famed English geologist Sir Charles Lyell and a series of reports from state geologist Michael Tuomey, generated optimism among financiers, industrialists, and relatively progressive planters.[12] As railroad development gathered momentum in the 1850s—linking the commercial hubs of Montgomery and Mobile to prominent cities like Atlanta, Chattanooga, and Memphis—the prospect of developing Alabama's coal and iron became irresistible. In 1860 the legislature designated the South & North Railroad to build a line through the heart of the mineral district. Public debate now turned less upon whether the district should be opened up than on how—that is, as a manufacturing center in its own right or (as many planters preferred) as a source of coal for domestic and railroad use and iron ore for export to manufacturing centers elsewhere.[13]

The Civil War arrested the railroad's progress and deferred resolution of the issues it raised. Nonetheless, as nine coal companies and sixteen ironworks sprouted up around the state to supply the Confederacy, Alabama's mineral prospects glimmered as never before. But the victorious Union forces annihilated the state's iron industry; by late 1865 the wartime mines and ironworks had vanished as quickly as they had appeared, most never to return. Physical destruction of the industry was compounded by the financial losses of worthless Confederate debts, a chronic lack of capital, inexperienced managers, poor transportation, and weak regional markets. The question of how the state's mineral potential might be realized would figure prominently in Alabama politics throughout the Reconstruction era.[14]

The shaping of that potential fell first to the planter-dominated government that arose in the aftermath of Appomattox. Although unwilling to indulge the emergence of an unfettered manufacturing class within a laissez-faire polity, leading political figures were not wholly resistant to what would crystallize as the New South gospel: railroad building and industrial growth, northern investment and agricultural diversification. Before its displacement by Radical Reconstruction in 1867, the planter government ushered through bonds for rail construction in the mineral district by certain Democrat-affiliated companies, including the South & North.[15]

The Radical Republicans brought to power a new regime in Alabama, purged of planter influence and more receptive to industrial development. Expanding on its predecessor's support for a railroad route through the mineral belt, the Radical government passed lavish bond issues to cover the Republican-connected Alabama & Chattanooga Railroad as well as the South & North. Over the next several years each enterprise raced to construct its line. Although the two routes were to intersect in the heart of the mineral district, they had differing plans for what should develop there. The Alabama & Chattanooga envisioned a region that would export coal and iron north to Chattanooga for manufacture there; the South & North preferred a manufacturing center within the mineral district. During the late 1860s and early 1870s the two lines—each represented by a cluster of politicians, land speculators, railroad officials, and advocates in the press—vied for favorable treatment from a succession of state governments. With byzantine twists worthy of the legendary railroad barons of the North, the protracted struggle for supremacy in the mineral district culminated by the mid-1870s in favor of the South & North—the outcome of its acquisition by the powerful Louisville & Nashville Railroad and the restoration to state power of the Redeemer Democrats in the elections of 1874.

BY THE MID-1870S, then, an essential precondition for the takeoff of Birmingham—the presence of a well-connected railroad line devoted to fostering a local manufacturing center—had fallen into place. And yet the prospects for industrial development now appeared bleaker than at any time since the Union army had demolished the state's coal and iron facilities. The optimism that had surged in the early 1870s was stopped cold by two nearly simultaneous jolts in 1873. In the summer, a severe cholera epidemic swept Birmingham; in September, the failure of Jay Cooke's banking house sent the nation into a five-year depression. During the mid-1870s Birmingham was virtually dormant. Most mines and ironworks either went out of business or were taken over by larger concerns, usually ones backed by northern or European capital. In the absence of fresh investment, Birmingham seemed to be dying an early death. "The cry of 'hard times' is heard on all sides," a correspondent to the Birmingham *Iron Age* observed in 1875. Only two years earlier, the paper recorded, "the saw and the hammer could be heard all over the city. . . . The future outlook was pleasing." But in one short month disease and depression had transformed the whole scene. "There is nothing wanting here," railroad engineer and industrialist John Milner wrote wistfully, "but capital and labor."[16]

Before the depression had run its course, however, signs of life began to reappear. Prompted by the discovery that better quality pig iron could be manufactured with Alabama coke than with charcoal, a group of enterprising industrialists launched an investigation in 1876 to reassess the productive capacity of the coal fields. Each of them would figure prominently in the rise of industrial Birmingham: Truman H. Aldrich, product of a well-established New England family, mining engineer, and owner of the Montevallo coal mines since 1873; James W. Sloss, a major force in the South & North Railroad and the Eureka Iron Company; Henry F. DeBardeleben, flamboyant son-in-law of Alabama textile manufacturer Daniel Pratt; and Milner. The successful experiments in pig iron production based on coke sent a fresh ripple of optimism through the district. The enthusiasm was not shared, however, by capitalists in the Northeast, who continued to show little interest in Alabama's mineral potential. It would be another three years before Birmingham would revive in earnest.[17]

Birmingham's prospects brightened in 1877, when Aldrich, Sloss, and De-

Bardeleben pooled their resources to buy up 30,000 acres in the Warrior coal field near Birmingham. The following year, as the depression lifted, they formed the Pratt Coal and Coke Company, which began shipping coal from the Pratt seam in February 1879. Although initial production at the Pratt mines was a modest 120 tons per day, its opening would long be remembered as a watershed moment. "Then, and not until then, was there any sign of life in the city of Birmingham," Milner later wrote. By the end of the year Birmingham boasted a number of manufacturing enterprises, including three foundries, three machine shops, several South & North Railroad shops, and an iron furnace under construction. A visitor pronounced Birmingham "the livest town I have seen in the South."[18]

As the New South industrial creed gained a following during the early 1880s, the mineral district underwent dramatic expansion. In 1880, Alabama coal production reached 323,000 tons, an almost 25-fold increase over the production levels of ten years earlier. The Pratt company remained the engine behind the boom, producing 1,500 tons of coal and 300 tons of coke daily in 1881. During the first half of the 1880s, however, a series of new coal or iron concerns cropped up, most of them launched by one or another of the founders of the Pratt company. Prominent among these were the Alice and Mary Pratt furnaces (started by DeBardeleben), the Birmingham Rolling Mill (DeBardeleben and Pratt), the Cahaba Coal Mining Company (Aldrich and Sloss), the Sloss Furnace Company (Sloss), and the Woodward Iron Company. The year 1884 brought the district's first million-dollar acquisition, as Tennessee interests led by Enoch Ensley combined the Pratt Coal and Coke Company (purchased in 1881) with local iron enterprises to create the Pratt Coal and Iron Company. Alabama coal and pig iron began to find markets beyond their initial network of Deep South towns (such as New Orleans, Columbus, Macon, Atlanta, Augusta, Decatur, and Birmingham itself) to reach such Upper South and Midwest cities as Nashville, Louisville, Indianapolis, Chicago, St. Louis, and Cincinnati. The Louisville & Nashville Railroad played a vital role in stimulating Birmingham's growth, and in the process its own. Rapidly becoming one of the South's premier railroads, the L & N laced the district with spur lines and belt lines, invested heavily in mines, furnaces, and land, and offered operators cheap freight rates to northern markets. "But a few years ago," a correspondent for the Cincinnati *Gazette* marveled in 1881, "this site was a cotton field, and not a profitable one either"; now, he continued, it was "a lively little city," where "some thirty trains a day arrive and depart, and the smoke of iron furnaces, foundries and rolling mills obscure the southern sky."[19]

By the mid-1880s the district had become one of the most celebrated industrial centers in the New South. Yet beneath this surge of activity remained a shallow foundation: the regional market remained feeble, quality of coal and iron marginal, technological capacity second-rate, northern capital scant, and returns on investment slight. As would so often be the case, Birmingham's industrial performance proved strong enough to encourage continued dreams of a "Pittsburgh of the South" yet not so strong as to ever actually deliver such an outcome.

Over the latter half of the 1880s, Birmingham made strides impressive enough to awaken concern in Pittsburgh itself. A depression in 1884–85, while slowing production, enabled the low-price iron of Alabama to penetrate northern markets. The recovery of 1886 spurred unprecedented growth around the district. As feverish land speculation seized Birmingham, consolidation and expansion swept the coal and iron industry. In 1886 the formidable Tennessee Coal, Iron, and Railroad Company (TCI) made its entry into the district, acquiring a majority interest in Ensley's Pratt Coal and Iron Company. No less ambitious, Henry DeBardeleben established the DeBardeleben Coal and Iron Company; backed by $4 million in London and Charleston capital, he founded *his* would-be Pittsburgh of the South, the town of Bessemer, ten miles southwest of Birmingham. Within a few years DeBardeleben's new industrial center would boast (with what C. Vann Woodward called "Biblical repetitiveness") seven blast furnaces, seven large coal mines, seven ore mines, 900 coke ovens, and 140,000 acres of the region's richest coal and iron lands. Not to be outdone, in 1887 Enoch Ensley laid out just west of Birmingham the town of Ensley, where TCI soon erected four furnaces. That same year the Pioneer Mining and Manufacturing Company, newly organized by the Thomas family of Pennsylvania, founded the town of Thomas just above Ensley; here too, new furnaces would soon go into blast. Meanwhile, Virginia and New York capitalists absorbed the Sloss Furnace Company into a newly formed Sloss Iron and Steel Company. TCI, DeBardeleben, Ensley, Thomas, and Sloss were names that would dominate Birmingham at the end of the nineteenth century—and, in some cases, well into the twentieth.[20]

At the end of the decade, Birmingham, foremost ironmaker of the South, radiated industrial vitality. Eight railroads traversed the district, bringing investors and inhabitants and carrying coal and pig iron to cities around and beyond the region. Scores of coal and iron mines and mining communities clustered along one or another of the long-distance rail lines or along the local Birmingham Mineral Railroad that looped through the coal fields. Together, Alabama coal mines were now producing over 3.5 million tons per year. The center of gravity of the district was of course the city of Birmingham, now a bustling center of ironworks, railroad shops, and an increasingly diverse range of craft and industrial manufacture. One boosteristic tract from 1888 gushed with unrestrained alliteration of "busy, bustling Birmingham, the city of unceasing activities and unwearied energies, invincible in progress, electrical in celerity . . . Birmingham, the best, biggest, brightest, boldest exponent of the New South." After a visit in 1889, Andrew Carnegie pronounced the overall district "Pennsylvania's most formidable enemy."[21]

But once again the threat to Pittsburgh failed to materialize. By the late 1880s, Pennsylvania steel was eclipsing pig iron, and Birmingham's prospects now turned on its potential to produce and market steel. Nature did not favor Birmingham in this endeavor: the local iron's high phosphorous content hampered conversion. Tentative strides were facilitated by the advent of open-hearth production, but that method was either insufficiently economical or insufficiently capitalized to afford a breakthrough. Lacking a profitable steel-making capacity, the Birmingham operators were forced to stake their survival on proximity to regional markets and, above all, on cheap production costs.[22]

Contracting markets at the start of the 1890s drove many iron companies out of business or into the hands of those strong enough to ride it out. In 1892, TCI achieved undisputed dominance in the district when it acquired the DeBardeleben and Cahaba companies, leaving only the Sloss company as a serious rival. The brisk consolidation of the Alabama iron industry between the late 1880s and the mid-1890s not only concentrated control in a few, more powerful hands, it also shifted the source of investment ever more decisively from southern to northern capital.[23]

The severe depression of 1893–97 further tested the Birmingham operators. With the collapse of the pig iron market, credit evaporated, iron stopped moving, production ground to a halt, and even the mighty TCI faced bankruptcy. A series of capital infusions from Wall Street averted the collapse of Birmingham. The operators responded to the depression in several ways. First, they intensified efforts to lower production costs, most dramatically through wage cuts, provoking a bitter miners' strike in 1894. Secondly, depressed demand for pig iron induced the operators to turn their sights to foreign markets, in such far-flung places as Canada, Latin America, England, Australia, India, and Japan.[24]

Suppressing labor costs and broadening iron markets bought the operators time, but the depression underscored a third imperative: the need to break into steel production. The *Engineering and Mining Journal* caught the mood when it argued that while southern states "are now producing pig iron more cheaply than [ever,] this is not enough. They must turn some of their raw materials into finished products, and without curtailing the output of pig iron change it into steel." For a time, this ambition continued to founder on the shortcomings of Alabama ore and coke, along with the reluctance of TCI's northern-based directors to expend serious capital in its southern holdings, which they preferred to milk for quick speculative gain. Only when the powerful L & N warned of economic disaster for the district and threatened to divert its resources to another steelmaker, did TCI respond. Following a successful series of open-hearth experiments in 1895, it built a steel mill at Ensley, which began production in 1899. The advent of Birmingham steel stirred a new round of millennial excitement. Sure enough, the achievement breathed new life into Alabama's coal and iron industry: "It saved us from going into the hands of a receiver," TCI president Nathaniel Baxter would later recall. But the 1900s would only begin to fulfill the overheated expectations that greeted the coming of steel production to Birmingham.[25]

As the economy revived in the late 1890s, Alabama's iron industry joined in two nationwide trends. First, it underwent another flurry of consolidations: TCI swallowed up sixteen companies, Sloss another dozen, while other large furnace operators, such as the Republic Iron and Steel Company and the Alabama Consolidated Coal and Iron Company, appeared on the scene. Second, the operators, seeking to regularize their long-turbulent labor relations, recognized unions for the first time. Industrial production soared, as Alabama pig iron continued to find new markets abroad; in 1898, Birmingham was reportedly the third largest shipping point for iron in the world. Between 1898 and 1903, coal production climbed steadily, from 5.8 million tons at the outset to 10.4 million at the end, while pig iron production rose from 1 million to 1.6 million tons. By the turn of the century, Birmingham operators had weathered a harsh depression, developed steelmaking capacity, consolidated their corporate resources, and rationalized their labor relations. The district now seemed poised on the threshold of industrial maturity—or, as the Birmingham *News* exulted, "the cynosure of the eyes of the iron and steel magnates of the world."[26]

Still, these advances failed to deliver the Alabama coal and iron industry from its ever troubled condition and regionally subordinate status. The pig iron produced by TCI found its way to the Carnegie interests in Pennsylvania for manufacture into steel, but it did far less to foster steel production in Birmingham itself. Even amid the renewed prosperity, TCI experienced a chronic dearth of capital, while its officers fretted regularly over "excessive" union power. When in 1907 New York financier J. P. Morgan acted to "rescue America" from an economic panic by absorbing TCI into the colossal United States Steel Corporation, Birmingham was swept up in another rush of optimism. But before long the merger's double-edged legacy sank in: while northern iron and steel would stabilize industrial Birmingham, it would also contain the district's growth, chiefly through discriminatory freight rates.[27] And so Birmingham continued on its Sisyphean struggle into the twentieth century. Amply endowed with all the minerals vital to iron and steel production, yet hampered by inadequate capital, backward technology, modest regional markets, an oppositional planter class, and competition from the North, the mineral district was locked in a perpetual state of becoming—its potential too great to dismiss and yet never fully realized.

Historians have long debated why and through what mechanisms the New South failed to develop on a par with the industrializing North.[28] For the study that follows, however, what matters is not so much the reasons for the New South's sporadic development as its effects upon the lives and struggles of Alabama's racially diverse coal miners.

Coal production had sprung to life in a rural, sparsely settled area. In social terms the mineral belt had been a tabula rasa, its inhabitants and their various cultures transplanted from elsewhere. The mixed composition of those who came makes the Birmingham district a remarkable setting for the study of race and labor in the New South. What gave rise so quickly to a biracial labor force? How did race shape, or conform to, the patterns of work and community that emerged there? How did it mediate ongoing conflict between miners and operators over the charged issues of livelihood and independence? How did it inform the miners' very sense of who they were and where their interests lay? Central to this study are the potent but often unpredictable effects of race upon the chronic clashes between miners and operators in the Alabama mineral district. We turn first to the process by which the diverse workforce of the coal fields came to be.

THE COAL DEPOSITS of Jefferson and Walker counties were first tapped commercially in the 1820s. Most early miners were recruited locally by enterprising landowners, and set to work during the summer extracting coal from the beds of local creeks and rivers. It was a process that would be unrecognizable to the coal diggers of a later era. It was strange even to contemporaries. Michael Tuomey, Alabama's first state geologist, reported "a novel process in the art of mining, namely, diving for coal." A flat-bottomed boat would be moored near the edge of an underwater vein of coal. From the boat, men would use mauls to drive wedge-shaped crowbars into the coal below, loosening two or so feet across a section of the seam. Next they would dive in, retrieve the loosened coal, and deposit it into the boat. Once the rains of fall and winter had rendered the waters navigable, river pilots would float and pole boats laden with many items—hogs, poultry, pile staves, corn, and cotton, as well as coal—through the rocky waters of the Warrior and Tombigbee rivers southward to Mobile, where the coal was sold. Requiring stamina more than skill, diving for coal was performed over a period of weeks by local people who followed other callings during the rest of the year. The extracting of coal in antebellum Alabama was an irregular, primitive endeavor. Cheap retrieval of easily accessible minerals was the overriding concern; existing knowledge and infrastructure permitted nothing more ambitious.[29] But even at the time Birmingham was founded, coal mining in Alabama remained to a large degree the humble activity of local inhabitants pursuing supplemental income. In 1874 the Birmingham *Iron Age* described how nearby "native citizens" would "simply dig the coal from the surface as they would clay or sand" and haul their yield by wagon several miles to Birmingham.[30]

Scarcely half a decade later, however, the picture had changed dramatically. New methods and conditions of mining had taken root in the district that would remain familiar well into the twentieth century. Alabama coal was mined predominantly by pick, under the age-old "room-and-pillar" system that prevailed in most bituminous fields at the time of Birmingham's emergence.[31] Mines were honeycombed with "rooms" separated by walls, or "pillars," of coal left standing to support the roof. A room was essentially a tunnel, the far end of which constituted the "face" of coal worked by the miner. It was typically around 24 to 30 feet wide and as high as the seam itself (anywhere from 2 to 10 feet). Each miner was assigned a room, which would customarily remain his station until it was worked out (approximately 350 to 500 feet). A room was usually worked by two miners, at times father and son, at times a skilled miner and a laborer he employed to help with the more muscular phases of extraction, and at times equal "buddies." The miner's knowledge and skill were passed on through an informal tradition of apprenticeship, commonly from father to son.

Equipped with the traditional tools of his trade—pick, lamp, shovel, pry bar, breast auger, needle, saw and ax, tamping bar, squib, oil, and black powder—the miner conducted a series of tasks. First, he spent two or three hours on his back or side picking a three- to four-foot wedge out of the base of the seam. Next, he bored holes into the face above the wedge with a five-foot auger. Drilling, like undercutting, was a taxing process: the miner cranked the auger, pressing with his body against its breastboard to push the bit into the coal. The miner then loaded the hole with black powder contained in a paper cartridge, in an amount he deemed necessary to produce chunks of coal small enough to load, but not so small that they would be rejected by the company. After tamping the powder with dampened dirt, the miner finally inserted and lit the squib (or fuse), raced for cover behind a pillar, and called out "fire in the hole" to alert others of the imminent blast. That was often the last act of the shift; the following day, the "shot" coal would be cleaned of slate and other impurities and loaded, either by shovel or by hand, into a railcar brought along a makeshift track to the face. Loading made up approximately half the miner's work. Once filled, the car was hauled by mule from the room, along the main haulage road, and finally to the mouth of the mine, where the coal was weighed and credited to the miner.

In addition to undercutting, drilling, blasting, and loading, miners were often required—especially where they lacked collective power—to prop up the roof with timber, lay tracks from the face to the entry, push empty cars from the entry to the coal, and other types of "dead work" (that is, uncompensated labor). Paid by the ton (or at times the carload), the miner (and where used, the laborer) stood at the heart of the productive process. Work at the coal face was supported by a series of tasks performed by "company men" (or "day men"), who were paid by the day. These ranged from lowly jobs, such as slate picking or "trapping,"[32] performed largely by children or over-the-hill miners, to such skilled positions as carpenter, mason, electrician, and engineer.

The pick miner enjoyed much craft autonomy. By custom, he (and his helper or "buddie") asserted a proprietary claim to the room where he worked. Paid by the piece, miners functioned somewhat as independent contractors, with often extraordinary latitude to come and go as they wished. Supervision in the decentralized workplace of a room-and-pillar mine was light; a miner might receive a visit from the mine boss once a day, if that. In the isolation of his room, he was called upon to blend heavy exertion with seasoned judgment. Upon a range of decisions demanding broad experience—when to timber the roof or pull down a precarious overhang, how deep to undercut the coal, where to bore the holes, how much explosive to use, and so forth—rested not only a miner's earnings but preservation of life and limb. No two seams—indeed, no two mines—were wholly alike, and the ongoing challenge encountered by even the most skilled miners injected their work with a pride of craft that made many loath to consider any other trade.

For many more, however, the romance of the "miner's freedom" was out-weighed by the harsh realities of life underground. Although conditions varied, the pick miner typically performed arduous labor in cramped quarters amid darkness, dampness, and myriad dangers, with the likelihood of long-term disease or debility, in exchange for meager (and often falsely calculated) pay. As tonnage workers, they routinely had to make the grim choice between a livable output and personal safety.

As mining via drifts, slopes, and shafts quickly displaced surface mining in Birmingham's early years, so the smattering of local citizens engaged in the gathering and selling of coal gave way to a convergence of wage laborers from far and wide. Observers were struck by their racial and ethnic diversity. In 1881 a visitor to Pratt Mines described how "hundreds of laborers of nearly all races and colors moved to and fro. . . . The heavy bur-r-r of the Cornish miner, the Irish brogue and Welsh accent mingled amusingly with the drawl of the native Alabamian and the broad lingo of the black." Twelve years later another traveler recorded an even broader diversity: "The fellow clad in stripes in the county prison camp at Coalburg, the diminutive Hungarian Slav, or Slavisch, as locally known, the industrious German or Frenchman, the native Afro-American in the majority, the honest Scotchman with his twang, all are here in one and the same mine. The native Alabamian of the mountains and experienced and raw material from every section of the United States are likewise on hand in no contemptible numbers."[33]

The diversity captured in these accounts was not atypical. In 1880, 42 percent of a significant sample of Alabama miners was black; in 1889, African American presence had climbed to more than 46 percent of the nearly 8,000 miners and quarrymen in the state; by the turn of the century, blacks made up approximately half of Alabama's 11,751 coal miners; in 1910 the figure was 54 percent (of 20,778 coal miners); in 1920, still 54 percent (of 26,204).[34] The white miners represented a broad, continually shifting mix of ethnic backgrounds. In 1889, 65 percent were native born, including many from the Alabama hill country and from coal fields to the north; most of the remaining 35 percent came from the United Kingdom, a large portion from coal districts there. By 1899, native-born whites comprised 81 percent of white miners; immigrant miners, whose numbers had remained almost the same while the workforce more than doubled, now included growing proportions of southern and eastern Europeans, especially Italians.[35] Most of the major coal towns or companies had racial mixes not greatly different from that of the district as a whole.[36]

The extent to which mine labor was arranged along racial lines varied over time and setting. While no comprehensive numbers exist, certain persisting patterns are discernible from localized and anecdotal evidence. African Americans and whites each made up a significant portion of the skilled miners.[37] Each also worked at various types of day labor. Overall, blacks figured predominantly but not exclusively among the laborers who worked under skilled miners. Many

African Americans, however, worked as skilled pick miners, hiring black laborers of their own. None were known to take on white laborers.[38]

In the eyes of the Alabama operators, the ideal labor force would be characterized by skill and efficiency, deference and reliability, low wages and high productivity, aversion to unions and receptiveness to the company store. Recognizable among employers everywhere, this agenda took on particular urgency in the Alabama mineral belt, where low costs—the common denominator for all these ideals—were essential to survival. As operators sought to cobble together a workforce with the optimum blend of qualities, they drew upon a rich lode of racial and ethnic assumptions. Although their preferences were constrained by the range of available labor, they remain significant enough to warrant close attention.

Alternately simplistic and convoluted, dogmatic and at times perversely insightful, the operators' social "wisdom" remained more or less constant throughout the late nineteenth and early twentieth centuries. "Native whites," drawn mostly from the surrounding farmlands and joined by a modest influx from the iron and steel centers of the North, were widely esteemed as reliable and educable.[39] Yet many southern whites, particularly those fresh from the countryside, were disdained for their inexperience and apparent aversion to industrial rigor. "Often they are the men who have failed at farming or other callings," one visitor observed in 1912. In 1886 the Birmingham *Age* melded these two points together, urging that the long-neglected peoples of the hill country be recruited and socialized into the New South.[40]

Depictions of European immigrants were likewise varied. Those from the United Kingdom or northern Europe enjoyed particular favor in the eyes of employers and superintendents (themselves often of northern European descent). They were also extolled for their work experience, as well as their reliability and overall strength of character. The English, Welsh, Scotch, Irish, Swedes, Germans, French, and American whites, the 1911 United States Immigration Commission concluded, were the best labor available in the district. From Birmingham's earliest days, industrial boosters viewed such immigrants as a vital source of labor for the coal fields. In 1871 the state's commissioner of industrial resources, in urging greater efforts to attract newly arrived European immigrants to the mineral district, derided the "fallacy" that "our climate is not suited to white labor."[41]

Immigrants from southern and eastern Europe fared less well in the operators' estimation. While Slavs were seen as sufficiently steady to meet Birmingham's industrial needs (especially as unskilled laborers), other southeastern Europeans—such as Bulgarians, Macedonians, and Italians—were widely viewed as wanting. Above all, southern Italians—a growing presence in the coal fields by the early twentieth century—elicited resigned contempt for their reputedly meager experience, "ornery" approach to work, and general tendencies toward "vice" and "disorder." The operators valued immigrants from southern and eastern Europe chiefly for unskilled work.[42]

But it was African Americans who drew the most extensive, often contradictory assessments from mine employers. On the one hand, the operators complained that blacks lacked the ambition, the training, and the social habits necessary to a stable labor force. "The principle trouble with the colored labor is its indisposition to settle down," complained Eureka Iron Works superintendent Giles Edwards in 1883. "Whenever the notion strikes them to go, they go. . . . No contract constrains them," remarked Sloss.[43] If the shortcomings of black workers were considered numerous, so were the virtues. Employers viewed them as brawny (and thus especially well-suited for unskilled labor), docile and less open to unionism ("It appears to be natural to them to submit to the white man," Edwards observed), open to work that whites found too disagreeable or dangerous, and most ready to plow their wages back into the company store. Ultimately, whatever their ambivalence, most operators shared the conclusion drawn by Llewylyn W. Johns, voluble mining engineer for the Pratt company: "We could not do without the negroes."[44] The substantial use of African Americans at the mines and furnaces—in the latter they were employed almost exclusively by the late 1880s as cheap, unskilled labor—extended a tradition rooted in industrial slavery.[45]

One of the miners' habits that the operators reflexively viewed as a "colored" phenomenon—the tendency to move about—was in fact a customary feature of coal districts everywhere. Miners of the hand-loading era often moved from mine to mine, town to town, district to district. This mobility reflected the miners' customary streak of independence. It was also a strategy of response to immediate circumstances: the idiosyncratic geology of coal mines, the irregular availability of work, and variations in pay, treatment, and conditions. Whatever the causes, it was not unusual for a miner's career to take him far and wide: from the Central Competitive Fields of Illinois, Indiana, and Ohio through the upper and central Appalachian coal districts of Pennsylvania, Kentucky, West Virginia, and Tennessee. Just as the launching of the Pratt company in 1878 extended bituminous coal production to the southern tip of Appalachia, so it expanded the geographical reach of miners' mobility.[46]

The first generation of skilled miners generally arrived in Alabama by way of coal fields to the north. Most were of northern European origin; more often than not they or their parents had come from some coal mining region of Great

Britain. These were the men of whom the operators sang praises and whom Birmingham's founders sought to attract to the district. As early as 1875 the Jefferson County Industrial and Immigration Society was hard at work advertising the splendors of the mineral district to capitalists and skilled workers around the northern United States and Europe. Subsequent civic boosters continued the effort. The local press frequently ran celebratory editorials. In the late 1880s Milner, by then a member of the Alabama Senate, campaigned for the establishment of a commission on immigration that might promote the state's resources to prospective arrivals. Most energetic were the railroads, chiefly the Louisville & Nashville and the Alabama Great Southern, which deployed agents in the North and across the Atlantic to induce people of hardy stock to settle along their routes. At times of labor scarcity, agents traveled through Kentucky, Tennessee, West Virginia, and states farther north in search of experienced miners.[47]

Small farmers from the surrounding countryside, predominantly white, comprised another source of mine labor. As was typical in coal regions, the line between farming and mining in the Birmingham district was permeable. A contemporary description of the Centreville section of Bibb County conveyed how intermeshed the worlds of agriculture and mineral extraction could be: "To the north is the mine, to the south is the farm. And here, on this spot, mine and farm dove-tail into each other." This proximity made farmers an easily available pool of labor, although, conversely, it lent them an easily available alternative to mining. Many retained ties to the land, periodically selling their labor at the mines to supplement their income. This pattern was conveyed in reports from the Alabama coal region in the *National Labor Tribune*, which alternated effortlessly between mining and farming news. Other whites from the nearby farms, however, crossed the line conclusively to become skilled miners.[48]

Most black miners came from the cotton belt 80 to 100 miles to the south. Some headed for the mineral district on their own initiative, seeking a better livelihood and a freer existence than that afforded by the world of tenant farming. Others were recruited by labor agents. Whatever the impetus, African Americans quickly became a major presence at the coal fields, to the point that planters began complaining about the effect on their labor supply. Using local contractors, the railroads brought in large groups of black workers from neighboring states, chiefly Georgia, Tennessee, and South Carolina. Induced by extravagant promises of high pay and handsome living conditions, most black newcomers were put directly to work on railroad extension building, at the foundries, or at the mines. Black agents were particularly valued by the coal companies. In the summer of 1899, for example, one local black newspaper editor toured communities of black miners and laborers in Georgia and Tennessee, passing out circulars and delivering speeches extolling the attractions of the Birmingham district.[49]

How black workers experienced life in the coal fields was as varied as the paths they took there. Some set down roots, bringing or establishing families, obtaining homes, becoming skilled miners. Others, preserving ties in the Black Belt, moved back and forth between mining and farming. Some, that is, had essentially become miners, returning to their home communities for the planting and harvesting seasons, for holidays, or when work at the mines was too scarce or unappealing; others essentially remained farmers, journeying to the mines during slack times for bouts of wage labor. Still others, mainly young men unencumbered by family and restless to explore new vistas, inhabited both worlds in relatively equal measure. Little remains of the voices of blacks themselves to illuminate how they sought to straddle these two worlds. Representatives of both agricultural and mining interests, whose views are more accessible, often lamented the porous line between these settings. "[The negro] gets dissatisfied so easily," a Mobile cotton trader observed. "The younger negroes . . . spend a portion of the year on one plantation, and then they want to go and see something of the world, so they start off to another plantation, or go to work on a railroad or in a coal mine." Operators complained of black miners' tendencies, in the words of one, to "scatter to the farms" as soon as they "feel the touch of hunger."[50] But if the countryside offered miners a respite from mine labor, it could also provide the operators with reserve labor in times of shortage.[51]

For all the movement back and forth between the farms and the mineral region, as the coal belt developed, many African Americans came to view it as home. "The movement of the negro from the farms and plantations, to the mines, to the lumber camps, to the railroad works," the state Department of Agriculture and Industries commented in 1906, "is little short of a race exodus." Numerous black workers moved not so much between the plantations and the mines as among different activities around the district. These were mostly unskilled hands who, in the broader tradition of common laborers, might feel equally at home—or equally alienated—working at coal or iron mines, at the furnaces or railroad construction sites, or as helpers in sundry craft or service trades. Sloss described how this "restless, migratory class" moved about: "They come down here, they work on railroads, they work on the streets, they work in the coal mines to some extent, and largely in the ore mines and about the furnaces and rolling-mills, and it is no terror for them to be discharged." His explanation focused, in typical operator fashion, on the congenital temperament of black workers: "They love to change, today at work, and tomorrow away or idle." Prevented by most employers from rising above menial, low-paying positions, many blacks existed on the margins of employment, drifting in and out of work around the district.[52]

Year after year, civic-minded whites fretted over Birmingham's "bad negro sections," such as Buzzard Roost and Scratch Ankle, where African Americans were said to indulge in all manner of drinking, gambling, crime, violence, and related dissipation. How to control the "idle negro" was a regular focus of public debate; some form of municipal or county vagrancy law, scarcely less draconian than the Black Codes following emancipation, usually materialized as the solution. Still, the operators, while always ready to join the chorus of alarm, had their uses for such people: they served as a convenient reserve of casual labor for a sporadic industry whose labor requirements were no more "stable" than the work habits of the "idle negroes." Asked by a senator how his business could function "with such a haphazard character of labor," Sloss replied: "Oh, I have got another set down-town that I can drum up whenever those who are at work leave. . . . We know where they congregate, and we have men go down and hunt them up." Coal companies seeking labor from the city also advertised in the local press. Thus the transiency of workers could mesh with the transiency of work.[53]

There remained one further source of mine labor—convicts leased to the operators.[54] Equally notable for its vital role in the state's mineral development and its notoriety in the public eye, the convict lease system maintained its grim presence in the coal fields from the early 1870s until its abolition in 1928. Although the practice had its genesis during Reconstruction, it was the Redeemer Democrats who turned the leasing of Alabama's predominantly black convicts into a large-scale enterprise. Bourbon governments found the system appealing on a number of grounds. First, it relieved a fiscally crippled state of the burden of maintaining its prisoners, and indeed generated a handsome revenue. It dovetailed with the Redeemers' dogma of retrenchment and economy in government, through which they audaciously contrasted themselves with their "corrupt" Republican predecessors. Ironically, corruption permeated the convict lease itself. The mechanisms by which employers bid for prisoners and by which the state regulated their conditions virtually cried out for the bribery of public officials, and the call was regularly answered. The arrangement also furnished industrialists with cheap and obedient workers without threatening the planters' supply of labor. Indeed, the convict lease functioned as a key component of the Redeemers' ever more elaborate legal apparatus—which also included vagrancy, enticement, contract enforcement, emigrant-agent, and criminal-surety laws—through which the planters worked to reassert control over the lives and labor of African Americans. In these ways the convict lease did much to reconcile the planter and industrialist wings of Bourbon Democracy.

State convicts were worked at the mines from the beginning. By the late 1870s the arrangement that would become generally standard had fallen into place at Pratt Mines and Newcastle: in exchange for the labor of the prisoners, the coal companies would maintain them at their own stockades, and pay the state a monthly rate per convict (graded by a hierarchy of classifications that reflected the convict's productive capability). From the late 1870s through the mid-1880s the Pratt company absorbed the lion's share of convict labor. In 1888 TCI, which had recently acquired Pratt, outbid ten other companies to win an exclusive, ten-year contract with the state for all convicts sent to the mines, under which the company would pay between $9.00 and $18.50 per month for each convict. (The remainder of the state prisoners were distributed between two Black Belt plantations.) In 1891, Jefferson County began leasing its own convicts, chiefly to TCI and Sloss; soon, counties all over the state were following suit. Between 1880 and 1904 the state's profits from the convict lease exceeded $2.3 million, approximately 10 percent of its accumulated budget for that period.

Convicts at the mines were overwhelmingly black. In 1880, African Americans made up 89 percent of the 734 convicts leased out by the state, most of whom went to the mines; ten years later, the same proportion held. At some places, they worked exclusively as helpers to skilled, free miners. At others, such as Coalburg and sections of the Pratt division of TCI, whole mines were worked solely by convicts. Whatever the arrangement, the operators benefited enormously from the use of convicts. An incomparably cheap source of labor, the use of convicts meant lower wages for free labor as well. Nor could they take days off or move from mine to mine, in the custom of free miners. Convicts enjoyed no autonomy at work; the mine boss had virtually uncontested authority. They could not effectively strike or join unions. On the contrary, they undermined organization among free miners. "They do their work well and regularly," TCI reported brightly in its 1890 annual report, "and in case of strikes they can furnish us enough coal to keep at least three of the Ensley furnaces running." Convict mines also served as training grounds for prospective free miners. Approximately half the convicts remained miners once their terms were up—miners who had learned their trade not under the guidance of a father or a "buddie" steeped in a tradition of miners' autonomy but rather under the harsh discipline of the penal warden. From the perspective of the operators—for whom profitability was seldom assured, free labor seldom abundant or readily controllable, and lower costs always imperative—the convict lease was indispensable.

The inevitable counterpart to convict labor was "free labor," and throughout this period the great majority of miners, black and white, fell into this category. Yet the "freedom" of the Alabama miner was not without its own coercions. Miners' efforts to realize an independent livelihood clashed continually with the operators' drive for stable, controllable labor, high productivity, and low costs. These conflicting class aspirations had always taken on a bitter edge in America's coal districts; equally so in the industries of the New South, a region marked by weak economic development and elaborate mechanisms of labor repression.

But if class tensions in Alabama's coal fields were clarified by the South's colonial economy, they were muddied by the region's powerful legacies of race. It was within the charged dynamics of class and race that the contours of the free miners' world—patterns of housing and consumption, religion and recreation, health care and cultural association, wages and work process, political activity and labor organization—took shape.

Pratt Mines, 1889: Several hundred white miners lay down their tools and leave the mines, with the go-ahead of the mine boss, to pursue a black miner accused of raping a white woman. Upon capturing the suspect, they lynch him.

Lewisburg, 1905: White miners and their families join black miners and their families along the riverside for a black baptismal ceremony.

Cardiff, 1891: A baseball match between the "white nine" of Cardiff and the "negro nine" of Brookside is canceled when an armed white man, hostile to interracial recreation, orders the black players out of town.[1]

2 The World of the Alabama Coal Miners

I N THESE EPISODES—one an act of rabid hostility, another a moment of social harmony, and the third an event blending elements of both—we can glimpse the gamut of racial sentiments that coexisted in the Alabama coal fields. The organized labor campaigns that black and white miners built together incorporated this variety of attitudes. How the Greenback-Labor Party, the Knights of Labor, and the United Mine Workers addressed the challenges of race in an ever more segregated region makes up the core of the ensuing story. This chapter explores the social terrain from which these campaigns, detailed in the following chapters, arose. The interracialism that made them so remarkable emerged out of the wide-ranging set of conditions that white and black miners faced in common.

The limits of this interracialism—imposed partly by external forces, partly from within the movement itself—reflected how deeply the culture of Jim Crow continued to shape the miners' world.

AT THE OUTSET of the 1880s, scarcely half a dozen mining communities existed around the Warrior and Cahaba coal fields. Each was associated with one or two mining companies.[2] With the surge of mineral production and railroad expansion during the 1880s, new mining settlements proliferated.[3] By the end of the decade, the coal fields encompassed a broad network of mining communities. The larger ones—such as Warrior, Blocton, and Pratt Mines—were generally known as mining *towns*. Although anchored in the coal industry, these had a measure of economic diversity, with independent shopkeepers and professionals. Their populations tended to run toward several thousand. Smaller and more remote communities, such as those at Blue Creek and around Walker County, were more commonly described as mining *camps*. Located miles from the larger towns, they seldom had populations much above 1,000. Here the miners were more subject to the operator's social and economic power, often compelled, by circumstance if not policy, to live in company housing, patronize the company store, visit the company doctor, and so forth. The material quality of the miners' lives had much to do with the degree of independence they enjoyed in their communities.[4]

The mining communities all shared certain features. Each had a web of social institutions that were familiar throughout working-class America. Churches were central: every locality had Baptist and Methodist congregations, and most larger ones had Catholic churches as well. Not that all families were devout. In 1891 the Birmingham *Labor Advocate* lamented that of approximately 2,500 whites in Pratt Mines, only about one-quarter were practicing Christians. Nonetheless, a resident of Graysville wrote in 1904, "The music of the church bells which is heard often during the week proclaims the religious status of our town." In addition to church attendance and the traditional marking of birth, marriage, and death, spirited revivals periodically visited the coal towns.[5]

Equally ubiquitous were fraternal orders, most commonly the Odd Fellows, Knights of Pythias, and Red Men. Beyond offering miners a sense of belonging, the lodges provided material and emotional cushions against the precariousness of their lives, ranging from dispensing death and sickness benefits to sponsoring balls to raise money for those maimed at the mines. Each community also had a cluster of associations devoted to recreation, culture, and self-improvement, including brass bands, baseball clubs, military companies, excursion groups, literary societies, mothers' clubs, Sunday schools, drama troupes, gun

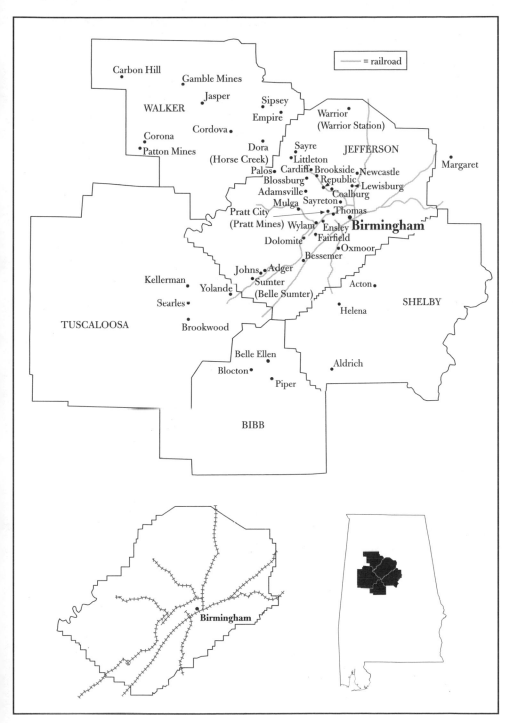

Map 2. Birmingham Coal Mining District

clubs, sewing circles, bicycle clubs, and ladies' aid societies. Such enterprises lent each mining town its own civic identity.[6]

All but the more remote settlements were enmeshed in a wider network of coal towns that had cropped up by the turn of the century. Common movement of miners and their families from one locality to another fostered social ties among the towns; the spread of passenger rail links facilitated them. Leagues of baseball and football teams from different towns reinforced the feeling of districtwide community, even as they sharpened local loyalties. Miners' families visited friends and relatives in neighboring towns, and special events such as lodge picnics or holiday celebrations in one place might draw thousands from around the coal belt.[7]

The mining towns did have their rougher side. Individual violence occurred frequently in the streets, the saloons, and the mines. Most incidents pitted white against white or black against black, although at times they were interracial. Often the cause was apparently trivial. Disputes over a shovel handle, a fifty-cent debt, a game of dominoes, or "a woman" were but a few of the episodes that led to fatal bloodshed among miners. Coal towns sometimes took on a Wild West quality. From Warrior in 1885 came word of "great excitement on our streets." Seven African Americans had been "knocked down," a Knights of Labor paper reported, and four or five pistol shots had been heard. The murky incident had been caused by two "desperadoes," who "defied the authorities and went home." Crap games, and efforts to suppress them, were common sources of violence; the unruly effects of alcohol even more so. One pay night in Wylam, the free flow of whiskey led to a commotion in which "chairs, bottles and missels commenced to fly through the air, and then some one commenced to shoot." This rough-and-tumble aspect became part of the way in which even hometown promoters painted the mining centers. One observer described happenings at Cardiff as a blend of "good, bad, moral and immoral. . . . If one goes to the church, lodge, or school, the [news] can be read by the refined with interest; but news obtained at the saloon will not suit the pious—for it will be too badly mixed with ribald songs, coarse jests, hiccoughs, and fighting."[8]

The coal towns nonetheless provided vehicles for material security, cultural affirmation, spiritual and recreational fulfillment, social independence, and, ultimately, a sense of collective identity. The identities articulated by the miners and their families were as multifarious as their institutions, such as churches and political parties, unions and baseball teams, mothers' clubs and sewing circles, Robert Burns societies and Bastille Day celebrations. Of course, the very terms "mining town" or "mining camp" highlight an occupational commonality suffused with class identity. At places where miners found measures of autonomy or material stability, they described their communities in tones of class pride. Thus, the labor press could reconcile searing indictments of company exploitation with sunny depictions of the mining towns. It came naturally to Knights of Labor leader Nicholas Stack, at times a caustic critic of the operators, to toast "Dear old Blocton" as a "lovely mining village among the hills." A portrait of Pratt Mines in the Knights local journal painted vividly the features of a "good" mining town: "Many . . . own their own homes and possess many comforts, not a few of them having cows and pigs and good gardens. Their hospitality cannot be excelled, and one sees evidences on every hand of neighborly feeling which finds vent in many acts of kindly expression. The local churches are . . . well attended. The hall which is used . . . by several orders would be a source of pride to any large city." All in all, the writer concluded, Pratt Mines was a town that "our mining friends can well be proud of from every point of view."[9]

A BROAD SENSE of common identity could never extinguish the effects of race upon the sensibilities of the coal towns' inhabitants. The color line bisected virtually all areas of the miners' lives: each town had black and white neighborhoods, black and white schools, black and white churches, black and white fraternal lodges, black and white athletic clubs, black and white brass bands. These mutually insular spheres seldom attracted comment, so naturally did they reproduce the prevailing practices—informal and, increasingly, codified— of the South in the late nineteenth and early twentieth centuries.[10]

The insularity was not absolute. Blacks and whites did interact on the margins of daily life. At times such contact could go beyond the purely incidental, to take on an aspect of genuine community. White citizens and their children might attend commencement exercises at the black school or line the river alongside African Americans for a "colored baptismal ceremony" (one instance was conducted "with a dignity and a simplicity of faith that had a very wholesome effect upon the white brethren present," the *Mineral Belt Gazette* reported). The funeral of a black miner's wife could occasion the shutting down of a mine, so that miners of both races might attend.[11]

Relations between blacks and whites were sufficiently varied to undermine any dichotomy between harmony and antipathy. The episode cited at the outset of this chapter involved no less a bastion of popular culture than baseball. An 1891 contest at Cardiff between the local "white nine" and the "negro nine" of Brookside was canceled at the initiative of a young white man named John Harden, who appeared with a Winchester rifle, announced that the white team "would not play with the negroes that day," and proceeded to chase the black players out of town. An integrated game between racially defined teams, a festive atmosphere dispelled by an act of sour intolerance—here was the range of racial attitudes that mingled uneasily in the mining communities.[12]

The prospect of racial conflict always hovered over the mining camps. Amid the more raucous forms of recreation, violent encounters were commonplace and could take on a racial dimension. In 1879 the governor dispatched the Birmingham Rifles to Helena to quell a so-called riot between black and white miners, originating, according to a press account, "at a negro bagnio, engendered by petty jealousy and plenty of mean whiskey." Liquor was also the explanation given for a racial melee on a Saturday night at Bradford Mines, where a white man named Cochran fought with a group of reportedly inebriated blacks, wounding several. The following evening a group of African Americans were said to have surrounded Cochran's house intending to lynch him; upon finding him gone, they fired shots into the house and broke the doors and windows.[13]

But such a scene was extraordinary; more commonly, crowd violence was directed by whites against blacks. Tensions could flare over chance encounters. One Sunday afternoon at Horse Creek, the *Labor Advocate* reported, a group of black and white children "got into a fuss." A black mother attempted to discipline some of the white children. Bart Thrasher, a white man, told her to desist. Allen Ford, a black man, came forward and "put in his say." Words were passed, Ford "made at Thrasher," who drew his pistol and shot Ford to death.[14] At Pratt Mines, word that a black woman had "insulted" the wife of a white man precipitated rumors of an imminent "race war," culminating in the deployment of state troops.[15]

The grisliest forms of violence against African American miners were reserved for those suspected of raping white women. When black Coketon miner George Griffin was so accused in 1880, the white miners mobilized a lynch mob. Griffin fled by foot into Walker County, and from there by rail to Tennessee, where he found work at the Sewanee mines. Nearly a year later he was apprehended and returned to Birmingham, where he was convicted of rape and hanged. For others, the end came more quickly. When news swept Pratt Mines in 1889 that a black man had raped Mrs. J. S. Kellum, a white woman, and murdered her son, 400 white miners left the mines—with the encouragement of the superintendent—divided into squads, and set out to find the "incarnate devil." Armed with squirrel guns, pistols, knives, and antique rifles, they combed the town and countryside on horse and by foot. Over the course of a long day, one black man after another was seized, questioned, brought before Kellum, and in the end released upon her determination that "he is not the man." "There is no dearth of men to go on every errand" the *Age-Herald* solemnly reported, "as the mines have belched forth their hordes of rough, honest men. . . . Of course there was no work in the mines yesterday, and there will be little or none today." The following day, the miners-turned-posse finally produced a suspect whose guilt Kellum was prepared to confirm—George Meadows, an African American newly arrived from the plantation district. After a perfunctory investigation, Meadows was lynched.[16]

Black miners did not always suffer mob attacks passively. A week-long hysteria triggered by the alleged rape of a white woman by a black man in Walker County revealed not only how deeply white views about the proper place of black men penetrated the coal fields, but also the boldness with which African Americans might defend themselves. On the evening of June 22, 1899, according to the Birmingham *News*, Mrs. Monroe Jones was accosted and raped on a country road outside the mining town of Corona by a black man—"very black and repulsive looking," in the inflammatory language of the *News*—armed with a Winchester shotgun and a razor. After the man, later identified as John Shepherd, ran off into the nearby woods, Jones staggered home and told her family of the ordeal.[17]

Before long, a posse of around 100 men was scouring the woods. Shepherd remained at large over the next few days. After he was reportedly spotted on a train near Brookside, the deputy sheriff there rounded up his own posse, and word soon came that another was on its way from Walker County. Alarmed that Shepherd faced imminent lynching, a group of around 50 local African Americans, mostly armed, proceeded to follow the Brookside posse. "The air was charged," the *News* reported, "and there were threats heard coming from the negroes that there would be serious trouble if the negro was caught."

Jefferson County sheriff O'Brien arrived at the scene, and quickly concluded that the black gathering and the white mob en route from Walker County put his men and the local posse "between two fires." After wiring a request for state troops, he approached the black group, shotgun in hand. "I am Sheriff of this county," he announced, "and if you do not disperse within twenty-five minutes I will put every one of you under arrest if it takes every white man in this county to do it." The blacks complied, sullenly; Ed Ellis, their reputed leader, reportedly muttered that he would "get even with the Irish son-of-a-bitch."

In the meantime, troops from Birmingham headed toward Jefferson. As their train passed through Brookside, they could see a large crowd, composed of blacks and whites, milling about quietly, if apprehensively. The scene that greeted them at nearby Jefferson was more ominous, featuring a large crowd composed of posses drawn from near and far. Soon a force of 150 armed white men, commanded by Sheriff O'Brien, was searching the woods for Shepherd; it was, the *News* remarked, "the biggest crowd of armed men that ever hunted a negro in this county, and the little army was an imposing sight as it traveled over hill and dale."

Imposing, but not effective. African Americans were believed to be harboring the suspect or at least alerting him to the whereabouts of his pursuers. Coming upon the black settlement of Trimble Hill, an exasperated O'Brien ordered a search of all houses. The sheriff's refusal to exempt the local church provoked an indignant response from the pastor. His protest was to no avail, but neither was O'Brien's search. He continued on to Blossburg, home of many of the black men who had gathered to monitor the Brookside posse. O'Brien again summoned local blacks and delivered another statement, warning them that he wanted no trouble from them and pledging that Shepherd would suffer no harm upon his capture. "The negroes had no reply to make," the *News* tersely reported.

The African Americans of Blossburg reacted less impassively that night, when around 125 whites from Walker County arrived to search the black quarters. Residents responded with gunfire, and in the minutes that followed over 100 shots were exchanged. Although only one man was injured, a pall fell over the area; saloons were ordered closed the next day, and few blacks showed up at the mines. That afternoon, two leaders of the previous day's black "posse" were mortally shot by snipers while walking along a road. Two other black men who ran to their aid were likewise struck down. By the next day, a number of African Americans were leaving the area, and shortly thereafter the story fades from the record.[18]

The episode reveals the explosive tensions that lurked beneath the daily relations of black and white miners. It further shows how, for all they had in common as miners, their status diverged sharply in the civil realm. And yet the story also suggests the limits that African Americans were prepared to set against extralegal assaults. The murders of the local black leaders and those who ran to their sides indicate the price black miners could pay for their militance; the trouble taken by a southern white sheriff to reason with (as well as talk tough to) an armed assemblage of black miners shows the collective boldness they could at times, with impunity, assert.

Although not an everyday occurrence, such spectacles of black miners' resistance to mistreatment were not rare. Nor was it directed solely against extralegal abuse. One Saturday night in 1899 the town of Dolomite erupted in a bloody conflict between black residents and law officers.[19] The trouble began when deputy sheriffs, together with two Woodward Iron Company officials, arrested three black men on charges of assault and battery, carrying concealed weapons, using abusive language, and resisting an officer. As word of simmering resentment among local blacks spread, the deputy sheriff of neighboring Bessemer led a party of officers to the scene. In the nearby woods, he encountered an estimated 250 African Americans, many of them armed. Not far off stood a smaller group of armed white residents. Somehow the imminent conflict was diffused, although not the tension. "The negroes at Dolomite have all quit work and nothing is being done at all," the *News* informed readers the next day. "The excitement remains at a high pitch."

Sheriff O'Brien arrived to calm matters. Summoning around 250 African Americans near the mines, he advised them to return peacefully to their homes, leave their arms there, and go back to work. If they complied, he said, he would guarantee their complete protection. He went on to detail the circumstances that had led to the three men's arrest, stressing how the incident had been triggered by an assault by one of them upon "another of your race." When the two others resisted, the officers had no choice but to arrest them as well. "The talk was well made and seemed to have a fine effect upon the negroes," the *News* observed. But its own elaboration suggested a more tepid reaction from O'Brien's listeners, grounded in resentment over rough treatment from law officers. "Say, boss," a member of the crowd interjected, "has a man a right to beat a nigger when he goes to arrest him?" The sheriff said no—such abuse, if proven, would be punished. Perhaps uncertain whether his assurances carried weight, he called on a black minister who had once worked for him to vouch for his honesty. The minister spoke as requested, and the group, having pledged to remain peaceful, was dismissed.

Sheriff O'Brien then proceeded toward a nearby hill to address approximately 60 armed white miners. He insisted that the law would be enforced equally against black and white, and appealed to their "superior intelligence" to assist in preserving the peace. But the whites had resentments of their own to express, recounting a series of offenses that had been committed against them by the blacks: one of them had been shot, others had had their homes fired upon. In the end, the white miners, like the blacks, accepted the sheriff's promise of protection and vowed to refrain from unlawful violence.

The peace did not last. According to the *News*, at three o'clock that morning a mob of around 250 black men opened fire upon "nothing in particular" from behind the coke ovens, "with the apparent object of intimidating the officers." Sheriff O'Brien and a posse of quickly deputized white men routed the blacks, and subsequently arrested 19 in a vigorous house-to-house search. By the next day the disorder was considered over; that night both whites and blacks dispersed to their homes, "though some of them slept with one eye open and with one hand in reach of a ready gun." The return of quiet to the mining town did not mean a return to calm. A number of African Americans had been wounded in the gun battle, and trials loomed for those arrested. Word spread among whites that the events of the past few days were not an isolated matter, that blacks in town had been preparing for such an outbreak for weeks and that more were on the way.

Confrontations such as these served as grim reminders that, however powerfully class relations shaped the miners' worldview, their sense of identity remained heavily affected by the region's racial climate. Nonetheless, open clashes were infrequent. The most salient feature of race relations in the mining towns was separation; blacks and whites experienced their community lives with a minimum of daily contact. At Sayreton, the *Mineral Belt Gazette* observed, "you will find the American, the Italian, the Slav and the colored contingent . . . so isolated from each other that not a shadow of a shade of social or racial friction is possible." The United States Immigration Commission reached a similar conclusion upon investigating the mining district: "With one or two exceptions, there has been very little friction among the races employed." In part because of the racial insularity pervading the mining communities, much of the contact that did occur on the margins of daily life was characterized less by tension than by some measure of neighborliness.[20]

Racial separation was not so pervasive at the mines themselves. It did exist: at many mines, rooms or whole sections were worked exclusively by black or white miners. In some cases the company designated entire mines either for blacks or whites. Still, as travelers' accounts cited in the previous chapter illustrate, black and white miners often worked together (although seldom as "buddies" in the same room). Interracial contact underground was often hierarchical in nature, particularly where white miners took on black laborers. But to a remarkable degree, black and white miners worked in the same mines on equal terms. This proximity below ground stood in marked contrast to the racial separation that characterized community life above ground.[21]

IN VARIED WAYS, then, relations between black and white miners reflected the congealing racial order of the wider region. But Jim Crow was not the sole defining feature of their world; they also inhabited the raw and precarious environment of a coal district. Chronic, intense, and wide-ranging conflict with the operators over material conditions and power relations focused the miners' consciousness in ways that could submerge the divisive capacities of race.

The miners' autonomy was circumscribed by the operators along a variety of fronts. Some communities, especially the more remote, were in every sense company towns—owned, run, and heavily dominated by the employer. The operators at Walker County's Kelly Mines "have things pretty much their own way," a visitor observed in 1885. "The Bank Boss, a big burly bully, is the monarch of all he surveys." His power extended well beyond the mines themselves:

> He keeps a boarding house and at a high price for board with him. They are compelled to eat just what he sees fit to give them. When pay day comes the men do not pay for their board, but the company hands the money over to the Bank Boss, and should there be any left the company condescends to give it to the men. If any of the men quit boarding with him he generally hunts them up and tries to intimidate and also to find fault with the person who takes them as boarders. If a miner quits work and wishes to leave, the company after hunting up all accounts against him, if there should be any money left, deducts 25 per cent from the amount and hands the miner the balance.[22]

A resident of Coalburg painted a similar picture: "The company owns all the land and all on the land, even . . . some of the men; that is, the convict part of them." After describing exorbitant stores, cramped housing, and dirty water, he concluded that "[n]early every man you meet has some complaint."[23]

The operators' social power was not always wielded to such dismal or unpopular effect. Employers sometimes subsidized white and black churches, schools, lodge buildings, and the like. Some would sponsor social occasions for the miners, such as barbecues and sporting events, excursions and holiday balls, as well as benefits for the injured or bereaved. Company officials often participated in fraternal lodges, churches, and other associations in which miners predominated. This kind of involvement could inject the mining communities with an expansive civic feel, able at its most effective to blunt the harsher edges of life. So too could relatively benign treatment from company officials. Miners were apt to show their appreciation for those they considered competent and honorable. "Joe Alexander is the efficient mine boss, genial and clever, full of push and energy," a Blossburg miner wrote. "He . . . looks well after the interest of his employer and yet . . . is very popular with the miners." When Richard Thompson, a popular bank boss, left his position at Johns, the miners there carried him on their shoulders. Upon his departure from Belle Ellen, Superintendent Llewylyn Johns received a gold watch chain and a gold Knights Templar charm from the miners.[24]

But even under the most benevolent company practices, the edges cut through. In an industry haunted by unpredictable demand, irregular profitability, and hence a yearning among employers for cheap and stable labor, paternalism was a luxury that even the most conscientious could afford only in limited measure. The bleaker side of the miners' world surfaced both in the precariousness of their livelihood and in the limits imposed by the operators upon their independence.

Dismal conditions were particularly stark in company housing. To be sure, its quality varied across the district. The *Alabama Sentinel* described the Cahaba company's living quarters as a model, where each family had its own house, "enclosed by a picket or other fence, with lots of room . . . for the cultivation of garden truck. The houses . . . are lathed and plastered inside, rents reasonable and fair, coal fuel $1.50 a ton delivered, and water for drinking and domestic use supplied by the Cahaba Company in abundance. This is as it ought to be." Unfortunately, the *Sentinel* continued, it was seldom as it ought to be elsewhere in the coal belt. "The prevailing idea with the average operator would seem to be, how cheaply dwelling-house accommodation for the mine-worker can be provided." Rents from company housing were a significant factor in the operators' business. In the coal divisions of the major furnace companies, profits from rents tended to make up somewhere between 10 and 25 percent of profits from the sale of coal itself. Cheaply constructed, overcrowded, and exorbitantly priced "box houses" were described regularly in the accounts of observers. At times the operators themselves quietly acknowledged the decrepit state of company housing. "Some of the houses are so old and out of repair as not to be worth repairing," a TCI official noted in a report on Pratt Mines. "The old log houses are gradually rotting down." In some towns large numbers of miners had the means to purchase their own homes and thereby avoid the degradation of living in company housing (or of being evicted during strikes). Others lived in independent boarding houses or boarded in private homes. But many of those who turned to the company for shelter as well as employment cited that circumstance among their chief grievances. Poor sanitation, and especially water quality, also inspired frequent complaint. Primitive disposal of sewage and garbage (at some camps, one observer noted, the only garbage collectors were the "pig, the buzzard, and the chicken") exacerbated a plethora of health problems common to the region, most notably pellagra, malaria, hookworm, and typhoid.[25]

Another locus of tension was social services. Many operators deducted monthly fees, generally ranging from fifty cents to a dollar per miner, for such services as health care and schooling. The miners, who seldom had a say in the hiring of doctors and teachers, perceived these deductions as an encroachment upon their independence. Involuntary fees could quickly become debts, which effectively compelled the miner to remain with his employer. Companies often recruited miners as much for the fees they could extract as for their labor. "The agent for Worthington keeps bringing in new men," a miner reported from Compton, "but they leave as soon as they can work out their transportation and other fees, such as 'house rent' (whether you occupy a house or not), insurance, school, doctor, etc. That is taken out of the first money earned." At times resentment led to protest. In 1887, for example, a group of miners, tired of having to support a school while having no say in who taught there, went on strike and closed it down.[26]

But nowhere in the mining communities was exploitation more blatant than at the company stores. In many towns miners were pressured to buy at the commissary—or "pluck-me" stores, as they called them—where prices were often higher than those at independent merchants and the selection and quality of goods inferior.[27] The boss at Wheeling Mines, a miner reported in 1888, had notified the men "that they would have to trade more in the Company's store or else—." "Or else" could mean being assigned the least desirable places to work at the mines; more often, it meant outright discharge, usually on one or another pretext. For example, twenty-five miners at Corona were reportedly fired in 1895 "for the heinous offense of not dealing in the Company store." Many thus discharged found themselves blacklisted around the district.[28]

Operators further coerced miners into buying from their commissaries by blocking access to independent merchants, often by buying up the surrounding lands to keep out rivals. "There is one bakery here, one postoffice, one little fruit stand, one barber shop and a company store," a Blocton miner wrote. "The company here have [sic] a monopoly of the land and won't sell a lot for love or money. I needn't tell why—pluck-meism explains it." Miners seeking alternative stores sometimes had to travel miles. Company railroads might compound the inconvenience by charging high fees for trips to the larger towns. Certain operators denied miners the use of dynamite not bought at the company store, even if the grade and make were the same.[29]

Coercion also flowed from the payment of wages in store checks, fully redeemable only at the commissary. Miners lacking the currency to stay afloat between paydays had to draw checks through the company store. Typically such advances entailed a deduction of 15 to 25 percent, during hard times as high as 50 percent. Low-paid employees were particularly vulnerable to this pressure. During June 1903, common laborers at Warner Mines drew nearly half (47 percent) of their collective earnings in cashed checks. Pay advances like these not only compelled miners to patronize the company stores; as debts, they signified a loss of independence to the employer, who could now constrain their ability to move elsewhere or to join a union. Store checks aside, low wages, infrequent paydays, and irregular production made it difficult for many miners to escape the debt trap. A Brookside miner described the situation in 1897: "Here we have been . . . trying to pay up our old debts and to keep from going in any further, but its [sic] impossible to avoid it. Debt claims nearly all of us, and as long as 37-and-a-half cents is paid will still retain us."[30]

Some operators deliberately widened the trap by recruiting an excess of miners. Not only did this strategy provide a larger pool of consumers for the commissary, but by spreading thin the available work, it pushed more miners to draw checks between paydays. The *Alabama Sentinel* denounced company recruitment of superfluous labor as a "lie perpetrated on the poor working man to bring him here to crowd us down, so we will not be able to have cash to draw on pay day, and then we will have to go to their company store, and then we are in their clutches."[31]

The operators defended the system, holding that the commissary was not coercive, that the quality and prices of its goods were competitive, and that many miners would object to its abolition. The claim that company stores were more competitive than coercive was in some cases valid, particularly where independent merchants were able to establish a presence. "We have five good stores," wrote a miner from Adger, "which gives us good advantages to buy our provisions and not trade at the company store." Local truck farmers could also provide miners with an alternative source of meats, vegetables, and dairy products. Especially in such settings, miners could find the commissary a fair and congenial place to buy. "Go to the company store to buy your goods," one wrote from Kimberly in the pro-union *Mineral Belt Gazette*. "Very polite clerks and good looking, too. Mr. Marshel Rickles to cut your meat and sell you cold drinks, too." But the volume of complaint among the miners suggested the widespread existence of company stores that were not so amenable.[32]

For the operators, cheaply constructed housing and coercive commissary practices were ways to alleviate sporadic profitability. Keeping wages down was another. In a region where mining remained highly labor-intensive, the price of labor comprised the largest portion of production cost.[33] Wages in the mining belt were fluid, reflecting variations in market conditions, mine geology, labor supply, profitability, and the balance of power between the miners and the operators. From the late 1870s into the early 1900s, mine wages hovered between around 60 cents per ton during the best of times and around 35 cents during the worst. A miner typically produced between three and five tons per day; his daily earnings might fall as low as $1.00 and rise as high as $3.00. Skilled workers such as carpenters, blacksmiths, engineers, firemen, and timber bosses earned comparable wages by the day. Unskilled laborers—those working underground as miners' helpers, or above ground as "company," or "day" men—usually made between $1.00 and $1.50 per day; often half or less of what miners earned. Starting in the late 1880s, Alabama miners' wages would rise and fall along a sliding scale linked to the selling price of pig iron—the only system of its kind in American bituminous coal production. Conflict between miners and operators arose over both the terms of the sliding scale and its very validity as a determinant of wage levels.[34]

Operators used additional methods to diminish their labor costs. They routinely deducted miners' pay for work-related equipment and services. The eighty cents per ton paid miners at Warrior, a resident wrote in 1884, sounded like good money, "but . . . [it] is not so much when the miners have to pay for sharpening tools, oil, dynamite, checkweighman." Controversy also surrounded the procedures for measuring the amount of coal a miner produced. Miners often charged that the company deliberately underweighed the coal; hence, their recurring demands that a checkweighman of their own choosing be present at the scales. Some companies docked miners for "dirty" coal—that is, coal deemed inadequately purged of slate—and required them to load their coal with forks rather than shovels, or pass it through screens before coming to the scales, in order to filter out slate and smaller pieces of coal. Miners viewed these practices as another form of "theft" from their rightful earnings. They felt further aggrieved by the operators' widespread refusal to pay for a variety of tasks external but essential to the actual mining of coal. Such "dead work" included going from the mine entry to the coal face and back, shooting rock, laying tracks and guard rails, timbering the roof, clearing away cave-ins, driving air courses between rooms, and hauling coal to the tram cars. Where the operators did not feel pressed for labor, the scope of dead work tended to widen. In his travels through the district during the bleak year of 1895, one observer found that miners had to invest unpaid time clearing fallen rock from their roadways and tramming the coal they had loaded from their rooms to be weighed. From a Walker County mining camp in 1908, a visitor described the spectacle of miners carrying timber on their shoulders from the woods to the mine. The men explained that the company paid only for coal produced; how the miners went about ensuring their safety underground was their own business.[35]

The question of earnings, then, was contested on a variety of fronts, at the points of consumption and production. But it made up only one realm of conflict between the miners and the operators. Tensions also flared over conditions, the arrangement and process of work, and power relations underground, as the miners' instinct to preserve cherished traditions of autonomy, egalitarianism, and collective mobilization—the essential ingredients of what Carter

Goodrich called the "miner's freedom"—clashed with the operators' quest for cheap and easily controllable labor.[36]

One point of dispute was the subcontracting system, under which miners would hire laborers. The extent and terms of subcontracting varied. Most typically, a miner took on a single laborer to work alongside him. A report from Patton Mines in 1891 conveyed the range of tasks a laborer might perform: "Some have the taking out of coal, some have the driving of the entries, others have the mining of the coal and the drilling of the holes, others have the blasting down of the same, while others have the blasting of the roof for road ways, etc." It was not uncommon, though, for miners to work two or three rooms and hire as many laborers to perform the bulk of the physical work, especially the loading and hauling of coal. A laborer worked on the "check" of the miner who hired him (that is, the coal produced by, or with the help of, the former was credited to the latter). Subcontractors might expect to earn three to four dollars per laborer per day, from which they usually paid each laborer one dollar, or marginally more. At some places, miners were known to hire as many as ten laborers at a time.[37]

Where the subcontracting system was in place, many miners were more than ready to embrace it. "Oh, yes, sir," exclaimed a black miner from Dora when asked if it paid to take on a laborer. "The coal is dirty and one man can't clean it by himself; it takes one man to load it and another man to pick it." Charles Counts of Jefferson told how he declined a friend's proposal that they work together as buddies: "There would be no sense for me to work with you," he explained, "when I can dig two tons to your one." Counts preferred to continue with laborers. For many miners, taking on laborers spared them much of the more dull and arduous toil.[38]

Others, particularly those steeped in a union culture, detested the practice, regarding it as subversive to both their livelihood and their mutualistic ethos. Central to each was what one miner called the "cardinal principle . . . [of] equal pay for equal work." Subcontracting, some suggested, brought exploitation into the relations among workers themselves, encouraging miners to "rob" others with less skill or fewer options. Subcontracting further undercut the miners' collective interests by glutting the labor supply with unskilled workers. "We create a vast number more of coal diggers," one miner wrote, "and have them do the greatest part of our work, and pay them half and sometimes less wages than we ourselves receive." Still more offensive, wrote another, was the preferential treatment that miners who hired labor received from the bosses when rooms were assigned. With the advent of union contracts in the late 1890s, subcontracting was abolished across much of the district and a system of equal partners, or "buddies," became the norm underground. Elsewhere, where the collective strength of the miners was weaker, subcontracting persisted into the 1900s.[39]

Miners also grumbled over the allocation of rooms and of cars for their coal, practices frequently marked by arbitrariness, if not active favoritism. "Some men in the shaft are getting all the cars they want, while a great many others are not getting half work," one miner reported from Pratt Mines. "I know a lot of miners which have paid to the mining bosses $20 or $30 to give them a good room," another reported from Blocton; likewise, he knew many who had quietly paid for access to cars. As with contract labor, the irregular distribution of cars threatened the miners' code of fairness, and so an "equal turn" emerged as a prominent issue in contract negotiations, one over which miners were at times prepared to down their tools.[40]

Mirroring one of the central themes of labor struggle in contemporary America, miners and operators often clashed over what each saw as their rightful prerogatives at the mines. The firing of workers was one such area. Miners generally conceded the operators' right to discharge employees, but only with reasonable cause and in consultation with their representatives. Likewise, the freedom of miners to work their rooms on some days and stay away on others, according to their own whims and priorities—a customary liberty in the coal fields of America and Europe—often ran up against the operators' desire for reliable and maximum production. Where employers held relatively uncontested power, miners might be disciplined or even fired for absenteeism; where the miners enjoyed more leverage, the right to stay away from work for up to three days, for any reason, was codified in their contracts. Miners also resented and, if they were able, resisted pressure by the operators to perform dead work. "Besides digging the coal," a miner from Coalburg muttered, "we have to play mule, pushing the empty cars in and the loaded cars out of the room." Where they could, miners imposed contractual limits on the bosses' power to have them do company work.[41]

Some operators enjoyed enormous power, which spilled into virtually every aspect of production. One miner from Carbon Hill wrote, in the vivid language of nineteenth-century labor republicanism, of the "bulldozing and tyrannical company" for which he worked. At the expense of "every right Almighty God has endowed us with, together with every privilege the Constitution of our country guarantees to every citizen," his "hydra-headed company make [sic] the rules and expect their employees to abide by them. . . . They assume the power to buy and sell alike. They dictate to us when we shall work, how we shall work, how long we shall work, how much work shall be done, and the amount they shall pay for labor performed. . . . labor has no rights they are bound to respect."

In less "tyrannical" settings, disputes were resolved between company superintendents and mine (or "bank") committees chosen by the miners.[42]

Working conditions further aggravated labor relations at the mines. A perilous trade under the best of circumstances, coal mining in Alabama was particularly notorious for the frequency of injuries and fatalities. Accidents occurred in many ways. Some were freakish and extraordinary, as when a rope hoisting an elevator cage snapped, or when a coal miner was caught in a coal washer or fell off a trestle or was kicked to death by a mule or fell down a shaft. Other accidents struck with numbing regularity: premature dynamite explosions, lethal inhalation of "black damp" (excessive gas), fire, gas explosions, and, most common by far, runaway tram cars and sudden falls of slate, timber, or coal. The steady occurrence of individual accidents was punctuated now and then by horrific mine disasters. On February 20, 1905, for example, an explosion triggered by a "windy shot" entombed and killed all 107 men working in Virginia Mines. Catastrophes on this scale would bring thousands of miners and their families converging to the scene, some to join in efforts to rescue the living and retrieve the dead, while the rest remained outside in a vigil, somberly awaiting the extraction of their often mangled friends or relatives.[43]

Although some danger to life and limb was inevitable, much injury and death was widely considered preventable. The operators, and some state inspectors, tended to attribute the grim number of accidents to the carelessness of the diggers. Inspectors admonished miners to avoid working under loose or improperly timbered roofs, to stay off the slopes when car trips were running, to keep powder away from their lamps, not to drill overly long holes, to tamp the dynamite with clay rather than coal, to shoot only one hole at a time, and not to check too hastily on an unexploded shot. Surely inexperience or lack of care on the part of the miner could heighten the risk of individual accident or even large-scale disaster. But, as miners were quick to point out, low wages made attention to unremunerative dead work like timbering, however prudent, seem a luxury. Low pay on a tonnage basis also tempted them to explode more coal than was safe. Long hours at arduous work dulled the miner's caution and alertness. Miners often emphasized the negligence of the operators, especially in the vital areas of ventilation and timbering. As if to acknowledge the point, some operators insisted that prospective employees sign documents relieving the company of legal responsibility in the event of injury or death at the mines. Miners mobilized continually for the passage and enforcement of mine ventilation and inspection laws.[44]

There remained a final, bitter grievance, one that concerned both work and community—the leasing of convicts to work at the mines. Free miners regularly decried the use of convicts as subversive of both the moral and material condition of the district, an assault on their livelihood, and an affront to the dignity of their craft. The convict lease, many argued, pitted free miners against cheap and powerless labor. When hard times left the free miners of TCI with only one or two days of work per week, a union official reported in 1893, the company's two convict-worked mines continued to run "full blast every day." "It is an outrage on any civilized community to have the produce of crime brought in direct competition with the produce of free and law-abiding citizens," the *Alabama Sentinel* declared in one of countless denunciations that filled the labor press and much of the daily press as well. The miners, and much of the wider citizenry, objected to the concentration, or "dumping," in the Birmingham district of prisoners—future ex-convicts—from all over the state. Convict labor also deprived local merchants of potential consumers, as the miners, seeking public support, never tired of pointing out. Opponents added that the convict lease compounded the hardships of the beleaguered farmer. "If the rich men and corporations who run the coal and iron mines were compelled to hire free labor," the Cullman *Advance and Guide* opined, "then many a poor man who has been ruined and broken up by crop liens and mortgages . . . could obtain employment at fair wages . . . and ere long return to his farming business."[45]

Many recoiled at the corrupting influence of the convict lease upon state government. The *Journal of United Labor* scathingly described efforts of the Pratt company, the largest user of convicts, to cultivate key members of the legislature with a luxurious visit to the mines. "They were entertained with champagne and wine flowing like water. This when the free miners . . . were barely eking out an existence." Opponents held that the convict lease perverted the administration of justice. The practice, observed the *Labor Advocate*, tended to "increase the number of convictions, to lengthen the term of the sentence." At times state convict inspectors voiced their own distress over the trivial misdemeanors "not involving moral turpitude"—violating prohibition laws, carrying concealed weapons, using obscene language, gaming—for which convicts were sent to the mines. The inequity of the convict lease, some critics noted, had a class dimension as well. State inspectors made the point vividly in a 1902 report: imagine two youths—one of affluent background, the other not—convicted of fighting in public. The former, able to pay the fine and court costs, would go free; the latter, unable to meet these expenses, faced hard labor at the mines, in association with "the worse class of criminals." "The vagrancy law makes it a crime for a man to be poor in Alabama," the Birmingham *Free Lance* astutely noted. "If a fellow can not get a job the state calls him a criminal and gives him one at Flat Top [a Sloss company coal mine]—as a convict."[46]

Finally, there was widespread revulsion over the convicts' conditions. The revulsion, to be sure, was not universal. Instead of finding "dirty, sullen, hopeless wretches," the state inspectors reported after a visit to Pratt Mines and Coalburg in 1884, "we find cleanly, cheerful, well-fed men, contented as men deprived of liberty can be."[47] But such upbeat portrayals were overwhelmed by innumerable depictions of cramped, vermin-ridden quarters and inedible food, the boss's lash and the warden's bloodhounds, and the numbing rates of disease and mortality. The death rate alone was scandalous. During 1895, for example, over 15 percent of the convicts held at Pratt Mines died; "very embarrassing," the board of convict inspectors blandly observed.[48] At the mines, one Knights of Labor paper asserted with a tone of pity hardly confined to that organization, the convict "has food scarcely good enough for a stray dog, is knocked about, cursed, abused, and in some places shut up in darkness, filth and misery. . . . [I]nhuman treatment for so long a time has made them fierce and furious, spitting venomous curses at all the world." Into the early decades of the twentieth century the practice continued to inspire moral outrage. In 1915 a legislative committee declared it "a relic of barbarism, a species of human slavery, a crime against humanity." Ultimately, both humanitarian concern and material self-interest converged to create a popular groundswell against the use of convicts underground. "Sooner or later," Board of Convict Inspectors president R. H. Dawson had prophesied in 1889, "public opinion and the real interest of the state will force us to abandon the lease system." But year in and year out for half a century, popular legislation to abolish it was stymied in Montgomery, usually at the senate level, by a powerful "penitentiary ring" working at the service of the operators and the Black Belt planters.[49]

African American slavery, it has been argued, functioned as a foil against which whites of humble origin clarified their own sense of freedom.[50] The neoslavery of convict labor could not so readily confirm the "miner's freedom." Race did not, after all, correspond as neatly to labor status in the Alabama coal fields as it had in the slave South; while the convicts were predominantly black, a small but visible minority (usually upwards of 10 percent) were white, and many free miners were black. Instead of confirming their own comparative independence, in the way slavery did for free whites, the convict lease manifestly undercut the miners' status, depressing wages, subverting collective power, encroaching on craft terrain. Ultimately, the operators' use of prisoners seemed to the miners less a contrast to their own treatment than an extension of it. The pockets of company paternalism that existed here and there in the coal fields were submerged, as previously described, by an overriding drive to extract income, productivity, and obedience from the miners. Conditioned by an ever-precarious profitability, this imperative encouraged exploitative and dictatorial practices. In the miners' communities, it meant coercive commissaries, high rent, and dismal living conditions; in the mines themselves, it was manifested in low pay, dubious weighing and screening practices, rampant deductions and dockage, unpaid dead work, subcontracting, arbitrary use of employer power, and dangerous working conditions. The convict lease was part and parcel of that environment, and the miners knew it.

ALABAMA MINERS came to terms with these conditions in a variety of ways. One response, mentioned earlier, was individual mobility: between mining and farming areas, between mining and other occupations, and among different mines around the district. Of those who remained with a single company, many routinely took days off. Employment records at Warner Mines from February 1903 suggest how extensive absenteeism could be: of a workforce of 768 miners, only 40 worked more than twenty days that month; half worked fewer than twelve. For most of the days that the mines remained idle, operators claimed, the cause was absenteeism. The coal companies had a number of reasons to lament absenteeism: not only did it deprive them of potential profit on the unmined coal, but fixed costs, such as daily wages for company men, were incurred no matter how many diggers appeared. Absenteeism also deprived furnace operators of coal for their coke ovens. An operator was often compelled to mobilize a workforce far larger than needed for the desired output.[51]

Continuing ties to agriculture—for southern whites, primarily in the surrounding areas; for African Americans, in the Black Belt—gave miners an alternative source of sustenance. Other industries and trades, especially at the iron and steel plants, provided further options. Diggers routinely shifted among the district's numerous coal mines, searching for the best prospects. And most miners, in the traditions of their craft and of southern rural life, preferred to intersperse wage labor with other endeavors. Hunting, fishing, and tending gardens and livestock offered relief from labor underground as well as a way to supplement the family's livelihood and circumvent the company store. "All the miners cultivate their own gardens, so plenty of fresh vegetables are on their tables," wrote a Cardiff resident. "Milch cows are numerous, so butter and milk is both fresh and good." Social functions drew miners away from work as well. "The output at the mines will drop below normal today on account of a great number of operatives taking in the Odd Fellows' picnic," went a typical item in the Birmingham *News*. Then there was the miners' monthly "payday holiday"—for mass meetings, sports, and other tasks and pleasures—as well as a monthly "trading day," for the purchasing of consumer goods; each of these days was often followed by a second day off. Rituals of mourning also superseded work underground. If a miner's wife died, operations would cease for the day of the funeral; should a digger be killed in the mine, it would remain closed until he was buried.[52]

The miners' tendency to integrate alternative activities into their working lives was reinforced by the notorious irregularity of coal production itself. Operations could cease for any number of causes beyond the miners' control: flooding, fire, bad air, disease, car shortage, strikes by workers elsewhere, or sheer lack of demand. Diversifying their daily activities enabled them at once to enrich the quality of their daily lives, to attain a more balanced and independent living, and to weather the vagaries of an unpredictable industry. Not least, mobility gave the miners a hedge against the various coercions of life in the coal fields. A more conspicuous response—labor organization—is the focus of the chapters to follow.

Prisoners sent to the mines adopted their own forms of resistance. The convicts were after all miners themselves, ensnared in their own substantially tighter web of constraints. They could scarcely respond as free miners, by withdrawing their labor, or moving about, or engaging in union or political organization. Nor would the convict miners of Alabama ever attain the opening provided their counterparts in Tennessee, where in the early 1890s free miners descended on the convict camps, released the prisoners, and torched the stockades. Still, the Alabama convicts found ways to confront their dehumanizing circumstances. Convicts might attain a measure of self-affirmation through activities sanctioned by the company, such as schools, moral and religious teachings, revivals, dances, troupe shows, and special dinners on holidays. They sought release through card playing, as moral reformer Julia S. Tutwiler archly reported after visiting the Pratt Mines stockades: "boys—country boys— who knew nothing about cards, become accomplished gamblers at the prison and go out to earn their living in this way." Where they could, convicts pooled their resources to ease the families of fellow prisoners through hard times; when, for example, the house of the wife of a convict at Pratt Mines' Slope No. 2 burned down in 1886, the men at the stockades worked extra to raise money for a new one. "Not a more orderly or well-behaved set of convicts is to be found anywhere than at Slope No. 2," the Birmingham *Age* commented.[53]

Convicts were not always so "orderly"; many resisted captivity and forced labor. Not all "disorderliness" aimed at the authorities. Some took steps to sabotage their own health, such as eating soap or rubbing poisonous substances into sores or cuts. Violence among convicts was endemic as well. An 1899 quarrel between two prisoners at Pratt Mines culminated in a duel, each wielding a pick, to the death. Guards intervened in such conflicts at their peril. When an assistant mine boss sought to break up a fight between convicts, one of them turned his fury on the interloper, knocking him down and driving a pick through his back.[54]

Some convicts simply refused to work. "God damn you I won't do it, and you can't make me do it either," announced a prisoner at Newcastle, brandishing a pick, when he was ordered by a guard to perform a certain duty. Two convicts were shot, one fatally, at the Loveless camp as they stood at the door of the barracks, picks upraised, refusing to go to work or to allow others to do so. At times convicts coupled such informal labor strikes with personal hunger strikes. Occasionally defiance became collective. One morning in 1885, approximately 100 Coalburg convicts struck, remaining stubbornly at the mouth of the mine upon learning that an unpopular former warden would be returning to that position. Twenty guards were promptly assembled, and after a tense several hours about half of the convicts returned to the mines, while the others were hauled off, the "ring leaders" to be shackled and lashed. Collective resistance could take on an edge of violence. On a spring afternoon in 1891, a group of convicts at Pratt Mines, emerging from an elevator cage, began hurling iron sprags at the guards, leaving one severely wounded. Several of the convicts had prepared for an armed response by placing steel coal shovels, their handles removed, under their clothing. (As the bullets ricocheted off their bodies, the *Age-Herald* mused, "they seemed to enjoy it.") Reports arose occasionally of attempted "mutinies." Such outbursts, observers noted, tended to occur at times when controversy over the convict lease was at a high pitch. John Milner attributed the recalcitrance of a Newcastle convict in 1882 to the emboldening influence of public debate: politicians "have got the matter into the camps, and some of the convicts are discussing it in a lively way." The establishment of a local Anti-Convict League in 1885 was said to have a similar effect. A reporter conveyed the feeling at the Pratt Mines, where a group of black convicts "plied [him] with questions thick and fast. . . . They believe that the anti-convict men have it in heart to . . . take the convicts out by force and set them at liberty."[55]

In lieu of an army of liberators, some prisoners placed their hopes in the mercy of the governor, and a steady stream of plaintive, semiliterate letters poured into the state house seeking pardons or at least relief from abusive treatment. "I Wish you Would please sir look Through my case," Pratt Mines convict Babe Ellis implored Governor Thomas Seay, in the despairing tone typical of such letters. "Ten years I have bin hear. . . . I never did kill any one. . . . have Mercy on me and see if you cant commute my Sentence are else my Wife and children are going to the destruction for ever more. . . . I am like a drowning man now. . . . I have two just as nice a boss men as men gets to be. . . . know one knows the trouble I see." "[M]y family is destitute of my absence," wrote Will Johnson to Governor Thomas G. Jones, adding that if he were pardoned, "I wont give the state of Alabama any more trouble."[56] Such appeals were not always quixotic: governors did on occasion issue pardons, particularly when physical disability could be demonstrated.[57]

Others pursued their liberty more directly, fleeing the mines individually or in groups. Convicts attempted escape in any number of ways, from stealthy to brazen: feigning illness, tunneling or dynamiting their way out of the mines, sawing their way out of the cell, starting a diversionary fire, swinging picks against slope gates, and simply overpowering the guard or hurling dynamite at pursuers. One audacious prisoner, assigned to train bloodhounds by pretending to escape, sought (unsuccessfully) to seize the opening by pretending to pretend. Most who fled were promptly recaptured, but a substantial minority succeeded, vanishing into the surrounding countryside, or the black "Buzzard Roost" section of Birmingham.[58]

ALONG A SERIES of fronts, then, the Alabama miners—black and white, free and convict—contested the power of the operators. One point on which they did not challenge the companies was the racial diversity of the labor force. The coal fields had been structured that way from the outset, and the miners never enjoyed much control over whom the operators hired or even how the racial lines at work were drawn. There is no evidence that the operators' recruitment of both blacks and whites represented any sort of divide-and-rule strategy. Rather, the availability of different groups of workers and their respective qualities as perceived by the employers seem always to have been the key considerations. Whatever its origins, this heterogeneity injected a powerful, and unpredictable, dynamic into the turbulent relations between miners and operators. If continuing struggles over conditions and power lent the miners an expansive sense of shared class interests, the divisive potential of race was never far from the surface. But while miners had little to say about the racial composition of their ranks, they had much to say about how it would inform their collective identities and shape their conflicts with the operators. Indeed, the extent and the meaning of the color line would become a crucial point of contention between the miners and their employers, and never more conspicuously than in the story of organized labor. From the late 1870s to the early 1920s, Alabama miners conducted several labor campaigns, under the successive banners of the Greenback-Labor Party, the Knights of Labor, and the United Mine Workers. The chapters that follow explore the vital, if ambiguous, meanings of race in these efforts.

3 The Greenback-
Labor Party and the
Knights of Labor

T HE BIRMINGHAM DISTRICT had scarcely come to life when
miners began to organize. Shadowy, short-lived, and long-
forgotten, the local Greenback-Labor Party (GLP) undercut
the image of southern working-class docility. Through its
expansive ideology and social following, it straddled the lines
between labor organization and political engagement, between farmer and industrial worker, and, most dramatically,
between black and white. In mobilizing against a Democratic
elite that had established itself on the grave of Reconstruction,
the Greenbackers of the mineral belt challenged the Bourbon
commitment to the supremacy of whites and the primacy of
capital. Greenbackism quickly passed from the coal fields, but
its standard was picked up in the 1880s by the Noble and

Holy Order of the Knights of Labor. Drawing from the ranks of the GLP, the Knights shared the Greenbackers' broad insurgent vision, although they anchored it more centrally in workplace struggle. During the second half of the decade the Knights built a strong following, rallying black and white miners around class concerns. Internal divisions contributed to the Knights' decline in the coal fields, but race was not the decisive fault line; rather, it was a divergence between the conservative trend of the leadership and an increasing militance among the miners.

If the interracial approaches of the Greenbackers and the Knights of Labor were impressive, they were not clear-cut. Ties of class on the one hand and of race on the other intertwined to give their racial strategies a complex, at times unpredictable character. Each organization featured the active collaboration of black and white miners, although within racially defined structures. Both the GLP and the Knights excoriated the operators' use of the color line as a device to weaken the labor movement, but many in each group retained the assumptions of white superiority. Just as the battles between miners and operators in this early period set the pattern for decades to come, so did the varied meanings of race in the miners' first labor organizations.

The Greenback current that surged through the nation's farmlands and industrial centers in the late 1870s extended a recurring theme in nineteenth-century American radicalism: the belief that economic want, inequality, and exploitation were rooted in the tight monetary policies promoted by avaricious bankers and bondholders. The chief vehicle of Greenbackism immediately following the Civil War was the National Labor Union (NLU). Underlying the NLU's program—expansion of paper money and government regulation of the terms of labor—was an ethos of egalitarianism, broad political empowerment, solidarity among the "producing classes," opposition to the spread of the "wages system," and a conviction that the American polity had become subservient to capital. Into its orbit the NLU drew trade unionists, feminists, businessmen, middle-class labor reformers, and black rights advocates. The diversity of its members lent vitality to the NLU, but it also generated a cacophony of competing agendas. Lacking internal cohesion or a record of tangible success, the NLU finally succumbed amid the devastating panic of 1873.[1]

While the depression decimated the labor movement, it also sharpened popular discontent. Western farmers revived Greenback politics. The formidable alliance of large capital and the federal government—dramatically confirmed by the crushing of the national railroad strike of 1877—rendered urban and industrial workers more receptive to independent politics. The bonds that developed among farmers, small businessmen, and workers during such conflicts encouraged the latter to fuse with local Greenback parties. Substantial electoral support for these efforts encouraged the formation in 1878 of a national Greenback-Labor (or National) Party, which found a prominent voice in the Pittsburgh-based *National Labor Tribune*. The GLP embraced a wide spectrum of backgrounds and ideologies. Yet in both agricultural and industrial settings, a broad "producer-citizen" self-identity enabled Greenbackers to contain, though seldom vanquish, the social and philosophical tensions among their adherents. By the fall of 1878, as the party joined forces with local Workingmen's or Labor Reform parties, Greenback candidates garnered nearly a million votes and elected fifteen congressional representatives. The party, it appeared, was approaching the stature of a mass movement.[2]

THE GREENBACK IMPULSE did not bypass the South.[3] In its more moderate form—bimetallism—it found support among the Bourbon Democrats. But a more radical version sprouted in the South as well, from varied constellations of independent parties, the Republican Party, and the Greenback-Labor Party. By railing against eastern capital and calling for the expansion of paper currency, these latter campaigns adopted the program of Greenbackers nationally. Yet their radicalism had a distinctly southern flavor. Composed primarily of upcountry farmers—but also embracing urban and rural wage earners, artisans, small businessmen, and renegade planters and industrialists—the southern independent movement rallied against the social order of an emerging colonial economy. In particular, it challenged the commodification of agriculture and the attendant decline in the yeoman's independence; the Bourbon policies of retrenchment, labor-repressive laws, and accommodation to eastern capital; the corrupt, one-party, "county-ring" arrangement of state and local politics, propped up by electoral fraud and intimidation; and the subordination of social and economic divisions—between planter and small farmer, laborer and employer, hill country and Black Belt, country and town—to the imperative of "white supremacy." In state after state around the South from the mid- to late 1870s into the early 1880s, independents, Republicans, and Greenbackers (on their own or in league) challenged the hegemony of Redeemer governments that were, in the words of C. Vann Woodward, "regularly aligned against the popular side of the struggle."[4]

The first stirrings of independent politics in Birmingham occurred in 1874. The elections that year marked the overthrow of Reconstruction in Alabama, as fraud, intimidation, and strident appeals to white supremacy overwhelmed the Republican and independent forces. The campaigns in the Birmingham district reflected this statewide pattern—neither miners nor the distinctive issues that would arouse them were yet a part of the picture. But the political environment that soon greeted the first generation of miners had roots in the watershed 1874 elections.[5]

The lifting of the depression in 1877–78 breathed new life into Birmingham's fledgling coal and iron industries. As a cluster of mining communities sprang up, local Greenback-Labor clubs quickly spread. Some twelve clubs existed in Jefferson County in August 1878; sixteen by late the following year. A band of roving organizers and periodic county conventions served to bind local clubs into a districtwide political community. The *National Labor Tribune* gave Greenbackers around the coal fields a sense of place in a regional and national crusade, one in which miners figured prominently.[6]

The Birmingham Greenbackers expressed the program and rhetoric of their counterparts around the country. Speaking before the Jefferson Mines club, Willis J. Thomas, a black organizer, read the entire platform of the National GLP. After attributing "the unparalleled distress" of the people to "legislation . . . dictated by . . . moneylenders, bankers, and bondholders," the document called for a reduction in the hours of work, the establishment of government bureaus of labor and industrial statistics, the abolition of contract prison labor, and the barring of imported labor.[7]

The outlook of Alabama's Greenback miners took on a southern cast as well. They assailed Bourbon Democracy and all that it stood for. Through their close ties to capital, they argued, the Democratic "rings" were key to the hardships of the region's common folk and the weakness of its labor movement. Bourbon rule reinforced the class deference that had so long denied them political power and prosperity. Southern workers were thus in a poor position to join in the nationwide revival of organized labor. "We, as a class, are hopelessly divided," "Olympic" wrote from Birmingham, "and it seems to me that capitalists are taking advantage of our helpless condition." Local Greenbackers were scarcely more receptive to the enfeebled Republican Party, which they regarded as more closely aligned with the "ring" style of the Democrats than with the Greenback brand of independent politics.[8]

Finally, local Greenbackers addressed their condition as miners, in tones familiar throughout the nation's coal fields. Correspondents to the *National Labor Tribune* complained of low wages and the lack of a uniform pay scale across the district. By acceding to uneven wage levels, "Olympic" asserted, miners were "blacksheeping each other out of work." Greenback miners also advocated state measures to enhance their conditions, such as the appointment of honorable and qualified mining inspectors and laws to ensure proper ventilation. But the topic most frequently raised was the notorious convict lease system. Anticipating themes that would long persist in the miners' campaigns, Greenbackers denounced the practice as an assault upon the dignity and livelihood of the free miners, a threat to the social stability of the district, and an abuse of the prisoners themselves. Together with other opponents, they attributed the system to an unsavory alliance between mining capital and the state. "Democracy upheld the slave oligarchy," Greenback organizer Michael Moran dryly remarked, and now it sought to replace slavery "with its improved convict system."[9]

Local Greenbackers insisted that social divisions must be subordinated to a broad popular alliance. One gulf they sought to bridge was that between miner and farmer. The barrier between agriculture and industry was quite porous, with farmers periodically selling their labor at the mines, and miners at times shifting to agricultural pursuits. Even long-term miners supplemented their earnings underground by tending their own modest garden plots.[10]

Greenback leaders strove to convince farmers and miners that they had adversaries in common. The depressed livelihood of the miner, they held, dragged down the farmer. "Times are very close here in this section for both farmers and miners," "D.J." of Jefferson Mines observed. The convict lease and company store systems had conspired to prevent miners from consuming the farmers' produce. The farmers and the miners were subjected to parallel schemes of exploitation—respectively, the "grab-all" system, under which a farmer who "happens to be in a little debt can't tell how soon the officer will come along, and grab all that he has"; and the "drive-all" system, under which the miner who "toils ten or twelve hours, and goes home with sore hands and aching bones and almost exhausted, [must tell his wife] 'Oh, I must work all I can, for I don't know how soon a convict will be put in my place, or how much of my coal will be docked, or how much of [my] day's labor will be taken off and given to the company. And if I contend for my rights, I may be shot down like a dog.'" "Dawson" distilled the plea for unity among farmers and industrial workers in a zestful pronouncement: "Miners, ironworkers, farmers, and every son of toil, must come to a general understanding. We are falling one by one, and amalgamation is the only thing that will bring us out of our present of servitude and tyranny."[11]

The mission of uniting miners and farmers required an ability to overcome their divergent circumstances. Prospects for the collaboration of black and white miners encountered obstacles of another order, rooted as they were in the region's pervasive sanctions against interracial organizing. The legacy of Reconstruction compounded the challenge. At its high-water mark, Reconstruction represented an astonishing assault on the culture of racial and class deference: it curbed the dominance of the planters over labor; it endowed African

Americans with civil rights and political power; and it rendered state and local government more responsive to popular aspirations. Eric Foner has aptly described this moment as a "massive experiment in interracial democracy." Yet no less dramatic than the rise of Reconstruction was its ignominious defeat, the product of social divisions among the southern Republicans, growing indifference from their northern counterparts, and a concerted campaign of terror, fraud, and racial hysteria by the "Redeemer" Democrats. How miners experienced and remembered this turbulent era cannot be presumed. It is unlikely, however, that it left unaffected the thinking of those who had so recently lived through it.[12]

If old traditions and recent history posed daunting challenges to labor interracialism in the mineral district, they were by no means insurmountable. On arriving at the coal fields, black and white miners encountered a range of common conditions. "Olympic" evoked the competing claims of racial ideology and shared class experience when he noted that local capitalists had two and sometimes three sorts of labor: "First, the poor white man, who is dependent on capital for his daily bread; second, the colored man, who is in the same fix; and third, the convict, who has no alternative but to go where his taskmasters choose to lead him in chains."[13] The GLP of the Alabama coal fields was an interracial enterprise. Mining towns typically had members of each race. Both white and black miners of the district wrote regularly to the *National Labor Tribune*. Public discussions of race by Greenbackers and their opponents further underscored the interracial character of the local GLP.[14]

Outside the South, the "race question" was seldom salient in Greenback campaigns. The 1876 platform of the Independent National (Greenback) Party (predecessor to the GLP) contained no reference to the racial issues that had convulsed the nation over the previous decade. Nor was there any mention of race in the more elaborate Greenback platform of 1880.[15] But in the Alabama coal fields, racial issues could never be so peripheral. How the local GLP navigated these treacherous waters would be essential to its viability in this industrial bastion of the New South. Here the party conceived of itself first and foremost as a labor organization, addressing the common class experience of all miners. Pay scales, the use of convict labor, mine conditions, and, more broadly, the bloated power of capital in the region and nation—issues such as these were raised by both white and black leaders for the consumption of white and black miners alike. It is noteworthy that in surveying letters from Birmingham Greenbackers published in the *National Labor Tribune* the race of the writer is frequently unclear. In a land obsessed with skin color, this vagueness was itself conspicuous.

And yet Greenbackers could not wholly avoid the race question. In the Redeemer South, interracialism could never be achieved simply by ignoring the issue; organizers had to make it an active project. Thus, they routinely urged miners to avoid racial divisions. Prejudice and bigotry, white organizer Dawson (no first name) declared, "is the lever that's keeping labor in bondage to capital." "D.J." (race unknown) picked up the theme: "[T]he colored people are all slaves, and the white laboring men are all in bondage." Capital, the white Moran reminded his readers, was unencumbered by racial sentimentality: "[Y]ou never hear those fellows fighting about their 'nationality' or their 'religion' when they meet. No! . . . The click of the dollar is the only God Shylock kneels to adore." The miners, then, could scarcely afford such a luxury as racial intolerance. In addressing white miners, Greenback leaders sought to dispel the impression that African Americans lacked the ardor and independence necessary to a labor organization. Ultimately, Dawson admonished, "we who are compelled to work side by side with [blacks] must drop our prejudice and bigotry."[16]

Their appeals did not fall on deaf ears. African American Greenbackers came to exert extraordinary influence among the miners, black and white. Most black leaders resided at Jefferson Mines and traveled the district organizing clubs. Easily the most influential was Willis J. Thomas, a "small, spare" young miner whom Warren Kelley (another black organizer) described as the foremost black Greenbacker in the area. Thomas's star within the local GLP rose rapidly. In mid-1878 Kelley related how Thomas, "a new colored member," had dazzled a public gathering with "a most brilliant effort," in which he "gave both of the old parties 'what Paddy gave the drum.'" Local blacks, Kelley added, "have settled upon him for a leader in the coming campaign." The *National Labor Tribune* was soon filled with references to Thomas. He was "doing a giant's share in the holy work," wrote a member from Helena. By August, Thomas had launched seven new clubs around Jefferson County, moving Kelley to call him "the leading spirit of the Jefferson club. . . . If Thomas is let alone he will turn the state of Alabama upside down in the course of twelve months." That month, Thomas was formally empowered by the central Greenback body of the district to oversee all the black clubs in Jefferson County, and was granted funds to sustain his activity.[17]

Part of Thomas's appeal lay in his oratory. It was claimed "by all who heard him," Kelley reported after a Greenback gathering at Warrior, "that such good words have never been heard from the lips of a poor coal miner before." A similar portrayal came from Haygood's Cross Road, where Thomas had just organized a club: "A colored man from [Jefferson Mines] brought some [issues of the *National Labor Tribune*] here and handed them around, and told us to subscribe for them. He seemed to be in a hurry. He made a speech here, and it beat anything we ever heard." Thomas had a charismatic charm. Kelley reported that the "ladies turned out in force and gave countenance to our young orator." And there was an aura of decency about him. "[A]ll accord the young man great praise for his speech and gentlemanly demeanor." Equally compelling was Thomas's ability to mobilize miners around a wide range of issues, some readily evident in their daily lives, others less palpable. An apparently uncomprehending report from Oxmoor illustrates how persuasively he could interweave local themes with national ones: "Do you know W. J. Thomas, a colored man? Our neighbors were roused up here last night by a carpet-bagger who had about a hundred newspapers, handing them around, and storming and crying something about greenbacks, public lands, banks, bonds, convicts, Chinese and poverty and starvation. He got eight or ten to join." Finally, Thomas confronted the political culture of the Redeemer South with vigor. "[H]e is not afraid to speak what he thinks," Kelley observed. White opponents of the GLP concurred. "[N]igger Thomas," one local Democrat said, "is too saucy; he don't care what he says to white nor black in his speeches." Nor was this courage confined to the substance of his speeches. His willingness to face down an ominous group of white Democrats at Oxmoor, detailed at the outset of this book, testifies to his boldness.[18]

To a remarkable degree, white Greenbackers grew receptive to the leadership of blacks in their organization. Certainly there were enough signs of this to unnerve the Bourbon element. One of the Democrats who accosted Thomas at Oxmoor had himself muttered that some whites considered Thomas "the best speaker in Jefferson County, white or black." And whites, he added, were coming to back the likes of Thomas. Before long, Thomas's "carpetbag full of newspapers" from the North would seduce away from southern white newspapers "all of the Negro subscribers and half of our white subscribers."[19]

If the conviction with which Greenbackers called on whites to "drop our prejudice and bigotry" reflected an openness to interracial collaboration, the frequency with which they took the trouble to make the point suggested the ingrained attitudes that haunted such a project. Also notable was the pragmatic language that framed calls for interracialism. Greenbackers presented racial division as debilitating more in a tactical than a moral sense. The practical imperative behind interracial unity inevitably confounds any effort to measure the extent of a "genuine," or moral, rejection of white supremacy. But seldom can the pragmatic and the ethical wellsprings of racial attitudes—harmonious, hostile, or somewhere in between—be cleanly divided. However practical the thrust of Dawson's statement, his condemnation of "prejudice and bigotry" clearly had a moral edge. There were further signs that collaboration between blacks and whites could inspire participants to reassess their premises about race. Most of the surviving evidence concerns the views of white members. Following a speech by Thomas to the central body of the district, a prominent white Greenbacker from Warrior rose to praise Thomas's club as the best, white or black, in the state, adding that he had never thought such work could be carried out by black people. When Thomas was nominated for a district-wide position, the white Warrior club declined to run a candidate against him, explaining that they all knew Thomas and preferred him to anyone else in the county. The two clubs agreed jointly to fund his organizing efforts. "Would to God every miner in the State had the same courage that [Thomas] embraces in his manly bosom," Dawson reflected a year later. The willingness of white Greenbackers to work with African Americans, however rooted in self-interest, had opened them to experiences that would alter their racial perspectives.[20]

The very existence of racially defined clubs conveyed the limits, no less than the presence, of interracialism—although here too, the biracial structure reflected pragmatic instincts as well as internalized attitudes. Strains surfaced in statements by black members. "A Close Looker" of Jefferson Mines drew attention to racial intolerance among white Greenbackers. After praising white leader Michael Moran as "a solid Greenbacker and a gentleman [who] treats every poor man alike," he lamented that not all whites shared this spirit: "I think Greenbackers that will not follow him are no friends to labor, and ought to be slaves." The author went on to identify politics as the Achilles heel of Greenback interracialism. Too many whites, he complained, could not bring themselves to break with the Democratic Party at election time, even as they urged blacks to stand by the Greenback ticket. A continuing alliance between black and white miners hinged upon the willingness of whites "deeply in love with the Democratic party" to stop sitting "astraddle of the fence." "[I]f the white men . . . want us to stand by them in the way of standing up for wages, and voting the Greenback ticket, and on many other things that they ask us to join them in, they must come out on the square, and stand up like men." This ambivalence in the relations between white and black Greenbackers, only hinted at in the evidence that remains, would reemerge in subsequent miners' campaigns over the decades ahead.[21]

AS THE Greenback-Labor Party was first and foremost a political organization, it focused its greatest energies on electoral campaigns. The party first tested its appeal in the elections of 1878. The Republicans, having never recovered from the debacle of 1874, could no longer serve effectively as the vehicle for popular disaffection in Alabama. Although it lacked the strength to field a full state ticket, the GLP ran candidates for the legislature from a number of districts, particularly the hill country of northern Alabama, the stronghold of independent sentiment.

Miners figured prominently among the Greenbackers of the Birmingham district. When the Jefferson County GLP convened in June 1878 to nominate candidates for the legislature, three of the six voting districts (or "beats") represented by delegates (Warrior, Oxmoor, and Jefferson Mines) lay in the coal belt.[22] Passionate denunciations of the convict lease reconfirmed the central role of the miners; while opposition extended well beyond the coal fields, it was there that the system's moral and material effects were felt most directly. Yet the Birmingham GLP embraced far more than miners. In fact, none of the Greenback leaders from the mining areas was included among the officials of the party's Jefferson County convention.[23]

Alabama Democrats subjected the GLP to a barrage of ridicule, blasting the insurgent party in tones familiar around the South. They depicted its calls for expanded currency, regulation of the railroads, and an end to convict labor as economically irresponsible. The Greenback activists were outsiders who brought with them "false and insidious doctrines and foreign money," in the words of the Birmingham *Iron Age*. But above all, advocates of Democracy played the race card. The Greenbackers were "powerless without the negro vote," the *Iron Age* asserted; their success would lead to "negro rule in Alabama again." The paper emphasized the high profile of African Americans in the GLP. Following the state election, it denounced one Jim Harper, "the most obnoxious negro about Birmingham, and the most extreme Radical," as a Greenback leader. The executive committee of the Jefferson County Democrats described the GLP as a stalking horse for the Republicans. They accused the radical party, "whose leaders have so long blackened the reputation of our country by sectional hate, glaring venality and unparalleled fraud," of having "their emissaries scattered, in disguise in our sunny land, urging people to join the National Greenback party."[24]

Local Greenback candidates for the legislature failed to win either of the seats for which they contested. But their support around Jefferson County was substantial; each candidate, W. F. Handy and H. J. Sharit, received 43 percent of the vote.[25] The election presented a checkered picture of political loyalties around the coal fields: in Oxmoor, the GLP reaped over 80 percent of the vote; in Warrior, 44 percent. Whatever may explain the variation, these figures suggest the popular appeal that the party had attained during its first year in the district. In Alabama generally, its following was concentrated mostly in the hill country, where William M. Lowe became one of the fifteen Greenbackers nationwide sent to Congress.[26]

Buoyed by this initial showing, Greenbackers turned their sights on the 1880 election. Presidential nominee General James B. Weaver and national organizer J. H. Randall toured the state. Building upon the party's national platform, the Alabama Greenbackers added planks protesting restrictive election laws, unfair taxation, poor schools, and the convict lease: it called, in the words of one Greenback paper, for political and economic relief for the "caucus-ridden, tax-burdened people of this state." Democrats, for their part, intensified their attacks, charging the GLP not only with fanning class conflict but, even worse, with being Republicans in disguise, an allegation lent credence by the open support the latter now offered the GLP. A pro-Democrat organ compressed the defamation into one dizzying epithet, describing the GLP convention as an "Independent-Greenback-Labor-Socialist-Radical-Sorehead conference." Lurking beneath such language was the inevitable bugaboo of racial equality. The insurgents often responded with disclaimers, accusing the Democrats of manipulating the black vote for their own purposes. "Wanted—A White Man's Party in Alabama," proclaimed the pro-Greenback (and possibly Republican) *Alabama True Issue*, in an acid reference to such "hypocrisy." Here was a notable contrast to the ways in which the Greenbackers of the coal belt responded to race-baiting.[27]

The election failed to fulfill the Greenbackers' dreams of ousting the Bourbon machine, but it did reveal a strong dissident pulse around the state. The party's gubernatorial candidate polled 23 percent of the ballot; much of this support remained concentrated in the northern counties, where Lowe was returned to Congress. In Jefferson County Democrats carried most of the offices, but the GLP and independents won several victories. Chief among these was the narrow election to the state legislature of H. J. Sharit of Warrior, who had campaigned extensively in the mining communities. The *Jefferson Independent* and Birmingham *Observer*, neither friendly to Greenbackism, conceded its popularity in Warrior, Oxmoor, and Pratt Mines.[28]

In 1882 the state Republicans threw their backing unambiguously behind the GLP, boosting at once the latter's strength and the virulence of Democratic attacks. The *Iron Age*, describing Sharit's efforts "to array one class of men against the other" and "to array the negro against the white man," concluded that his politics represented "nothing less than communism, radicalism, and niggerism." In the end, however, the GLP and allied independents made gains around the state, raising their presence in the Alabama House of Representatives from five to twenty, and their gubernatorial tally to 32 percent. Local results showed continuing support for the GLP in the coal towns; in Pratt Mines and Warrior, the two for which polling figures survive, Greenbackers and their allies prevailed up and down the ticket, although they all (including Sharit) lost the county generally.[29]

The Greenback challenge to Bourbon hegemony faded rapidly around the state following 1882. In part this trend mirrored the broader regional pattern; in Alabama, it was hastened by the sudden death of the charismatic Congressman Lowe.[30] By this time the GLP had waned in the coal fields. While the mining district continued to vote for independents during the early 1880s, Greenback activity seems to have ceased; most tellingly, correspondence to the *National Labor Tribune* tailed off after the turn of the decade. The reasons can only be guessed at. The petering out of the GLP around the nation no doubt undercut its momentum in the mineral district. Moreover, however notable its electoral performance, the party could never match the overheated expectations of the late 1870s. Finally, even at its height the GLP was hampered by a persisting wariness in the coal fields over collective action. Organizers often lamented the miners' apparent indifference to their own interests. "It seems as if the lowest grade of Workingmen in existence are entrenched in the coal mining districts," reported "Olympic" from Jefferson Mines. "[W]e can never get men to stand up for organization." They blamed this on the grinding effects of destitution, together with an abiding superstition that left many miners distrustful of labor organizers.[31]

Just as the GLP's popularity belied monolithic depictions of southern politics, so the race question revealed, and sharpened, important tensions in Greenback thought. Outside the district, blacks did not rally in significant numbers to the party. Greenbackers elsewhere in Alabama defiantly rejected Democratic claims that their movement was beholden to African Americans; on the contrary, they responded, it was the Democrats who, through either fraud or persuasion, reaped the benefits of an ever pliable black vote. In this light, the open endorsement of interracialism among Greenbackers in the mining district becomes all the more notable. In the tradition of upcountry Republicanism, small white farmers who predominated in the state GLP viewed blacks as pawns in their contest with the Black Belt elite. In the coal fields, the Greenback perspective was markedly different. Rather than an abstraction, an object of dispute, African Americans were central actors in the party; interracial cooperation was a compelling imperative in the miners' challenge to the operators and to Bourbon rule.

Black and white Greenbackers in the coal fields articulated their struggle above all as miners. But interracialism had its limits. The powerful holds that the

Democratic and Republican parties had come to exert over southern whites and blacks respectively during the 1870s died hard even in the mineral district. The separate social worlds that black and white miners occupied were reproduced, it would seem effortlessly, in racially distinctive Greenback clubs. The frequency with which Greenback leaders of both races stressed the importance of black-white collaboration highlighted the continuing significance of racial differences, just as it did a determination to overcome them. This uneasy alliance would reemerge in the 1880s under the banner of a larger, more prominent crusade.

THE FADING OF Greenbackism coincided with the rise of an organization that would by the mid-1880s pose an even greater challenge to the new industrial order—the Knights of Labor. As early as 1879, local Knights assemblies were established at Helena, Warrior Station, Newcastle, Pratt Mines, and Jefferson Mines. Over the following decade, dozens of assemblies would crop up around the district, some short-lived, others lasting and vital. The Knights drew upon the Greenbackers' critique of the iniquities of Gilded Age America and the Redeemed South, but adopted a new set of structures and strategies. Like the GLP, the Knights met with formidable obstacles, internal and external. Even at its peak the Order scored few major victories, either at the mines or in politics. In workplace battles it played an ambivalent role, at times initiating collective efforts, at times simply promoting struggles that had begun on their own, and at times striving to rein in actions, especially strikes, that it deemed excessively militant.

The Knights of the mineral district mobilized around such issues as wages and the terms of work. Through its organizers, its newspaper, and official gatherings, the Order gave miners a voice for their immediate concerns and a sense of connection with a broader movement. Above all, it kept alive an ethos of labor mutuality that tested the dual traditions of employer paternalism and white supremacy. On all of these levels, even in its eventual decline, the Knights laid the groundwork for a more enduring organization, the United Mine Workers.

The Knights of Labor was the first working-class enterprise in America to achieve the status of a mass movement.[32] Established in 1869, it was intended as an association of "producing" citizens that could blend the format of the fraternal lodge with values of broad-based labor solidarity. Through the depression of the 1870s the Knights remained a secret, ritualistic brotherhood concentrated in Philadelphia and the coal fields of northeastern and western Pennsylvania. In 1878, as the economy recovered and prospects for organization brightened, the Order assembled at Reading for its first national convention.

The Knights refashioned labor republicanism to address what they saw as the baleful trends of industrializing America. Like the National Labor Union and the Greenback-Labor Party, the Order assailed the corruptive influence of large capital upon the republic. But it traced the source of popular distress less to arrangements of finance than to those of production. It sought the abolition of the "wages system," which had come increasingly to subvert the dignity and independence of labor. Toward that end, the Knights adopted a platform that combined such standard union demands as the eight-hour day and the abolition of contract and convict labor with calls for public regulation of transportation and communications networks, arbitration of labor disputes, monetary and land reform, and self-education and self-improvement among workers. Above all, the association advocated "co-operative institutions, productive and distributive," through which workers could enjoy humane economic relations reflective of the public good. While condemning "the recent and alarming development and aggression of aggregated wealth," Knights leaders (most prominently General Master Workman Terence V. Powderly) eschewed both the rhetoric and tactics of class struggle. Grounded in a traditional republican respect for property, their mission was not to overthrow capital or pursue what they saw as quixotic strikes, but rather to restore equality and harmony to the relations between labor and employers. They discouraged independent political action, regarding it as fruitless (as the GLP experience seemed to confirm) and divisive. Education, arbitration, and economic cooperation were the Order's primary agenda.[33]

During the late 1870s and early to mid-1880s the Order grew dramatically. Assemblies proliferated nationally, and membership rose from 6,000 in 1877 to 30,000 in 1880 and 100,000 in 1885. In 1886, as labor organizing of every description flourished around the country, the Knights' ranks rose to a phenomenal 750,000. By the mid-1880s, they had attracted a highly diverse following, crossing lines of occupation, skill, nationality, gender, race, and even class (excluding only such unworthies as lawyers, financiers, saloonkeepers, and employers who performed no labor). While the national leadership scrambled to ascertain just what sort of following it had acquired, a "movement culture," or rather a network of local movement cultures, spread across the nation, cemented by lodge meetings, public lectures, newspapers, social events, and local campaigns.[34] As membership swelled and labor gained a heightened sense of power, the impulse among Knights both to strike and to pursue independent political campaigns (notwithstanding Powderly's aversion to such tactics) became irrepressible. The varied ideological and strategic approaches competing within the sprawling Order would eventually hasten its decline. In 1886, however, the rise of the Knights seemed to many to herald an end to the era of unrestrained acquisitiveness, and a triumph of the cooperative ethos.

LIKE THE Greenback-Labor Party, the Knights of Labor established a presence below the Mason-Dixon line.[35] Confronting the dogma that southern workers were too deferential and individualistic to embrace the labor movement, Knights campaigns sprouted in the port towns of Charleston and Mobile, the sugar plantations of Louisiana, the textile towns of the Piedmont, industrial centers such as Richmond and Atlanta, and the coal fields of central and southern Appalachia. The Birmingham district rapidly emerged as one of the hubs of southern Knighthood.

The first general assembly of the Knights of Labor empowered General Master Workman Uriah S. Stephens to appoint organizers to spread the gospel throughout the nation. Of the seventy-nine he named, four were southerners. Among the latter was Michael Moran, a white resident of the Alabama mining town of Helena, who set about organizing local assemblies across the district. During 1879, lodges were chartered in such key mining towns as Warrior Station, Pratt Mines, Jefferson Mines, Newcastle, and Helena. A number of these early Knights were active in the Greenback clubs as well. Moran was one; so were James Dye, Willis J. Thomas, and H. J. Sharit, who now resurfaced as leaders of Knights local assemblies at Jefferson Mines (Dye and Thomas) and Warrior Station (Sharit). In the fall of 1880, the local assembly at Jefferson Mines had nineteen members; the Newcastle assembly, twenty-six; the Pratt Mines assemblies, sixty; the Warrior Station assembly, twenty-five; and the Helena assembly, twenty-seven. Conforming to the Order's national approach, local Knights kept a low public profile during the late 1870s and early 1880s.[36]

But the resurgence of the Knights nationally in the mid-1880s engulfed the South, and actually crested there well after the organization had begun to recede in other regions. By 1886, the number of local assemblies in Alabama had risen from fifteen to forty-nine. Early the following year, representatives from forty-four of them met in Birmingham to form an Alabama state assembly. By the end of 1887, the Alabama Knights listed ninety-nine assemblies. The greatest expansion occurred in the Birmingham district. Pratt Mines and Warrior Station had the largest assemblies, but others appeared at Coalburg, Dolomite, Cardiff, Henry Ellen, Wheeling, Blue Creek, Bessemer, Blossburg, Patton Mines, Cordova, Carbon Hill, Corona, Aldrich, Helena, and Blocton. An estimated 7,000 to 9,000 people participated in a Knights of Labor parade in Birmingham in May 1887, and the Order claimed as many as 4,000 members in the district.[37] The Knights' presence was not confined to the coal fields. In Birmingham itself, ironworkers, woodworkers, carpenters, cigar makers, tailors, and coke workers formed their own assemblies, while several "mixed assemblies" combined workers from a variety of trades.[38]

The Knights of Labor aimed to foster a distinctive labor culture among the miners. Assemblies held regular meetings, where routine business was conducted, social ties cemented, and the general principles of the Order rehearsed. They conducted funerals and published death condolences, staged parades and holiday celebrations. The people of Warrior passed Independence Day of 1885 with what one observer described as a "bounteous picnic" of spring chickens, cakes, and "choice palatables," culminating in a Fourth of July oration by W. J. Campbell, a local black organizer for the Knights of Labor. The miners' perception of the Knights assembly as an important community institution was conveyed by a Blocton miner who described his hometown as boasting "a flourishing lodge of Odd Fellows, a large assembly of Knights of Labor, a lodge of Knights of Pythias, an athletic organization, two Base Ball clubs and one lawn tennis club."[39]

On another level, the Knights operated unambiguously as a labor organization, mobilizing local mine committees to craft demands, promoting labor legislation and candidates for mine inspector, advising which mines to avoid due to strikes or overcrowding, raising money for striking miners in other districts, establishing cooperative businesses in the mining towns, running tickets for state and local office, and conducting or mediating strikes. A local branch of the Knights of Labor's miners' division, National Trade Assembly No. 35, also functioned. The circulation of the *Alabama Sentinel* and the *Journal of United Labor* and periodic visits by state and national organizers reinforced members' sense of place in a districtwide, even national, movement.[40]

However local their immediate concerns, the Birmingham Knights were attuned to the broader themes of the Order. Banners at a large 1887 parade conveyed its essential tenets: unchecked accumulation amid widespread poverty was perverse ("If nobody has too much everybody will have enough"; "Beneath the shade of the mansion is the hovel of distress"; "Vanderbilt left his money behind"); a decent livelihood went hand in hand with citizenship ("Good wages and clothes make good citizens"); morality, not material acquisitiveness, should be the organizing principle of society ("Manhood, not wealth, must rule"; "Ye cannot serve both God and Mammon"; "Water only the live stock"); labor and capital should meet on terms of mutual respect ("We seek to unite labor, capital and honesty"; "Labor creates wealth. The Created should not dictate to the creator"); labor should mobilize independently ("Go with no party unless it is going your way"; "Organization! Organization! Organization!") within the ethos of broad mutualism ("An injury to one is the concern of all"); and only through such initiative could working people reclaim their dignity ("A contented slave is lower than a beast"; "Liberty or bondage— take your choice").[41] Countless variations on these themes were sounded in Knights of Labor newspapers, meetings, and demonstrations.

Embracing the social vision of the national Order, the Alabama Knights also responded to the particular circumstances of the district and the state. While the association enjoyed a presence throughout Alabama, its center of gravity lay in the Birmingham district and, within that area, in the coal fields. The miners' situation rendered them especially receptive to the Knights' program. Harsh and unsafe working conditions, coercive commissaries, low and infrequent pay, dubious weighing procedures, arbitrary supervision, long hours, and, above all, the employment of convicts; together, these represented an assault of singular scope on the values that the Knights expressed so eloquently—the livelihood and dignity of labor. The central role of the miners was revealed in the state assembly's 1888 legislative demands: of the ten listed, half were concerned primarily or exclusively with the coal fields.[42]

The Order cast the miners' plight in the classic language of late-nineteenth-century labor republicanism. "The emancipation of slaves in the United States commenced in 1863," wrote Edward Ronald Harris of Warrior, "and to-day there is a system of slavery carried on by capital that if properly brought to light would make our free institutions shake." Local Knights drew upon broad moral currents in demanding a greater reward for their labor. "We are willing to enrich others by our toil," wrote a miner from Coalburg, "but, by the beautiful canopy above us, are we not entitled to more of the wealth we produce?" The mining community envisioned by the Knights was a haven of stability and sound habits, economic self-sufficiency, and social interdependence.[43]

Picking up where the now-defunct Greenback-Labor Party had left off, the Knights mobilized politically. Despite the uneasiness of Powderly and his associates, the burgeoning Order of the mid-1880s gave rise to independent political campaigns around the nation. Historian Leon Fink identifies several forces behind this trend: the sheer momentum of the Knights' growth bolstered political confidence; a distinctive worldview lent such campaigns a sense of moral gravity; defeats at the workplace, often at the hands of increasingly hostile employers and courts, prodded the Knights toward political engagement; political power could enable the group to obtain favorable labor legislation; and political action could itself become a medium for the development of a "movement culture."[44] In Alabama, where labor conditions and state and local law were so tightly interconnected, the Knights were drawn almost inevitably into politics. Indeed, given the overlap between the memberships of the Knights and the GLP, this tendency might better be described as a *return* to politics.

Much of this activity was channeled through the regular party structure, as the Alabama Knights lent support to candidates who best promoted their program. Prominent on the Order's legislative agenda were demands emanating from the coal district: the abolition of company scrip, the institution of biweekly pay, the appointment of mine inspectors, health and safety provisions at the mines, and abolition of the convict lease. The Knights ran a series of local tickets around the state, some in the mineral district, with a measure of success. The organization turned to independent politics with the formation of a statewide Union Labor Party, an amalgam of local assemblies, the Agricultural Wheel, the Farmers' Alliance, Granges, and trade unions, the latter based mainly in the Birmingham area. By mid-1888 a new cluster of labor clubs around the coal fields were running slates in local elections. Despite initial fanfare, however, the Union Labor Party achieved little success, and quickly faded.[45]

The other locus of organization was, of course, the mines themselves. On the first of March 1879 a cut in miners' wages at Warrior triggered a spirited strike. Local sympathies were evenly divided between the miners and the coal companies. In mid-March, when the Jefferson Mines company announced that it too would reduce wages, miners there launched a strike of their own. The strikes were ultimately defeated, due in part to their localized character, in part to the companies' stockpiles of coal, and above all, to an influx of strikebreakers who, in the words of one participant, had "flocked in here from every direction."[46] These were the first sustained strikes in the Birmingham coal fields. In both the issues they raised and the forces that resolved them, these disputes anticipated many of the confrontations to follow.

Throughout the 1880s the coal belt witnessed periodic labor skirmishes. These were mostly isolated, short-lived affairs, without the benefit of a district-wide union. The Knights of Labor provided a structure and a rhetorical vocabulary that emboldened miners in these battles. The Order's leadership, however, wavered between involvement and aloofness, reflecting an ambivalence about direct conflict with employers.

The miners and the operators clashed over the range of issues surveyed in the preceding chapter. Strikes broke out most frequently over wage levels. Reductions were usually the catalyst, although at times it was the demand for higher wages.[47] Miners struck as well over the methods by which the company measured the amount and quality of the coal produced. They charged the operators with undervaluing their output in myriad ways—using faulty scales, placing scales at the mouth of the mine (along the way to which lumps of coal tended to drop off the trams, at the miner's expense), defining a ton at figures above 2,000 pounds, increasing tram sizes for miners paid by the car, screening the coal before weighing it ("slack" coal which fell through the screen was discounted, even though the company would sell it), and at times simply diverting coal from the car without crediting the miner who had produced it. Miners also walked off over the practice of docking their pay or discharging them for "dirty" coal. Resentment of mining bosses deemed arbitrary or dictatorial led to a number of strikes, as did the operators' refusal to recognize miners' bank committees. The conditions and arrangement of work—extending from matters of safety to the line dividing the miner's and the company's duties to the distribution of rooms and cars—were further points of conflict. High prices and coercive practices of the company store were prominent among strikers' grievances. Curiously, the use of convict labor, one of the miners' chief complaints, did not occasion stoppages; here miners confined their protests to the political arena.[48]

Perhaps most notable is a matter that did *not* arise as a source of conflict—the racial composition of the labor force. In fact, the only known instance in the Knights of Labor era when miners directly challenged the employment of a particular type of labor came during 1884 and 1885 at Warrior, where the miners, black and white, fought a protracted battle against the introduction of Italian contract labor. As a window on the miners' self-identity, the episode merits close attention.

The trouble began in August 1884 when the Warrior Coal and Coke Company brought several groups of Italians from New York to work under contract at reduced wages (65 cents per ton, down from the previous rate of 80 cents). By early September upward of one hundred Italians—virtually all unattached, inexperienced in mining, and unfamiliar with English—had arrived. "A remarkably fine body of men and excellent workers," the Birmingham *Iron Age* held. The miners of Warrior felt otherwise. Determined to uproot what local Knights leader Edward Harris called "that terrible curse upon American soil," over a hundred miners walked out, ordered the "Italian paupers" to leave the mines, escorted a number of them to the depot, and directed them north toward Muscle Shoals, where new employment arranged by citizens of Warrior awaited them. The Warrior company finally acquiesced, returning all but five of the remaining intruders to New York and reinstating the wage of 80 cents per ton. The miners' concern over the specter of cheap imported labor continued; at the general assembly of the Knights of Labor in Philadelphia that fall, Harris called on that body to "adopt some means by which the influx of pauper labor to the Western and Southern States be stopped."[49]

The peace at Warrior did not last long. Labor conflict reignited the following spring, when the Warrior company announced that henceforth it would pay the miners only once per month and require them to lay the iron tracks in the mines. When the miners walked out, the company agreed to suspend all changes for three months. But sour feelings persisted, as the miners grumbled over what they saw as petty oppressions at the hands of company officials. "Operators make agreements with the miners and break the agreements at their own sweet will," Harris complained. Tensions came to a head in mid-August, when one of the mine's five Italians returned from a visit to New York City with around twenty new Italian recruits, most of them fresh from the province of Naples, their passage paid by the Warrior company. Their arrival was hardly inconspicuous, as they emerged from the railroad depot firing guns in the air, a performance they repeated upon reaching the mining area. The newcomers later explained that they were simply saluting each other and celebrating the end of their journey, but local miners did not greet their arrival with such good cheer.[50]

Invoking the need to protect their livelihood, the miners voted to suspend work as long as the Italians remained. In the classic response of the times, the company cast the dispute as a question of who was ultimately in charge. "We are going to run the mines," Superintendent Kelsey declared. "It is absurd. The idea of 150 men dictating as to what we shall do! We are going to take a more vigorous stand, and will see who rules the mines." Pursuing an unusual strategy, the strikers prevailed upon local officials to arrest twenty-one of the Italians on charges of carrying concealed weapons—an improbable source of outrage in the nineteenth-century South—and shooting across the public road. Detained overnight, the Italians were then brought to Birmingham and marched in twos to jail. The prisoners, bewildered at their treatment, vowed to write to their compatriots in New York and warn them to stay away from Warrior. All the same, they were arraigned in county court and held for trial. The Warrior company, meanwhile, locked out the miners and took on the Italians' defense. On August 22, Salvator Fileno was convicted on a count of pistol carrying and fined $50. The Warrior company paid his fine and made bond for the remaining defendants, whose fate in court is not known. After several weeks, the strike was resolved by a compromise in which the Italians would be confined to one entrance of the mines.[51]

The Warrior battles of 1884–85 provide clues about the miners' self-definition. It did not go unnoticed that Italian contract labor had been introduced by a newly arrived, northern-based company, while the four older companies in the area, all southern-owned and -operated, had declined to join in the Warrior company's initiative. This distinction between indigenous and outside capital had meaning for the people of Warrior, a perspective no doubt reinforced by the appearance of foreign labor (not only from Italy but, perhaps just as alarming, from New York!) so soon after that of the new company itself. Even more intriguing was the readiness of black and white workers to unite against the encroachment of Italians upon *their* terrain. There is no evidence that such an alignment—so notable for today's students of race and labor in American history—provoked any discussion among participants at the time. Like all groups, the Warrior miners sharpened their sense of identity against the foil of "outsiders." What stands out is where they drew the line between "us" and "them." In this instance, the color line was subordinated to the lines between regional and northern capital, American and Italian workers, independent and pauper labor. This multilayered sensibility was captured in the Huntsville *Gazette*, a black newspaper, following the first departure of the Italians: "This failure of a Northern firm to cut down the wages of colored laborers and to substitute imported Italian paupers is due in a great degree to the noble stand in behalf of the miners taken by the old Southern companies."[52]

Nonetheless, race figured centrally in the story of the Knights of Labor. Most mining communities had a biracial system of white and black local assemblies. Although there exist no figures on the racial composition of the Knights in the Birmingham district, scattered descriptions show that at some places white members outnumbered black, while elsewhere the reverse held true.[53] In the atmosphere of the New South, the demography of the coal fields guaranteed the prominence of the "race question"; the conflicting imperatives of labor solidarity and white supremacy ensured its complexity.

The national organization publicly rejected racial exclusion. "We should be false to every principle of our Order," the *Journal of United Labor* declared, "should we exclude from membership any man who gains his living by honest toil, on account of his color." The Alabama Knights Assembly affirmed the interracial creed in its preamble, urging white workers "to set aside all race prejudices against the colored laborer . . . so that the toilers of both races may be successful in their struggles with trusts, monopoly and organized capital." In the Birmingham district, the Order commissioned members of each race to organize local assemblies. W. J. Campbell of Warrior was particularly prominent among black organizers, touring the coal fields as well as other parts of the state to build "colored" assemblies. At a Birmingham rally in 1887, black and white leaders spoke from the same platform to a racially mixed crowd of 5,000

Knights. White and black assemblies, while separately organized, cultivated close working relationships. "The fraternal visits to each other are productive of much good and encouraging to our members," a Warrior Knight reported. Contrary to the widespread pessimism of white organizers, many black miners came to view the Knights as an integral part of their world. Thus, a black resident of Aldrich described his town as "the garden spot of Alabama . . . we have one assembly of K of L and. . . . a good church and a good school . . ." He went on to describe the burial of Granson Richson, "a true knight," by 75 Knights and 300 others, at the Order's expense.[54]

Reminiscent of the Greenbackers, the Knights' racially inclusive approach blended ethical and practical considerations. Its aversion to the color line was conditioned by an understanding that the powers that be manipulated race at labor's expense. When the Democratic Party raised the specter of black domination to discredit the Union Labor Party, the *Alabama Sentinel* rejected this "appeal to prejudice" as "an insult to the citizens of Jefferson county—both white and colored." The Knights accused the Democrats of seeking gratuitously to sow racial discord among them. "That we of the Knights of Labor have any desire to raise a race issue, we indignantly deny," the *Sentinel* insisted; "on the contrary this issue is being deliberately raised by the Bourbon Democracy in a vain endeavor to create disruption in our ranks." Likewise, Knights leaders called on African Americans to disregard what they suggested were vacuous Republican appeals. "[T]he colored people are slowly learning," the *Sentinel* observed in 1888, "that all the Republican party of this State require of them is their vote, and have no other use for them. . . . [The black's] only hope is to vote with his white brother, who is painfully working out the problem of the working man's salvation." The Republican Party freed you from chattel slavery, but it left you in a condition of wage slavery, was the *Sentinel*'s message to its black readers. The Knights, meanwhile, propose to break the shackles that bind us all. "Should the negroes draw away from the whites," Harris held, "it would only be a short time before drivers with long whips would be in demand. There are men here . . . who would to-day enslave the negro and also the poor white man if it were possible." Ultimately, then, whether in the arena of politics or the workplace, the Knights stressed the need for white and black workers to avoid mutually debilitating divisions of race.[55]

If such appeals were grounded in pragmatic concerns, they were often expressed in unambiguously moral terms as well. At times the Order voiced a genuine endorsement of civil equality between the races, as when the *Sentinel* called on southern whites to "insist that the negro shall be protected in the enjoyment of his rights of citizenship—even in voting—with the same determination they now insist upon their own." When such a moment arrived, it argued, African Americans would "owe a lasting obligation to the grand order of the Knights of Labor." At a time when the harrowing imagery of black misrule still resonated in white Alabama, and when massive black disfranchisement was not many years off, the Knights' endorsement of black political rights represented a bold challenge to the prevailing Bourbon gospel.[56]

Equally remarkable was the association's approach to race relations underground. Here too, participants worried about racial manipulation, focusing on efforts by the operators to displace white labor with black.[57] Yet in an era when much of the labor movement, including many Knights, responded by barring African Americans from both membership and (if possible) the workplace, the Knights in the coal fields chose instead to bring miners of each race into the fold. Why did they favor the logic of inclusion over that of exclusion? Much of the answer lies in the origins of the encounter between black and white miners in the Birmingham district.

Unlike so many other settings in industrial America, where the initial contact involved an "invasion" of white turf by African Americans (often as strikebreakers), the Alabama coal fields were racially mixed from the beginning. The line dividing skilled from unskilled, moreover, never corresponded cleanly to that of race. Whatever their sentiments, miners had little say over the racial arrangement of the labor force in its formative stage. By the time the Knights of Labor had established itself, the presence of African Americans in skilled as well as unskilled occupations was a fait accompli.[58] Thus, any notion among white miners of a racially defined territory had little chance to take root. To be sure, the miners did come to assert a proprietary claim at the workplace, a claim that drew urgency from employer efforts to increase labor competition through the leasing of convicts and the overrecruitment of labor. However, as already noted, the line dividing "legitimate" miners from "intruders" was not a color line. The former included black and white miners. "Outsiders" were defined in other, nonracial terms.

One group of "outsiders" identified by miners of both races were the convicts. While the great majority of convicts were African American, the miners posed the issue not as a contrast between white and black, but rather between free and unfree. Their disinclination to fuse white with free or black with convict was conveyed neatly in the statement by "Olympic" cited earlier: that local capital had three sorts of labor—white, black, and convict. Strikebreakers constituted another "outsider" group against whom long-term miners, black and white, asserted territorial claim. As with convicts, the fact that many strikebreakers were black did not render the two interchangeable in the miners' eyes.

Too many regular miners were black, and too many strikebreakers white, to easily sustain such an association. Here the vital difference lay between honor and dishonor. The one instance in which social background flavored the miners' notions of territory came with the struggle of black and white miners at Warrior to thwart the importation of Italian contract labor. In this case the miners' move toward exclusion reflected instincts of ethnic xenophobia mingled with material self-preservation.[59] Thus, white and black miners drew the distinction between the realm of "legitimate" workers and intruders upon that realm not along the color line but rather along the lines between free and unfree, honorable and dishonorable, and, at times, "American" and Italian.

This perspective belies the juxtaposition of blacks and organized labor that has so often framed historical study of race and labor in industrial America. How, the question traditionally goes, did the unions approach black workers—with open arms or a cold shoulder? (Or, conversely, how did black workers approach unions?)[60] Such a framework may capture the situation in some localities and trades. But in the Alabama coal fields, where the labor force was multiracial from the outset, blacks and whites combined naturally to *become* the labor movement. In light of the particular conditions of that setting—the common lot of black and white workers, the absence of a "white" occupational territory upon which blacks might intrude, the incapacity of miners' organizations to regulate the composition of the labor force anyhow, and the miners' keen sensitivity to divide-and-rule tactics—the logic behind interracial collaboration became powerful indeed.

If the receptiveness of black and white miners to interracial organizing owed much to the early foothold each had attained in the coal fields, it was reinforced by the Order's broad, intensely moral standard of self-improvement. An 1887 piece in the *Alabama Sentinel* showed how this vision could press against the color line: "The man—*white or colored*—who enters [the Knights of Labor's] sanctuary with honest purpose and sits under her wise and pure teachings week after week, cannot fail to show to the world about him, in his conduct as a citizen, a husband, a father and a neighbor that his progress, intellectually and morally, has been upward. He cannot fail to win respect and fair treatment from his neighbors." Coupled with this moral thrust was an awareness of the shared concerns of black and white workers. African Americans, the *Sentinel* reminded white readers, "are workers with us in a common cause." This sense of mutual interest was illustrated by a resolution introduced at an 1887 convention of the Alabama state assembly, which reported that "there have been two of our colored brothers discharged from the employ of Duke, Bivins, & Co., for no other reason than that they belong to the Knights of Labor, and the said Duke and Bivins have in their employ convicts, which is against our interests as laboring men." Again, it was convicts, not blacks, who lay beyond the sphere of legitimate labor.[61]

The bonds among black and white Knights in the mining district were reinforced by the experience of collaboration. Some whites came to reassess their assumptions that African Americans could not be trusted as allies in the labor movement. "The Knights of Labor in Alabama have had but little cause for complaint against the colored Assemblies," the *Sentinel* observed in 1887, "much less than was feared by some of our white brothers." The black community, meanwhile, showed signs of warming to the Order. "The Knights of Labor," the *Negro American* said, "are doing a noble work for humanity in trying to elevate and protect the rights of the working classes. It also relegates the color line." An 1885 report from Warrior by black organizer Campbell showed how this alliance could transform race relations in the coal towns. The burial of Haywood Davis, a black Knight, had served as a "splendid example of friendly and christian feeling between our white and colored citizens . . . in our community." When two white members had been buried only a few days earlier, the black local assemblies had turned out "to pay their last respects to the departed brothers." Upon Davis's passing, "[e]verybody was looking to see whether or not the white organization would turn out to the funeral of a deceased colored Knight of Labor. Well they did." In fact, the funeral procession for the black member included fifty-one white and twenty-six African American Knights. Campbell was moved. "It was a rare scene . . . one of immeasurable benefit to our laboring masses."[62]

Periodic strikes were a significant index of interracialism in the coal fields. The Birmingham *Iron Age* reported that African Americans were "among the most persistent and determined leaders" of an 1882 walkout at Pratt Mines. During a strike in Walker County five years later, the *Alabama Sentinel* noted that a number of black miners had "left and joined their brothers in the strike." Most notable was the joint effort of black and white miners at Warrior during 1884–85. It is indicative that the interracial funeral for the black Knight described by Campbell took place in the thick of that protracted battle.[63]

Certainly, white Knights sometimes perceived events through their own racial prism. During an 1887 Walker County strike, the *Sentinel* complained that black strikebreakers had been brought in "while the old miners have been forbidden to go to the postoffice in the company's store or take a bucket of water from the well or even walk upon a railroad track." The race of the strikebreakers only accentuated the paper's indignation. Likewise, race-based notions might frame how whites explained a weak strike effort, as in William

Kirkpatrick's analysis of the Walker County conflict: "A strike is a strange thing in this part of the country, as it is not very long since the boss applied the whip to his men." Former slaves, he implied, were not the most likely exemplars of labor militancy. (Although here, it might be noted, historical background rather than innate racial characteristics were invoked to explain their limitations.)[64]

Nor, in moving beyond the juxtaposition of black workers and organized labor, should we dismiss it outright, for the distinction shaped the thinking of white Knights themselves. Even those most receptive to black members tended to view them as outsiders, or allies (real or potential), as opposed to fellow *subjects* of the Order. Consider the *Sentinel*'s assertion that "[blacks] are workers with *us* in a common cause" (emphasis mine), rather than "colored and white workers share a common cause." The implied "other-ness" of black workers, expressed even by outspoken interracialists, was often voiced in patronizing tones. The blend of inclusionism and racial condescension that permeated the national Order was illustrated in a report from the South by national organizer Tom O'Reilly to Powderly: "Colored Assemblies are the most faithful in the discharge of their obligations, and are the most perfectly disciplined. . . . The poor niggers believe that 'Massa Powderly' is the man born to lead them out of the house of bondage." Such attitudes thrived among local Knights as well. Note the *Sentinel*'s statement, cited above, that "colored people are slowly learning" the shortcomings of the Republican Party, and that "[the black's] only hope is to vote with his white brother, who is painfully working out the problem of the working man's salvation." No less revealing was the *Sentinel*'s claim that through "the lessons of our order the colored man"—no mention of the "white man"—"has exercised an elevating influence over those of his race who are not yet in the order."[65]

These supremacist assumptions were expressed most forcefully at moments when black members challenged them. At a meeting at Wolf Den to elect a checkweighman, black miners proposed that one of theirs be chosen, maintaining that whites had held the position long enough. When their point was brushed aside and another white selected, all but a few of the blacks declined to contribute to his salary. Their defiance drew a sharp rebuke from a white participant who called himself "A Wolf Denner." Rehearsing the grievances that black and white miners held in common—dead work, bad air, inadequate props, "pluck-me" stores—he chided the African Americans: "[S]hame to the colored man. Just because you cannot run the thing your way, you will not help to do good." "A Wolf Denner" urged them, "meet oftener with your white brethren for concert of action," arguing that, after all, "two heads are better than one." The black miners had to learn that "what is for white man's benefit is for the benefit of the colored man also." But his calls to class unity between black and white "brethren" left no doubt who should "run the thing"—"I for one am going to take the stand for white supremacy." African Americans had "too great a lack of intelligence and education for me to put [them] in a place where one hundred and fifty men's daily bread is at stake." They were too docile, susceptible to the "threats or bribes" of the "white monopolist." "A Wolf Denner" counseled the black miners to "follow the steps of your brother white men, without wanting to have all the offices." For him, labor interracialism was viable only within a scheme of white control. The story he recounted and the pains he took to make his case, however, suggested that many black miners thought otherwise. And, in prefacing his "stand for white supremacy" with "I for one," he hinted that not all white miners shared his formulation either.[66]

Ultimately more revealing than any particular glimpse of race relations among the miners was the overall dearth of public references to race during strikes and other affairs, from either participants or the newspapers that covered them. As discussed above, notably absent among the variety of issues miners raised during the 1880s was the racial composition of the labor force. Nor were miners' defeats often attributed, as they frequently were in other settings in industrializing America, to racial disharmony or (by whites) to the native unreliability of black workers.

But the Order's determination to subordinate the dictates of white supremacy to those of labor solidarity was not unqualified. Like the Greenbackers, the Knights ventured only so far beyond the racial orthodoxies of the New South. They too organized into racially separate structures. No discussion of that policy survives in the records; it is unlikely, given the dominant assumptions of the day and the separate social worlds that blacks and whites inhabited, that members of either race dwelt much on the matter at all. At times, the ethos of white supremacy arose unambiguously. A Fourth of July celebration at Powderly, for example, was open to "all white members of the Order." And more explicitly, in 1888 the *Alabama Sentinel* countered race-baiting of the local labor party by insisting that "this is a white man's fight and the Labor party is the white man's party." The workingman's candidate for sheriff "will see that the workman shall be protected, he is a white man and will have none but white men to serve as deputies."[67]

The ambivalence with which the Order tackled the "race question" was never more fully encapsulated than in the *Sentinel*'s response to allegations that an 1887 parade had broken the unwritten rule against "social equality." A commentator in the *Negro American* had complained, according to the *Sentinel*, "of whites and blacks mingling in inconceivable confusion, marching side by side, jostling each other, desecrating the sacred plane of 'social equality', etc., etc." The *Sentinel* testily denied the charge. To the contrary, the doctrine of racial separation had been scrupulously observed: black Knights "did not march side by side with the whites, nor did they seek to intrude themselves into improper prominence." If at the picnic that followed African Americans mixed "somewhat indiscriminately" with whites, this was "more the result of an oversight than of any desire on their part to do so." A black speaker had himself declared that "no self-respecting colored man would seek to intrude himself socially upon the whites any more than a self-respecting white man would seek social relations with the colored people." For good measure, the paper insisted that it was the black participants who had cheered this dictum the loudest.[68]

Having done what it could to placate respectable sensibilities by denying any yen for "social equality," the *Sentinel* attempted to diffuse the allegation by watering down its widely accepted meaning—broadening that meaning, improbably, beyond race. "Social equality," either between *or within* races, was impossible. "Social life is an artificial product dependent upon wealth, custom, tradition and many other things, and there is just as much inequality among the various elements of the white race as there is between the white and colored races. And this is equally true among the various elements of the black race." Thus, the paper converted "social equality" from the plane of race to that of class (although whether in the spirit of accepting or deploring class inequality is not clear). Elsewhere the *Sentinel* ventured another way to drain "social equality" of racial meaning: "There is no such thing as 'social equality,' never was, and never will be. When the declaration of independence said, 'all men are created equal,' it did not mean physical, intellectual, moral or social equality."[69]

Here was an audacious sleight of hand. However vague in the abstract, the racial thrust of the "social equality" charge was lost on no southerner. The indirectness of the phrase only heightened its demagogic force, leaving it room to accommodate all the shocking images it could conjure in the white mind, images so recently crystallized in the "dark era" of Reconstruction: noxious familiarity with a "barbaric race" no longer restrained by traditions of deference, no longer respectful of white social space, above all, of white womanhood. The Knights knew better than to flaunt this primal fear. They were surely mindful of the previous year's controversy surrounding the Order's Richmond general assembly, whose public breaching of racial etiquette had scandalized the white South, and provoked a flurry of charges about "social equality."[70] Even the *Negro American*, which at times applauded the Knights' willingness to challenge the color line, distanced itself from any suggestion of "social equality." Challenging all norms of southern race relations had never been a part of the

Knights' agenda; many, especially whites, would have found such an endeavor personally unpalatable, and all would have viewed it as politically suicidal. Like the southern labor movement generally, the Birmingham Knights regarded the "social equality" charge as a cynical diversion, designed to encourage divisions within its ranks and hostility without.

And yet for the Knights simply to disclaim "social equality" would have been to remain on the defensive, implicitly conceding that the race question, not the labor question, was the salient issue of the day. Hence the *Sentinel*'s creative, if disingenuous, effort to stretch the meaning of "social equality" beyond relations *between* the races, to include relations among the "various elements" *within* the races. Hence its effort to broaden not only the *parties* to these relations beyond race, but also the *content* of these relations beyond social contact, to encompass comparative levels of wealth, status, and physical, moral, and intellectual attributes. Hence the claim—one that simultaneously mocked and reassured the Knights' accusers—that "social equality" could never be. In constructing "social equality" so broadly, the *Sentinel* sought to render it innocuous, to strip it of its visceral terrors and thus of its capacity to distract attention from the labor question.[71] This elaborate response illuminated the outer reaches of the Knights' challenge to the southern racial order. To retain legitimacy and unity—and, for many, to be true to their values—they had to deny the charge. To keep it from overwhelming their mission, they sought to redefine and thus dilute it. To regain the offensive, they sought to subordinate it to the issues *they* considered paramount. In the end, however, the Order did not dwell on the race question very often. Instinctively its advocates understood that whenever race came to the fore, the cause of labor receded. It was an instinct that would persist in the labor movement of the Birmingham district for decades to come.

HOWEVER HEDGED and qualified, interracialism was the Knights of Labor's most remarkable feature. It did not arise in a historical vacuum, nor did it flow from an abstract devotion to fellowship across a color line already etched in the Alabama coal towns. The alliance, or even bond, that developed between black and white miners under the aegis of the Order arose from their ongoing struggle to alleviate working and living conditions that they experienced largely in common. The material basis for miners' interracialism came to light in the labor conflicts that flared, on a gradually widening scale, during the Knights of Labor era.

Few of these conflicts lasted long or spread beyond one locality. During the early 1880s, collective action and the Knights of Labor waxed and waned together in the coal fields. The largest strike of this period—a month-long walkout of 300 Pratt company miners to oppose a wage cut—took place in the spring of 1882, during the initial wave of Knights activity.[72] The organization's presence in the district diminished after 1882, to revive only in 1886. Likewise, strike activity ebbed during these years. But by the late 1880s, miners' struggles were becoming increasingly large scale. The Knights offered a vital framework for this trend. As time passed, however, the rising militancy of the miners ran up against the conservative instincts of the Order's leadership.

In February 1887, over 1,300 Walker County miners ceased work at Patton Mines after the Virginia and Alabama Coal Company refused to increase wages from 65 to 75 cents per ton.[73] The strike soon took on a cast that would become familiar in the coal fields. The company deployed a raffish crew of guards, evicted strikers from company housing, and brought in strikebreakers; the strikers called on miners elsewhere to stay clear of Walker County and, with some success, implored strikebreakers to join them. Material support flowed in from around the mineral belt.

If the battle revealed a broadening solidarity across the district, it also brought to light the ambiguous role of the Knights. Several weeks into the strike, State Master Workman Nicholas Stack broached the possibility of an arbitrated settlement with the Virginia and Alabama company (which demurred). But Stack had little credibility among the strikers, who considered him out of touch and overly eager to terminate the strike. Stack had "often wanted to settle," strike leader John Marsh later recalled, "but we would not acknowledge him." When the strikers rebuffed Stack's appeal for arbitration, he prevailed upon the state executive board to revoke their charter. Rather than quell their mutinous behavior, his move fueled it. "[C]heered on in their work by many members," Stack reported, the Walker County Knights went about denouncing "in bitter terms the action of the executive board, and especially myself." Nor was this heresy confined to the Knights of that area. Stack reserved special irritation for Local Assembly 2950 of Coalburg (Jefferson County), which he claimed had sent a circular to all miners' assemblies "treacherously" condemning the state executive board and urging that generous aid be given to the Walker County miners. After more than five months, the strike came to a murky conclusion.

A much larger-scale strike the following year reconfirmed the ambiguous role of the Knights.[74] On May 1, 1888, the miners of Pratt Mines walked out in a wage dispute with TCI. Soon afterward, for the first time, a miners' strike spread to other parts of the district, engulfing the Sloss company mines at Blossburg and Coalburg and the DeBardeleben company mines at Blue Creek. Over much of May and June a body of miners' representatives from around the coal belt convened in Birmingham to create a general wage scale, collect and distribute strike funds, and coordinate strike activity. Thus, the first strike in Birmingham coal fields to expand beyond one locality arose in tandem with the first districtwide miners' organization.

These developments both reflected and accelerated the alienation between the Knights leadership and the increasingly militant miners. Half-hearted overtures to the Knights from the miners' convention drew a prickly response. The *Sentinel* had admonished the miners not to strike, arguing that their interests and those of the company were "identical." When the miners disregarded this counsel, tensions sharpened. Stack, a self-described "bitter enemy of strikes," unleashed withering critiques of the strikers and operators alike, castigating "the heartless greed and legalized oppression of capital and the ill-advised and reckless work of labor." Never before had the Order so harshly condemned strikers in the Alabama coal fields.

The strikes were settled one by one in late June, on terms that pointed toward a more coordinated system of payment around the district. The Blue Creek miners returned first, accepting a proposal under which mine wages would rise and fall with the selling price of pig iron. A similar agreement followed at Pratt Mines. The miners' convention promptly ratified each settlement. The Sloss company miners were last to return, grumbling that they had been sold out by the others. Whatever the justice of their complaint, it conveyed the uncertain direction of miners' unionism even in the aftermath of their first districtwide mobilization. These uneven prospects were reconfirmed that fall by a failed strike of 1,400 Blocton miners in opposition to a wage cut. The *Sentinel* traced what it depicted as labor's failures that year to miners' "forsaking" the Knights in favor of the "loud-mouthed men" of the miners' convention.[75]

In the following two years, the Knights leaders and the miners continued to diverge. Stack's unwavering antipathy to strikes was perhaps more acceptable to those in the urban crafts. In the coal fields, where exploitation was so stark and broadly encompassing, strikes seemed to many an essential part of the miners' arsenal. As the fortunes of the national Order dissipated, the Alabama leadership endorsed Powderly's move toward a working partnership with the Farmers' Alliance. The miners balked at the Order's growing insistence that they subordinate workplace battles to electoral campaigns. Indeed, recent experience and current conditions reinforced their determination to expand local labor actions into districtwide affairs. From these efforts, and especially the spring 1888 strike, the miners derived a heightened sense of interreliance, militance, and organizational autonomy.

The contest between the conservative Knights leadership and the more aggressive miners took organizational form in July 1888, when a gathering of miners' representatives in Birmingham voted to create an Alabama branch of the National Federation of Miners and Mine Laborers (NFMML). The founding convention nominated a slate of officers for the approval of the district miners at local meetings. The new federation sent delegates to that year's NFMML convention in Indianapolis and lobbied for mining legislation in Montgomery. Although it inherited the miners' mounting disaffection toward the Knights, the federation enjoyed only a fleeting presence in Alabama.[76]

But new organizations would arise in its wake. In 1889 the NFMML was supplanted by the Miners' and Mine Laborers' State Trades Council of Alabama, which came to be known as the Miners' Trade Council (MTC). Stressing such familiar concerns as convict labor and company stores, the MTC added its own goal of a uniform period for mining contracts around the district. At root, the MTC represented a renewed effort to reconcile the Knights of Labor and the mainstream of Alabama miners. The ethos of Knighthood retained broad influence in the coal fields, even where its strategies did not. "It was thought by [the] delegates, as well as many other miners," explained J. L. Conley, the MTC's president, "that by giving up former names and affiliations and meeting on neutral ground, we could better consolidate our forces[,] . . . pool issues, bury past differences, and so increase our abilities to secure justice and oppose encroachments of monopoly." Toward this end, the delegates determined that the MTC would remain loosely affiliated with the Knights but retain an independent status, enabling the miners to control their own affairs. On these terms, the officers of the new organization, together with state leaders of the Order, launched an organizing drive around the mining district.[77]

By that time, the Alabama Knights of Labor had lost most of its members. In part, this decline reflected the diminishing fortunes of the Knights nationally, as the Order collapsed under a welter of problems—the social diversity of its members, its inability to deliver on the dazzling promise of 1885–86, internal dissent over strategy and purpose, and a concerted employer offensive. In the South, such challenges were compounded by the relative weakness of the union tradition, the power of Democratic rule and New South boosterism, and the distracting effects of race. The Alabama coal fields could scarcely have withstood this broader decline, although, as with the local Greenback movement, racial division seems not to have contributed appreciably to it. Local circumstances—defeated strikes, clashes between Knights leaders and militant miners, and the emergence of a rival organization—only hastened the Order's demise.[78]

With the coming decade, however, prospects for a renewal of miners' unionism appeared to brighten. Within a year of its launching, the MTC would be absorbed into a new, more promising force—the United Mine Workers (UMW). In the decades to come, the evolving fortunes of interracial unionism in the mineral district would be linked to the rises and declines of the Alabama UMW.[79]

4 The United Mine Workers in the Populist Era

INTERRACIAL UNIONISM confronted new challenges during the last decade of the century. The crash of 1893 and the four long years of depression that followed sapped the collective power of the Alabama miners. State government proved less friendly than ever, not only continuing to lease convicts to the mines but dispatching troops to quell the largest strike the coal fields had yet seen. The operators, for their part, adopted bold new strategies to divide the miners along racial lines. Such developments severely tested prospects for a labor movement of any sort, not to mention one that encompassed both black and white miners.

The miners' impulse to unionize nonetheless persisted, even as their power waxed and waned over the decade. Although their organization, the United Mine Workers, never achieved recognition during the early and mid-1890s, it did provide miners on both sides of the color line with a vehicle for mobilizing around common concerns. Despite increasing regional hostility to interracialism—not to mention racism and distrust among their own ranks—black and white miners continued to organize collectively. Their collaboration drew upon a continued sense of shared experience as miners, coupled with an awareness that the color line functioned, indeed was deliberately used, at their expense. Interracialism in the mineral district also mirrored the institutional approach of the nation's now foremost labor body, the American Federation of Labor (AFL). Although after the turn of the century the AFL would harden into a bulwark of segregationism, at this point its leadership still officially endorsed the recruitment of all workers regardless of race, and excluded unions that refused to comply.[1] Miners drew encouragement as well from the Populist crusade that arose to challenge the New South elites in the early 1890s. But perhaps the greatest single impetus came with the founding of the most ambitious union America's miners had yet seen.

IN JANUARY 1890, 103 delegates from the Knights of Labor's National Trade Assembly No. 135 and 87 from the National Progressive Union (an offshoot of the National Federation of Miners and Mine Laborers) convened in Columbus, Ohio, to put an end to five years of bitter rivalry. Out of the gathering emerged a new organization, the United Mine Workers of America (UMW), charged in its preamble with "educating all mine workers in America to realize the necessity of unity of action and purpose, in demanding and securing by lawful means the just fruits of our toil." At its founding, the UMW represented an estimated 17,000 miners, divided into twenty-one districts.[2] That April in Birmingham the Miners' Trade Council voted to reconstitute itself as District 20 of the UMW, under the continued presidency of John L. Conley.[3]

An air of expectancy greeted the founding of District 20, and the leadership moved promptly to capitalize on it. During May and June, Conley and National Executive Board member Patrick McBryde built locals across the mineral district. That spring's electoral campaigns heightened public attention to issues such as convict labor and mine safety, and UMW organizers invoked them readily. That winter and spring the coal fields were further enlivened by a wave of local strikes and disputes: at Carbon Hill over wages and the introduction of screens; at Milldale over low pay; at Kitchen Mine over the miners' demands for mules to draw their carts and for rules by which they might (in the words of the *Alabama Sentinel*) "govern themselves"; and at Blue Creek, over a wide range of work rules.[4]

It was at Pratt Mines, ever the bellwether of trends in the district, that the union sought to establish its influence. There, as the miners' annual contract with TCI approached expiration that June, District 20 held its first convention. The delegates, comprised of fifty-one whites and twenty-one blacks, adopted the Alabama section of a national wage scale formulated at the UMW's founding convention. The union invited all operators to negotiate in joint convention.[5] As the contract expired and District 20 prepared to reconvene, the mines fell idle. Only three minor operators had accepted the union's overture. Judge Bond of TCI had offered to receive a delegation of employees, but would not meet with "outsiders" connected with the union. It was a sobering rebuff: TCI, easily the most influential company in the district, had declined either to recognize the union or to grant any of its proposed terms.[6] District 20 concluded that present circumstances did not favor a strike, and the miners glumly renewed the existing contracts. But the operators' intransigence rankled, encouraging talk that, come fall and higher demand for coal, the miners should defy their contracts and fight for their wage scale.[7]

In November another District 20 convention called for a five-cent wage increase. Again virtually all operators refused to meet with the union. "You cannot afford to make any concessions to the men now," TCI general manager Alfred M. Shook admonished the company's Pratt Mines superintendent, "for the reason that you cannot afford to recognize their right to break a twelve months contract at the expiration of six months. . . . No matter how unjust or how onerous the present contract is, it should be carried out." Nonetheless, on November 29 nearly 3,500 miners went out; by the following week, the figure had risen to 6,000–8,000; by the end of that month, the number idle reached 15,000–20,000. The iron-producing operators banked their furnaces. The strike, the largest the coal fields had yet seen, stretched into the new year.[8] But by mid-January the union—unable to overcome mass evictions, strikebreakers, repression by armed guards, and a hostile press—conceded defeat. The effect of the miners' loss was most vividly illustrated at Warrior, where returning strikers were required to sign ironclad contracts sharply curtailing their right to raise demands and conceding to the company broad powers over hiring and firing, discipline, and the organization of work.[9]

During the strike, the operators used tactics broadly familiar to veterans of the coal fields. At Blue Creek, however, the DeBardeleben company introduced a new inflammatory element, with repercussions that would shape the fortunes of miners' unionism for years to come—the large-scale importation of African Americans. Henry F. DeBardeleben made his move at the outset of the strike. "Five hundred colored miners wanted at Blue Creek mines," announced advertisements circulated far and wide. The notice promised earnings of a handsome $3.25 per day. DeBardeleben's recruitment of African Americans from farther south was vigorous enough to cause concern among Black Belt planters over their labor supply.[10]

Not all the operators were comfortable with DeBardeleben's provocative approach. TCI's Shook confided to an underling his fear that "you are going to have trouble if you force negroes into the mines." But before long the strategy was bearing fruit. On the morning of the strike's first day, seventy black men appeared at Blue Creek's Johns mines, where they were personally greeted by DeBardeleben and his superintendent. In the coming days African Americans continued to arrive at Blue Creek, while strikers—predominantly whites, who tended to have wider employment alternatives—began leaving the area. By early January DeBardeleben was renting the houses they had vacated to black strikebreakers. The slopes at Blue Creek were by then in "full blast," the *Age-Herald* reported, and company mine bosses were "jubilant over the success they were making with the colored miners." Even Shook, discarding his initial apprehensions, advised that TCI follow suit.[11]

The mass introduction of black strikebreakers tested as never before the miners' capacity to withstand the divisive potential of race. Prestrike statements by Blue Creek whites had already thrown that capacity into question. When 80 whites and 200 blacks announced their intention to remain at work, the *Age-Herald* reported that white unionists "immediately raised the color question." One white vowed to make the company's anticipated use of black strikebreakers "the issue of the strike"; the resulting indignation, he predicted, would "have the effect of bringing all the white men who do not want to strike over to us." His words suggested how ripe Blue Creek was for a divide-and-rule strategy.[12]

The UMW worked to keep black miners in the fold, pledging material assistance to strikers of each race. "We have no fight to make on the negro," declared national organizer McBryde. Strikers were counseled to curb any hostility toward black strikebreakers and to seek instead to win them over. Thus, an appeal from the *Sentinel* to the African Americans working at Blue Creek: "We beg leave to assure them that no matter what opinion we may have had toward them since the strike began, if they are now prepared to assert their independence we will extend to them the right hand of fellowship. . . . Shoulder to shoulder we will march together in our struggle of right against might."[13] Even the antiunion *Age-Herald* acknowledged that the strikers were leaving the black strikebreakers unmolested: "They seem to think that they will do more good to their cause by forbearance."[14] The strategy paid off, as large numbers of black miners rallied to the union's call. "[T]o the honor of the colored miner," the *Sentinel* said, "he is as a rule devoted to union principles and dreads the stigma of blackleg as badly as does his white brother miner, as witness his gallant stand at Blocton, Pratt Mines, Warrior and elsewhere." Regularly the *Sentinel* described spirited meetings of black and white strikers.[15]

The union sought in a variety of ways to deflect the racial wedge. First, it resisted the equation of "colored" and "blackleg," emphasizing the presence of white as well as black strikebreakers. The *Sentinel*, for example, allowed that there were "a few pitiful, sordid souled white and colored skinned creatures practicing . . . this most despicable of all callings." It took pains as well to distinguish between black "practical miners," who even at Blue Creek had sided overwhelmingly with the union, and the black "green hands" who had taken the strikers' places. Likewise, the UMW leaned over backwards to deny any suggestion that whites enjoyed a primary claim on work, at times even hesitating to urge the ouster of black strikebreakers lest it stoke racial divisions. Finally, at a number of points the union took exception to the language of white supremacy. The *Sentinel* highlighted DeBardeleben's reference to his black strikebreakers as "niggers." The paper also reminded African Americans of the contrast between the union's and the operators' views on race. "Our coal operators don't want to treat with the United Mine Workers of America, we hear, because of their colored membership. Hallo, colored men of Blue Creek and elsewhere, who are working, what do you think of this?"[16]

Yet statements like these do not establish the absence of strains and wariness between black and white miners; if anything, they hint at the contrary. Nor does the dearth of positive evidence: day-to-day racial tensions would likely have eluded the notice of even those most eager to amplify them. One Birmingham paper, however, did describe an interracial "sensation" occurring on New Year's day at Adger, where the DeBardeleben company had begun renting houses to black strikebreakers. The white miners' wives "got rampant when they saw the colored folks sandwiched among them" and confronted them with frying pans, tongs, "and everything else imaginable," causing the frightened blacklegs to take refuge in the mines. When the women spotted DeBardeleben himself, a "score or more" surrounded him and unfurled a banner depicting a black miner, the word "scab" inscribed underneath. In a puckish act of derision, they set about tying a white apron around the company president. Whether the women were implying that his tactics were unmanly or reminding him of the effects on their domestic lives, DeBardeleben reportedly "took it all good humoredly and had more fun out of it than did the women."[17] The strikers' reactions to the importation of black labor did not usually involve such levity. There were reports of whites shooting and otherwise intimidating the offending newcomers.[18] What remains impossible to gauge with any precision—despite the widespread assumptions of contemporaries, not to mention historians of race and labor—is the extent to which such actions were "racial" in nature; that is, where the reaction to "blacklegs" ended and that to "blacks" began.

Reminiscent of the Greenback-Labor Party and the Knights of Labor, the interracialism of the United Mine Workers was uneven and qualified. White unionists actually contested the operators' manipulation of race from both within and outside the tradition of white supremacy. While they challenged that tradition openly and often, they also succumbed to it—even embraced it—by charging their adversaries with betraying it. Discussing DeBardeleben's resort to black strikebreakers, one contributor to the *Sentinel* asked, "[I]s it possible that the numerous editorials in the *Age-Herald* against negro supremacy have no meaning? Is it possible that the said paper will advocate the displacement of honest, good-citizen white labor by the negro it so violently antagonizes?" Similarly, miners at Warrior referred to the poststrike ironclad contract as the "Warrior force bill," a sardonic allusion to the 1890 Lodge force bill that would expand federal supervision of state elections, to the consternation of the white South.[19] In likening the operators' repressive measures to the specter of renewed northern intervention, many miners borrowed from, and reinforced, the spirit of the Redeemers.

Beyond noting the operators' racial hypocrisy, the *Sentinel* offered appeals of its own to white supremacy. Pointedly it asked how many Birmingham merchants wished to see the coal fields filled with "negroes, dagos and Hungarians?" It uncritically reprinted a warning from the *Age-Herald* about the growing presence of African Americans brought about by the strike. In the long run this would spell "a great setback to the industrial life here. If the negro is unprofitable in the cotton belt it is not because he does not produce cotton, but because he is not a good citizen." One union supporter, W. E. Russell, dwelt upon the docility he viewed as intrinsic to African Americans: "The negro has no independence, no manhood, no ambition beyond the moment—he will be as the potter's clay in the hands of the operators." Should African Americans become the majority of the miners, he asserted, "you may rest assured that the continually-reducing wages that they receive, will find its [*sic*] way to the commissaries . . . and the negro dives that will grow and fester in the community." This broad vein of racist thought residing in the union came to light as well when white UMW advocates held strikers of their own race to a higher standard of loyalty. A writer from Pratt Mines castigated "the man that has a white face and straight hair that . . . goes and blacklegs or scabs." For the "chicken-hearted" whites of Blue Creek and Blocton who had stayed at work, the *Sentinel* reserved "the most supreme contempt." But even some who argued the superior capacity of whites for labor solidarity remained hopeful that African Americans might come around. The operators "had better not be so sure that the Negroes will consent always to be slaves," warned the *National Labor Tribune*.[20]

By playing the race card the DeBardeleben company had brought conflict between miners and operators onto new and explosive terrain. Each side could view the outcome with a blend of satisfaction and disappointment. For DeBardeleben, the recruitment of black strikebreakers had served to restore production and dampen strikers' morale. But his divide-and-rule tactic did not visibly undermine their unity. The strategic imperative of broad solidarity, the egalitarian ethos that historically suffused miners' unionism, and the recent experience of collaboration across the color line all combined to blunt the racial wedge. On the other hand, there was the ambivalence of the union's public stance, as it strained to negotiate the conflicting claims of labor unity and a racist white public—and in turn to bring these contradictory purposes into line with the strikers' own varied sentiments. The charged yet ambiguous dynamics of race in the strike of 1890–91 introduced patterns that would shape miners' struggles ever more markedly in the coming decade.

IF THE EFFECTS of DeBardeleben's strategy were mixed, the outcome of the strike was not. The miners had suffered a sharp defeat. District 20 remained formally intact; indeed, within a week of the strike's termination its executive committee was meeting to review current mine legislation. But despite brave words from its leaders, the union languished over the next several years. Faced with a declining market for pig iron, the operators reduced the cost of extracting a ton of coal from 70 to 60 cents, a reduction facilitated by the wholesale substitution of common laborers for miners in the loading of coal.[21]

The Alabama UMW remained essentially dormant during the early 1890s. Periodic efforts were made to revive organization, under a charter from either the national UMW or the Knights of Labor or as a new entity such as the short-lived Alabama Miners' Protective Association. There was talk of establishing a miners' Relief Society.[22] But from mid-1891 to the fall of 1893, collective action was scarcely in evidence. Not even the startling Tennessee miners' revolt of 1891–92 against the convict lease could break through the torpor that now covered the Alabama coal fields.[23]

The one area in which the miners continued to find a public voice was in politics—and that was a voice suffused with populist sensibilities. The bleak working environment of the coal fields quite naturally opened the miners to political avenues for relief. They found such avenues in a series of broad-based anti-Bourbon alliances that arose in Alabama during the late nineteenth century.

In the late 1870s the Greenback-Labor Party had offered the miners a potent if fleeting vehicle for popular protest. The mid- to late 1880s brought a succession of efforts to mobilize agrarian and laboring interests against the planter-industrialist axis. These efforts crystallized in 1887 with the founding of the Alabama Union Labor Party (renamed the Labor Party of Alabama a year later), an amalgam of the reformist Agricultural Wheel, the Knights of Labor, and assorted trade unions. The Labor Party blended the concerns of southern agrarianism with those of the labor movement, advocating such reforms as better pay and working conditions, government ownership of communication and transportation systems, fair election laws, and an end to convict labor. It stirred interest among the miners but, like the GLP, ultimately failed to check the power of Bourbon Democracy.[24]

Into the vacuum moved the Southern Farmers' Alliance.[25] Founded in 1875 in the frontier country of Texas as an offshoot of the Grange, the Alliance expanded dramatically in the following decade. To alleviate the debt peonage afflicting much of the rural South, the Alliance promoted a network of buying and selling cooperatives through which farmers might circumvent their onerous reliance upon creditors. It adopted an ambitious program of political education, made up of local weeklies and a lecture bureau, to spread its neo-Greenback assault on monopoly and hard currency. By the end of the decade the southern Alliance enjoyed a large rural following, black and white, and had forged links with its counterparts in the corn and wheat belts of the West, and with the Knights of Labor.

When its cooperative enterprises proved no match for large private concerns, the Alliance began moving beyond its avowedly nonpolitical orientation, calling by 1890 for federal measures to rectify gross inequalities of wealth and political power.[26] In the strongholds of western agrarianism such as Kansas, Nebraska, and South Dakota, state Alliances organized independent political parties around these demands. The southern Alliances, not yet prepared to break with the Democratic Party, sought instead to capture it from within. In exchange for Alliance support, Democratic candidates were required to "stand up and be measured" on its demands. The results were impressive: Alliance-backed candidates in the South captured four state houses, eight legislatures, three seats in the U.S. Senate and forty-four in the House.

But this newfound influence proved illusory. Alliance-supported officeholders failed to deliver on their pledges and drifted back to the Democratic mainstream. Before long many Alliance activists were sufficiently disillusioned to throw their support to a third party. Such an option materialized in February 1892, when National Alliance representatives met in St. Louis with delegates from a wide range of reform organizations to lay the groundwork for what would become the People's (or Populist) Party. In Omaha that July, the new party adopted a platform reaffirming the Alliance principles concerning currency, credit, transportation, and land. But the Omaha platform went further: it addressed more directly the interests of labor (the eight-hour day, immigration restriction, and opposition to the use of Pinkertons in labor conflicts), embraced more far-reaching measures for fair elections, and called for government ownership (not merely regulation) of transportation and communication systems. Populist (or allied) tickets contested state and local elections throughout the South. Recalling the Greenbackers of a generation earlier, the issues raised by the Populists—indeed, the very fact of their challenge to the Democratic rings—resonated widely. Yet the Populists also met daunting obstacles: chiefly, white reluctance to go against the Democratic Party, the rigging of elections by Bourbon machines well practiced in that art, and, as ever, the divisive powers of race. These forces combined to preserve Democratic hegemony through most of the South in 1892, although not without notable gains for the Populists. The Alliance's first foray in Alabama politics in 1890 achieved mixed success. Alliance Democrats claimed nearly 60 percent of the state assembly. However, the movement's candidate for governor, Agricultural Commissioner Reuben F. Kolb, lost the Democratic nomination to the conservative Thomas Goode Jones, some charged through extralegal means.[27]

During that year's primary campaign, the Birmingham labor movement—the miners' union and the Knights of Labor, the Birmingham Trades Council and the various craft unions, the *Alabama Sentinel* and the recently founded Birmingham *Labor Advocate*—rallied to "Workingman's" Democratic tickets. Spanning their agenda was the range of issues that were cementing bonds between agrarian radicals and large numbers of workers; thus, they endorsed candidates who backed a fair election system, an expanded currency, the development of public schools, and legislation for proper mining conditions, semimonthly payment in cash, and abolition of the convict lease system. From Pratt Mines, the *Alabama Sentinel* described how sharply support for congressional candidates diverged along class lines: the anti–convict lease candidate, R. J. Lowe, had the support of "one thousand free miners"; J. H. Bankhead, himself a convict contractor, enjoyed the support of "the Mayor of Pratt Mines, one capitalist, one civil engineer, two doctors, seven merchants, an engineer, a dentist, a book-keeper, a druggist, six clerks, two market men, an editor, and one mechanic." Following the defeat of the Alliance and "Workingman's" forces at the May Democratic convention, the *Sentinel* cried foul. At Warrior and Pratt Mines, the "trained politicians, tricksters and hired under-strappers of corporated and capitalistic power" had employed all manner of fraud to brush aside popular support for the "Workingman's" tickets.[28]

In the elections that August, only the Republicans remained as an alternative to the Bourbon Democrats. Voting returns from the mining communities varied widely: in Brookside and Newcastle, the recorded Republican vote was virtually nil; in Warrior, Republicans carried one third of the vote; in Pratt Mines, over 40 percent; at Henry Ellen, nearly half; at Johns, 55 percent. The Republicans prevailed in the Walker County mining towns of Carbon Hill (56 percent) and Cordova ("by 79 votes") but lost handily to the Democrats in Bibb County's Blocton (35 percent) and Shelby County's Helena ("by 39 votes"). Official counts, it must be noted, were notoriously inaccurate, usually to the benefit of the Bourbons. Still, even these suggest continuing disaffection among miners with the prevailing political order. After Jones went on to win the governorship, the *Sentinel* observed: "Workingmen will not forget the lesson that has been taught them, that the Democratic party will not recognize them nor give them representation in its ranks. . . . What action they will take looking to their political enfranchisement in the future, remains to be seen."[29]

Over the next two years Governor Jones did little either to antagonize or to win the hearts of the miners.[30] The first test of the miners' loyalties would come in December 1891, when Jefferson County voters were to elect delegates to the following June's Democratic convention. The campaign brought a rematch between the Kolb and Jones wings of the party. Kolb and Jones clubs appeared in all the major coal towns, and each candidate crisscrossed the district. The contest galvanized the public. "Parades, mass-meetings, a few free fights and one or two shooting affrays made matters interesting," noted one observer.[31]

As in 1890, the results produced a checkered picture of miners' loyalties. At Blue Creek, the Jones ticket carried Adger and Sumter by substantial margins, but ran virtually even with Kolb at Johns; at both Pratt Mines and Warrior, support for Kolb hovered near 40 percent; at Blossburg, the race was even; and at Brookside, Coalburg, Bradford, and Mary Lee, Kolb prevailed with large majorities. Once again, the results demonstrated the continued presence of "Kolbite" sentiment in the coal fields, but also the ongoing influence of the Bourbon machine—or, perhaps more accurately, of the operators. The *Sentinel* expressed disgust for the "miserable cusses" who voted "largely against their own interests and for the candidates of the 'grinding capitalists,' against whom they complain and charge with unnumbered oppressions."[32]

By 1892 a groundswell was sweeping Kolb/Alliance circles for a state People's Party, and such an enterprise was formally established that February. Kolb himself, still reluctant to break fully with the Democrats, challenged Jones in the general election as a "Jeffersonian" Democrat, although with a distinctly Populist platform. After a bitter campaign, Jones won another victory amid renewed charges of fraud. In the major mining beats, the governor won Blue Creek, Bessemer, and Warrior, while the Kolbites carried Blossburg, Brookside, and Pratt Mines.[33] The outcome reconfirmed that neither the operator-Bourbon pole nor the reform-organized labor pole enjoyed uncontested sway in the coal fields.

THIS UNEASY BALANCE was disrupted with the panic of 1893. The mineral district was hard hit: declining demand sent state coal production plummeting 72 percent that year. As work at the mines grew scarce, talk of wage reductions filled the air.[34] So too did resentment over the continued use of convict labor in the midst of widespread distress. Recent electoral contests had kept alive a current of opposition to the coal companies. The coming of underemployment and wage cuts infused this critique, and intensified it. Out of this atmosphere came a gradual reawakening of miners' unionism. Signs of life flickered that summer, when the miners of the Tennessee and Sloss companies mobilized against anticipated wage reductions.[35]

That fall, unionists conducted an organizing drive throughout the mineral district, the first of any consequence in three years. On October 14 a convention of thirty-nine delegates in Bessemer established a new organization, the United Mine Workers of Alabama. Over the remainder of the year local councils of the Alabama UMW were established around the coal belt. The union retained the District 20 structure and sent periodic reports to the *United Mine Workers Journal (UMWJ)*, although it did not formally reaffiliate with the national organization. The atmosphere of a labor crusade was once again palpable. One miner described a series of "great revival meetings" conducted across the district by a "celebrated Georgian evangelist" who lambasted the large manufacturers as "mean, contemptible creatures" for reducing their workers' wages. "Soulless sinners, you are going to hell at express speed!" the preacher would thunder.[36]

The newly established UMW of Alabama adopted the qualified interracialism of earlier campaigns. In the larger communities (such as Adger, Sumter, Coalburg, Blocton, Dolomite, and Corona) separate black and white councils appeared, following the biracial pattern of organization that had prevailed in the coal fields since the Greenback-Labor era. Union leaders drew upon past experience to promote interracialism, stressing the demonstrated commitment of black miners. "On Thursday night previous [at Blocton]," the *Labor Advocate* announced in typical fashion, "a strong local Council of colored miners were [*sic*] formed of nearly a hundred members and it is thought that within the next few weeks every colored workman will be enrolled under the union banner." From Corona came word that the "colored" council, named the Eagle of West Alabama, had outshone the white council, the Morning Star. "The Negros of Corona never allowed the K. of L. to die here," an observer wrote. "Can the white man say as much?"[37]

In like spirit, union leaders urged white and black miners to reject the color line, arguing that it diminished the cause of labor. First, it undercut the miners' leverage. A Corona man described local miners' failure to resist a pay cut: when black miners demanded ten cents, they were refused because white miners failed to follow suit. In addition, race distracted miners from the real sources of their oppression. The *United Mine Workers Journal* made the point flamboyantly, after describing "race prejudice" as the leading obstacle to unionism in the Birmingham district: "Capital and its paid diplomats, in the persons of wily, scheming underlings, or desperadoes whose fitness for favor and place consists of braggadocio or desperadoism, spies, cut-throats, traitors and the rest seem to fade into insignificance as barriers to the great desideratum when compared to the 'color line.' " Capital, meanwhile, "knows neither color, race, nation or creed." Beyond dividing and distracting the miners, the color line legitimated principles of natural distinctions that might shift from the plane of race onto that of class. With that in mind, the *United Mine Workers Journal* warned members not to "set up among themselves distinctions of birth-merit." And racial division enervated the miners morally as well as tactically. Pratt City organizer W. S. Hannigan described the failure of workers to "come all together, never mind the color," as "degrading, distrustful, unholy and unmanly." The color line, moreover, defied common sense. The *United Mine Workers Journal* considered it "preposterous in the face of present day civilization" that a miner should decline to stand "on a plane of equality" with another simply "because that man's face is black." Race simply did not bear on one's character. On the contrary, it was those who raised "the evil spirit of race prejudice" who were "the most infernal," followed by "those who are so ignorant, so bigoted and peevish as to be moved by such empty, inane, foolish and false notions." Finally, it was incumbent upon white workers to recognize the capacity of African Americans for unionism. "A great many white men will say, 'what are you going to do with the nigger?' " the Corona miner wrote. "Let me tell you something. The best men we have today are negros." After a successful local strike in early 1894, he likened "our colored brothers" to a "stone wall," adding that "any man may well be proud to take a Corona darkey by the hand and acknowledge him to the world as a brother."[38]

White praise for the union loyalties of black miners was often blended with a patronizing habit (echoing the Knights) of discussing them as outsiders to be brought into the movement rather than as an intrinsic part of it. "Excelsor" lauded the UMW for "placing a strong reliance upon the true unionism, courage, and devotion and sacrifice for principle of the colored members when once they can see its benefits as seen by us." "Don't leave our colored Brother out in the cold for a bone of contention," he urged elsewhere. "Take him and share and share alike. Educate him on the line. Take care of him and he will do you good." And, as in earlier campaigns, the presumptions of white supremacy were invoked for rhetorical effect. That black miners at times organized before whites, the Corona unionist said, "brings a blush of shame to my cheeks." Or there was the resort to racially charged imagery, albeit with an ironic twist, in the *United Mine Workers Journal*'s denunciation of racial division: he who would promote the debilitating color line was "a fiend of the blackest hue."[39] Ultimately, then, the revival of miners' organization breathed new life into the tortured "race question." Around the corner lay the next great encounter between the miners and the operators, one that would clarify—and intensify—the tensions within the union's racial approach.

AS THE DEPRESSION persisted into 1894, hardship grew around the district. Few mines operated more than three days per week, some as few as one, while the operators continued their efforts to cut costs by reducing wages. Growing numbers of miners departed the state. In early spring the operators presented the miners' union with a proposal for a general reduction of 22½ percent. The miners agreed to a cut, but of no more than 10 percent, and then only if coupled with various concessions from the operators. At this the operators balked. Accusing them of seeking the "virtual enslavement of the miners and the turning of our nineteenth century civilization back into its barbaric past," the Alabama UMW declared a districtwide strike.[40]

The response was impressive. On April 14, 6,000 miners came out. Morale was bolstered a week later by the launching of a massive strike called by the national UMW in response to wage cuts ranging from 10 to 30 percent. The strike drew out 100,000 of the country's nearly 200,000 bituminous miners; in the coming weeks another 80,000 more had joined in. Although the Alabama strikers were not formally linked with the national strike, the latter gave their efforts a heightened sense of purpose. The breadth of solidarity in the Birmingham district was demonstrated on April 23 when 4,000–5,000 strikers, evenly composed of blacks and whites, marched through the city, carrying banners of the various coal towns. By the end of the month 9,000 miners were out.[41]

Henry F. DeBardeleben, now a vice president of TCI (which had acquired his company since the last strike), took charge of the operators' campaign with vintage flair. Extra convicts, along with new mining machinery, were brought in to replace the miners; strikers were evicted from company housing; furnace operators leased out whole mines to contractors; the pro-company press, most prominently the *Age-Herald*, embarked on ritual denunciation of the strikers as misguided and lawless.[42]

But the centerpiece of the operators' strategy was the mass importation of black strikebreakers. Pioneered in the strike of 1890–91, the tactic came as no surprise now. "The end might be the displacement of white men with negroes," the *Age-Herald* predicted as the strike began, and "our North Alabama mineral region may become the Black Belt."[43] Again, DeBardeleben chose Blue Creek as his first target. But this time he pursued his divisive paternalism with a boldness only hinted at three years before. Early in the strike he issued a public statement touting Blue Creek as a potential "Eden" for black workers. He described it in visionary tones: "This is a rare chance for all first-class colored miners to have a permanent home. They can have their own churches, schools and societies, and conduct their social affairs in a manner to suit themselves, and there need be no conflict between the races. This can be a colored man's colony." His offer was also a challenge. "Colored miners, come along; let us see whether you can have an Eden of your own or not. . . . You can . . . prove whether there is intelligence enough among colored people to manage their social and domestic affairs by themselves in such a way as to command respect of the people at large . . . without the aid or interference of the white race."[44]

His biblical solicitation gained instant notoriety among the strikers, who promptly dubbed its author the "serpent" in the Garden. Sarcastic references to his vaunted paradise filled the labor press. Thus, a portion of "The Maiden's Prayer," carried in the *Labor Advocate*:

What will become of Alabama, Pa,
If things go on this way?
There are hundreds of colored men
Scabbing every day

I hear there's a man in Birmingham,
Or not very far away,
Who's trying very hard to make
A negro Eden pay.

In a distressed economy, DeBardeleben had little trouble making his negro Eden pay. TCI officials recruited trainloads of black laborers, some from the plantation region and others from urban areas around the South. By late May the Blue Creek mines were at full production, while the Sloss company had adopted the DeBardeleben strategy at its mines. By June, TCI had turned its sights on the heart of miners' unionism, Pratt City.[45]

The coal fields bristled with resentment over the influx of black strike-breakers. "[O]ur enterprising city can now boast of a flourishing *slave trade* in our midst," one striker wrote. Pinkertons employed by Governor Jones reported mounting agitation among white strikers. They appeared "very bitter against the negro 'scabs'," detective "JHF" wrote the governor, "and that is about all their conversation." That white miners would ultimately be displaced by African Americans became a regular refrain—at times indignant, at times resigned—at saloons and street corners across the mining district. One striker confided to detective T. N. Vallens that the companies were (in Vallens's words) "throwing the white strikers out of their houses and replacing them with ne-groes, and when the strikers lost[,] many of the white men would never be allowed to go to work here again." But anger at the use of black strikebreakers was not confined to whites. "[T]he negro strikers are very bitter against them," Vallens observed, "and threaten them at every opportunity."[46]

Beyond weakening the strike's effect on output, the importation of black labor tested the cohesion of the miners themselves. Would the white miners' resentment of black strikebreakers lead them to turn on the black strikers? In such an atmosphere, would black miners—many of whom had first arrived as "blacklegs" in 1890-91—remain committed to the union? The stakes were high: Vallens crisply advised the governor that if the black miners returned to work, "the strike is broken." Union leaders, thinking likewise, sought to block efforts to divide the miners along racial lines. Once again the imperative of interracial unity found expression in the labor press. "Poverty is the odium here, no matter what the color," wrote J. H. Bergen of Coalburg. With explicit reference to DeBardeleben's recruitment of black strikebreakers, the *Labor Advocate* printed the formal policy of the AFL eschewing discrimination based on color. Union leaders reminded white miners that the abuse of black miners affected all. The *United Mine Workers Journal* detailed how the agents of "a certain coal company" had taken to lashing with "black snake" whips African Americans who refused to work. "This may not cause the blood to boil in the veins of our Caucasian brothers in the South . . . but . . . the blows of the 'black snake' fall indirectly on your shoulders; the shoulders of labor and the arm of plutocracy alone are involved in this shameful incident."[47]

It is difficult to know just how far such sentiments permeated the strike district. The Pinkerton reports, however, offer unique glimpses.[48] Certainly the inflow of black strikebreakers affected how white miners viewed African Americans. At Fox's saloon in Birmingham, "JHF" heard white strikers vow that they would "never trust a negro nor ever work with them again." Nonetheless, the Pinkertons detected a remarkable, if casual, camaraderie between black and white strikers. At the miners' saloons where they plied their trade so assiduously, the governor's moles would find miners of both races drinking, smoking, and discussing the latest developments together. In Pratt City, "JHF" encountered "15 or 16 strikers both negro and white and all about half drunk in Wilson's saloon." On another visit he came upon black and white strikers "all talking together," agreeing with stony determination that they would not return to work for 35 cents per ton.[49]

Signs that the bonds of unionism submerged the racial divide appeared more publicly as well. In late May an estimated 300 Pratt City strikers, black and white, descended on the mayor's office to protest the arrest of a black striker charged with shooting into the home of a black labor agent. A parade of 2,000 miners from Pratt City to Birmingham (followed by a day of speeches and brass band music) was comprised equally of blacks and whites. Public displays of interracialism were not confined to the miners themselves. In Blossburg, detective Vallens observed a "large procession" of wives and children, "both white and black." At the head was a banner reading, "No blackleg."[50]

The interracial spirit emerged in all its potency—and fluidity—at a miners' meeting on the eve of the strike at the all-black coal town of Johns. Company officials who came to address the meeting, reported "Pendragon" in the *Labor Advocate*, "were no doubt surprised to see it half painted black and the other half white, as the meeting was supposed to be for colored men only. But be that as it may, the whites remained. The men of [neighboring] Adger and Sumter, both white and colored, were well represented, and the men of Johns, who are all colored, appreciated the presence of their neighbors too much to ask them to leave. I may say there was quite a discussion as to whether the meeting should be a joint one or not. Anyhow, the colored men moved by a unanimous vote that it should, and that settled it." The event epitomized the textures of race relations in the strike district: the southern culture of segregation was not strong enough to prevent a joint meeting of black and white miners, but it *was* strong enough to compel "quite a discussion" among them. And yet to inject the issue was not to ensure a change of course; in this case, the airing of the race question concluded in a reaffirmation of interracial engagement at the initiative of the black miners, whose endorsement put the issue to rest. Equally revealing was the relish with which the episode was described. "Pendragon" could not resist an impish parting shot at the company officials: while they may not have been "altogether pleased" by the black miners' vote, the fact remained that the racially mixed group of miners from Adger and Sumter "had come to stay. . . . If we hurt the feelings of anyone by being so obstinate, we humbly beg pardon and promise on our word of honor it will not occur again—until the next time."[51]

Despite the interracial bonds reflected in such moments, black and white miners did not experience the strike in identical ways. A combination of past experience and present circumstances left African Americans with more to lose by striking. White strikers enjoyed broader employment alternatives than black strikers, both within and beyond the district. The latter were likely to have fewer material resources to sustain them. And the state treated blacks on the wrong side of the law with special harshness; indeed, the racially lopsided demography of the convict population suggested an extensive criminalization of blackness itself. In this setting, African American strikers must have viewed the deployment of troops by an unsympathetic governor with particular trepidation. Many black strikers felt intimidated by the prospects either of staying out or of returning to the mines. From Pratt City, Vallens reported that they were "divided as to going to work . . . those who want to work are afraid to do so or say so, but some of them declare they are starved out and must do something pretty soon."[52]

Throughout the conflict, however, black miners remained widely committed to the strike. "We, the colored miners of Alabama, are with our white brothers," read a sign at a mass rally. Vallens privately confirmed the impression: "The negro strikers are standing firmly with the white men," he wrote the governor in early June, though he took pains to explain their union loyalty in the only way he could make sense of it—"probably from fear." White and black local union councils functioned throughout the coal fields. Word came regularly that miners around the district, black and white, were "as firm as ever," "still solid," etc. "Pendragon" gleefully described the dismay of TCI officials when their all-black labor force at Johns joined the strike. The company had "always had the idea that because there was [sic] only colored men employed at Johns . . . they could do anything they like with them. . . . It has dawned upon their somewhat clouded vision that the colored men down here are just as wide awake to their own interests as the white men are." Union leaders were careful to distinguish between loyal black miners and strikebreakers, whom, they noted with equal care, were both black and white. "We have nothing to fear from the colored man in and around Pratt City," strike leader Hannigan wrote. "[T]hey are true and noble; it is the mean scrubs of white men that are our enemies." The union's executive committee brooded over the "employment of negro 'scab' labor to the exclusion of white and the better class of colored labor."[53]

The "better class of colored labor" received substantial respect within the union. Racial slurs were reserved for the black strikebreakers, described as "DeBard's coons" or "DeBard's pets." The black strikers were "solid," "true," and "noble"; DeBardeleben's "coons" and "pets" resembled the fabled Sambo—docile, ignorant, manipulable. Pratt City union leader W. J. Kelso noted that the black strikebreakers had been recruited and trained by white "gulls": "Now, can we blame the ignorant negro for working, when men that call themselves white lead them[?]" Ready to assign blame, however, a miner from Brookside breathlessly denounced newly arrived blacks as "a benighted, bestial, brutish and criminal horde who, in their besotted ignorance and voracious cupidity are a disgrace to the genus 'hog.' " In these ways the prevailing white caricature of African Americans was brought to sharp focus on black strikebreakers.[54]

At times the caricature was expanded to include white strikebreakers as well. The Brookside writer who considered black strikebreakers something less than "hogs" thought little more of the white guards, whom he designated "cattle." A white writer to the *United Mine Workers Journal* alluded to "white-faced, black-hearted leading scabs." With a similar twist on the standard terminology of racism, a black miner described having recently gone to Pratt City "and lo, and behold! all the black-legs working there were not 'niggers,' from the color of their skins at least. So-called white men of the lowest grade of humanity are working there under guard more humble and time serving than the lowest bred African slave before the war, and yet these black-leg white pimps have the audacity to call us 'niggers.' "[55] In such ways did strikers redirect the contemptuous language of white supremacy from skin color to behavior.

This painstaking distinction between black strikers and black "scabs" extended even into the most sensitive corner of Jim Crow thought—the relation between black men and white women. Union leader William Mailly described a "disgusting episode" one Saturday night at Sumter, in which a conversation between two white women, Mrs. George Davis and Mrs. Lewis Owens, in the home of the latter, was disrupted by the appearance of a black strikebreaker who had recently moved into the other side of the house. The intruder pointed a revolver into Mrs. Davis's face and "used a vile expression." The women screamed and the man fled. Most astonishing to Mailly, no efforts were made by the deputies to arrest the man. "It would be interesting to know what would have happened to the man if he had been a striker, instead of one of the 'gentlemen' with whom Mr. DeBardeleben intends to found his Eden." No genuine black miner, he claimed, would ever breach the sanctity of white womanhood in this way. "In all the past three years I have lived in Blue Creek I have yet to hear of one white woman being insulted by any of the colored men who have been working here and who are [now] on strike."[56]

With no small irony, themes of white supremacy were thus invoked to defend an interracial union. Cramped by a racial atmosphere that they could neither weather unchallenged nor reject outright, the strikers often turned to irony as their most viable rhetorical strategy. Much of the grist for this approach lay in the fact that many "blacklegs" were white, and many strikers black. Yet the large-scale displacement of whites by African Americans remained too large a trend to deny and too threatening to the white miners' social and material status to ignore. Inevitably, white strikers came to view the matter through the lens of race and to express their outrage in the language of white supremacy. And here there was no trace of irony working to recast or subvert its meaning. In an open letter to the "merchants and citizens of Birmingham," the union's executive committee pointed out that "most of us white miners" had come from else-where, at the invitation of the operators. "We accepted their offer, broke up our homes, came to Alabama with the intention of staying," built houses, joined churches and fraternal lodges, "and at the same time assisted to build the 'material development of the sunny south.'" And now their position was threat-ened by the dual curse of starvation and black displacement. "Gentlemen . . . can you stand to see your white brother driven from the state[?] . . . By saving us you save yourself; our interests are mutual." Referring to the present exodus of whites from the mining district, the *Labor Advocate* bewailed the loss of "our best citizens." It was one thing for whites to respect, and even collaborate with, blacks in common cause, quite another to accept the mass displacement of the former by the latter. "The colored people of this country have rights which should be conceded to them," the Warrior *Index* editorialized, "and they will be sure to get their rights as long as they behave themselves. But this country does not belong to the negro, and white people will not be driven out of this country." If DeBardeleben was so insensitive to this point, the paper continued, let him "go to some . . . country with negro supremacy."[57]

AS PRODUCTION resumed and prospects for the strike dimmed, the union showed no sign of giving in. In the mining communities, feelings intensified. One evening in early May a group of strikebreakers at Wylam, six white and thirty black, were greeted by two dozen women beating tin pans, shouting "scab" or "blackleg," and imploring the transgressors to quit. When one of the strikebreakers pleaded that he had a family to feed, he was subjected to a volley of sticks and stones. The first serious violence occurred on May 6, when 200 Horse Creek strikers marched on Price's Mines and dynamited the operations. Mass evictions at Coalburg, Blossburg, Brookside, and Blue Creek provoked heated battles. Rumors flew that the strikers were preparing to descend on the Pratt City stockades and dispatch the convicts from the district, in the manner of the Tennessee miners during 1891–92. Tensions were further inflamed by the introduction in Pratt City of black strikebreakers in mid-May. At Bud Hanley's saloon, the only talk detective "JHF" heard was about "so many negro 'scabs' going to work at Slope No. 5." Some of the more agitated patrons muttered that, as "JHF" conveyed it, "the mines ought to be blown up and some of the 'black-leg' negroes killed"—to the dismay, he added, of the "more conservative" strikers. Over the course of May 16, Chat Holman, a black labor agent for TCI, was fired upon, arrested, and jailed for carrying a concealed weapon, and threatened with lynching. Matters came to a head on the night of the 20th, when Walter Glover, a black strikebreaker, was shot to death through the front door of his home. Local sympathies were suggested when the two men charged with the murder, one white and one black, were acquitted despite strong evidence of guilt. Less indulgent, Governor Jones ordered Alabama's Second Regiment to nearby Ensley, where they soon conducted wholesale arrests of Pratt City strikers.[58]

From the outset, the strikers enjoyed considerable public support. The notorious convict lease system was an especially powerful source of anger: "everyone here sympathizes with the miners on that particular point," detective Vallens observed. On May 26, 2,500 citizens met in Birmingham to consider how to induce the operators to conclude a fair settlement. But the operators, expressing confidence that they could maintain production with their present workforce, declined to make any concessions.[59]

June brought little to improve the strike's prospects. As "blacklegs" continued to pour in, hardship mounted; at Fox's saloon in Birmingham, "JHF" heard miners talk of "starvation." Wholesale arrests by raw young troops armed with gatling guns stirred high indignation in the mining communities; military excursions here and there in response to false alarms only heightened the tensions. As strikers' desperation grew, railroad trestles became the targets of fires and dynamiting. By mid-June the national strike wound down to an ambiguous resolution, and the Alabama miners were fully on their own. Union leader Hannigan privately concluded that the strikers had "little chance to win."[60]

Early in July the Walker County miners returned to work. Strikers elsewhere felt a ripple of hope when the yard workers on the Louisville & Nashville and the Kansas City, Memphis & Birmingham lines joined the nationwide Pullman strike led by Eugene V. Debs's newly formed American Railroad Union. If the yardworkers could induce the trainmen to join the strike, scab coal would not move. Well aware of that possibility, the governor had troops clear Birmingham's Union Station of strikers and guard railroad property. The trains kept moving. In less than a week the railroad strike was over in Birmingham.[61]

Sensing an opportunity to end the conflict, TCI offered the Pratt miners 35 cents per ton, in cash, twice a month. The union held out for 45 cents. Tensions returned to a boiling point. In Pratt City, a striker delivered what one paper called an "anarchistic and incendiary" speech, in which he described the strikebreakers as "damned niggers" and asked if the miners "proposed to stand it any longer." The answer came on the 16th, when about one hundred strikers marched to the mines, armed with Winchester rifles, and began shooting at "blacklegs" as they emerged. "Strikers are killing my negroes at Slope No. 3," Superintendent Jones G. Moore wired the sheriff. Troops were duly dispatched. What became known as the "Pratt Massacre" left three black strikebreakers and one white company guard dead. Dozens of miners, thirty-four white and twenty-two black, were arrested for murder. At Lockhart Mines in Walker County, an explosion of a powder keg caused another fatality. More arrests followed and, despite vehement disclaimers by the union leadership, public sympathy swung sharply against the strikers.[62]

The strikers placed their final hopes in that year's gubernatorial election, to be held August 8. The race pitted the Jeffersonian Democrats, led once again by Reuben Kolb, against Jones's anointed successor, William C. Oates. Festering controversy over the legitimacy of the prior election, combined with dire hardship across the state, intensified Kolbite passions.[63] By early summer, the battle in the coal fields had become a central issue in the campaign; conversely, the elections began to appear pivotal to the strike. "The political significance of the strike and the bearing of politics on the strike," as the *Age-Herald* put it, "is more and more apparent daily."[64] With convicts working in the place of strikers and state troops patrolling the district, these connections were not hard to see. The Kolbites championed the strike and lambasted the use of convicts and troops at the expense of law-abiding citizens. They denounced the importation of black strikebreakers, which they felt resembled the political manipulation of African Americans practiced so artfully by the Bourbon Democrats. The Oates campaign, for its part, pointed to strike-related violence as a vindication of the governor's use of troops.

The elections became the talk of the district, as newly formed Kolb and Oates clubs held rallies in the mining towns. In this campaign Kolb drew the support of a clear majority of miners, who clung to the hope that a Populist would turn the tide in their struggle. Opponents of the strike lined up behind Oates. Perceiving this balance of loyalties, the Tennessee and Sloss companies offered in late June to pay the expenses of any miner who chose to leave the state—a transparent effort to diminish Kolbite support in the district. Union leaders, convinced that the miners and Kolb would rise or fall together, scrambled to persuade strikers to remain through the election. The results were mixed: some miners took up the operators' offer and left for other regions, Vallens informed the governor, "but very few of the strikers either white or black are going to work." Few miners at Corona, the *Age-Herald* reported, deny "that they will continue to hold out until the race between Kolb and Oates is settled. All the miners in this camp who are out on the strike are for Kolb." "Kolb politics," Pinkerton "JMP" found at Pratt City's White Elephant saloon, was "the main topic of conversation. . . . They seem to have completely forgotten the strike and think of nothing but politics." Many joined in a broad concern that a corrupt Democratic machine would find ways to deny a genuine Kolb victory. "If Kolb is elected we ought to put him in if we have to take up arms to do it," a striker told "JMP."[65]

The loyalties of African Americans, who made up over one-third of the registered voters of Jefferson County, were of prime concern to each side. In the key mining communities they figured even more prominently: of the registered voters of Warrior, Coalburg, Pratt City, Bessemer, Woodward, and Dolomite combined, over 45 percent were black. Both the strike and the election campaign brought to light sharp divisions among African Americans of the district. Urging black miners to back Oates was a small army of local black middle-class leaders. The Kolbites, they argued, were no friends of the black worker. On the contrary, claimed black Democrat E. E. Carlisle, they were infected with "negro haters." The miners' union fared little better in their estimation—its efforts to prevent blacks from working during the strike, a group of Democratic black ministers suggested, was an assault on their civil rights. The operators, meanwhile, had offered opportunity to African Americans who refused to strike. "The Tennessee Coal and Iron company," Carlisle said, "has already manifested its appreciation of the negro's loyalty by giving those at Blue Creek the better dwellings, church and school house, formerly occupied by white men, and this is but the beginning." And in this regard, he argued, blacks had been protected by a vigilant governor. For African Americans, these circumstances dictated a clear course of action: support the Oates ticket and abandon the strike. Statements like these were denounced by black Kolbites at rallies around the district, along with the labor press, which stressed the need for unity between black and white miners on the overlapping terrain of unionism and politics. There is little evidence, in fact, that the conservative preachings of black professional figures wielded much influence in the coal fields at this moment, or indeed during the late nineteenth century. (It was not until the 1910s, when the spread of corporate welfare programs provided it with a far more visible platform, that the black middle class found a significant voice in the mining district [see chapter 6].)[66]

On election day Oates narrowly carried Jefferson County, but all the major mining towns backed Kolb. Most black miners, rejecting the advice of black Democrats, showed a Populist leaning comparable to that of white miners. At places where black miners were a majority of the voters, such as Bessemer (57 percent) and Woodward/Dolomite (62 percent), Kolb carried the vote by substantial, even overwhelming, margins. This is particularly noteworthy in light of the Populists' ambivalent stance on the rights of African Americans. Long exasperated by Bourbon race-baiting and fraudulent use of black voters (dead or alive, registered or not) to provide their margins of victory, the leadership of Kolb's Jeffersonian Democrats called for the disfranchisement of African Americans altogether in 1894. Only when that gambit failed did the Alabama Populists compete with Oates for black votes.[67] It would therefore appear that the black miners' preference for Kolb reflected more what they felt they had to gain as miners than as African Americans. Whatever the motivations, these results suggested the degree to which common cause among the miners had during this election season submerged the racial divide, in political as well as collective labor action. But statewide, Oates prevailed. In the end, the elections had no more revived the fortunes of the miners' strike than had the railroad strike.

In mid-August, as the strike approached its fifth month, TCI and the union reached a settlement. The terms included 37½ cents per ton (to rise when iron prices rose), reductions in the prices of company housing and mining supplies, semimonthly pay, and a commitment by the company not to blacklist returning strikers. These concessions notwithstanding, the miners could hardly claim victory. The largest miners' walkout yet in the district had failed to keep down production, attain an adequate wage, or gain union recognition. Nor was the strikers' harmony left wholly intact. Sloss miners complained that the union had sold them out by signing only with TCI, leaving them without the same benefits and protections; TCI's Blue Creek miners, for their part, grumbled that the union had conceded too much.[68] Once again, a daunting range of obstacles— inexhaustible reserves of convicts and strikebreakers, a hostile governor, a depressed economy, an ambivalent public, and the varied agendas of the miners themselves—had conspired to thwart a districtwide strike.

"The Agony is Over," the *Labor Advocate* proclaimed; "Everything . . .

Looking as Bright as Can Be!" But such bravado could not answer the somber realities that awaited the miners after the strike. As the depression dragged on through the mid-1890s, employment remained sporadic; miners sometimes worked only one or two days a week. Some operators, most prominently the Sloss company, deliberately recruited excess labor, either to increase the pool of consumers at the company store, or to dilute the militancy of the "old-timers" with more docile "green-hands."[69] Wages were scarcely adequate to the miners' needs.[70] Reports mounted of coercion to patronize overpriced company stores.[71] Operators flexed their muscles at the workplace as well, broadening the scope of dead work, denying miners the right to checkweighmen, and docking their pay in arbitrary ways.[72] Union officials surveyed this deterioration of conditions in early 1895: "Since the late strike last summer certain Companies in Alabama have been subjecting their employees to treatment fitting only for slaves; by depriving them of the right to meet in public under penalty of discharge; by refusing them the right to have checkweighmen on the tipples and dumps; by docking cars of coal for small quantities of slate; by discharging men for not dealing in Company Stores and 'pluck-mes;' by forcing employees to sign agreements or 'iron-clads' under threats of discharge; by keeping up a systematic method of blacklisting . . . by lengthening the already long hours of work from 7 A.M. to 5 P.M. to 6 A.M. to 6 P.M.; and other outrages too numerous to mention."[73]

"Miners, we call you to arms!," the Alabama UMW announced as it launched a fresh organizing drive just two months following the strike.[74] The operators moved vigorously, however, to squelch a revival of unionism. At Adger, TCI denied miners the right to hold union meetings on its premises, compelling them to repair to the woods. At Corona, miners were warned that anyone calling for a mine committee would be discharged.[75] Amid this atmosphere, the miners' response was mixed. Large crowds greeted organizers at Warrior, Coal Valley, Stockton, and Cardiff. Perhaps most gratifying in light of the late strike was the warm reception extended organizers visiting Blue Creek. "Under the trees by the baseball grounds at Johns," President John G. Smith and Secretary-Treasurer William Mailly wrote, "with the November sun casting its rays through the leafless branches on white and colored alike as they grouped around, surely it was a sight long to be remembered." Elsewhere the response was less memorable. At Pratt City only around forty miners showed up; "We could hardly have believed that in such a large camp . . . so few men would turn out to hear their own interests discussed." Anemic turnout was reported at Coalburg and Blossburg as well. Organizers attributed much of the indifference to politics; the fall election, they suggested, had eclipsed the union as the miners' key sphere of action. While the union leaders may not enjoy the politician's gift of oratory, Smith and Mailly wrote, "what we preach is of more benefit to the miners of Alabama . . . than all the political issues of the present day. We could go on voting and voting and voting, but the result would never benefit us one iota if we do not organize."[76]

But beneath electoral distractions lay a deeper explanation—the demoralization that blanketed the district following the strike's defeat. "There are men who can take a great interest in religious matters, politics, sports and all the other many topics of the times," an Adger resident wrote, "but the instant the subject of labor organizations is brought up, a sudden chill seems to strike them, a stony silence follows, and the sudden remembrance of another engagement wafts them swiftly from the scene." Poststrike repression had fostered what Mailly called "cold-blooded indifference." The *Labor Advocate* described it as a "stupor."[77] Organizing efforts did keep the miners' union alive. In February 1895, for the first time in four years, the Alabama UMW elected to reaffiliate with the national UMW, generating a ripple of optimism that miners' unionism might revive.[78] The great majority of Alabama miners, however, remained unorganized. Ultimately, unionism neither blossomed nor vanished during the mid-1890s.

Relations between the miners and operators grew equally varied, a sign of diverging labor practices among the major companies. TCI adopted a relatively benign paternalism. To be sure, the company's miners at Blue Creek, Blocton, and Pratt City continued to air a full catalogue of grievances: low wages and infrequent pay, "pluck-me" stores and myriad deductions, irregular work and unfair dockage, suppression of organizing and arbitrary firing, subcontracted labor and the contracting out of mines. But their resentments were tempered by company concessions that surpassed those of other operators.[79] Three successive 2½-cent wage increases during summer 1895 (dictated by the rising price of iron) defused tensions further; enough, indeed, to worry organizers that the miners' taste for unionism might abate.[80] On a number of occasions TCI addressed miners' demands in a conciliatory manner. When a delegation representing Pratt Mines, Blocton, Blue Creek, and West Pratt waited on company officials to ask that the wages of company men and the price for mining narrow seams be raised, the officials assented.[81]

Other operators pursued a harder brand of paternalism. In the year following the strike, miners' complaints arose largely from the Corona Coal and Coke Company in Walker County and the Sloss company mines at Cardiff, Brookside, Coalburg, and Brazil.[82] During 1895, labor tensions at these places erupted in a series of battles. Two hundred miners at Corona struck that April in response to a wage cut. The strike continued for half a year, generating the now customary spectacle of evictions and "blackleg" labor. This time the Corona miners won, as the company restored the original wage level.[83]

But by 1895 it was Sloss that had become the chief symbol of employer intransigence in the coal fields.[84] In late July, 300 Sloss miners at Cardiff, Brookside, and Brazil struck after Superintendent T. C. Culverhouse first rejected their choice of checkweighman, then fired the committee that brought him news of an alternative selection. Only by walking out, Mailly declared, would Sloss appreciate that "their workmen are free-born citizens of this Republic and not servile slaves." The strike lasted scarcely a week, ending when the company agreed to reinstate the fired miners' committee and deal with future ones, to recognize the miners' chosen checkweighmen, to pay for dead work, to establish consistent rules for docking for dirty coal, to sharpen the miners' tools at reasonable intervals, and to implement the Pratt scale.[85] Within weeks, though, fresh allegations of mistreatment were in the air. Sloss officials, a Cardiff resident wrote, "scare those they can, dock those whom they have a grudge against, and tell them if they don't like it to lift their tools and go. Rules are ignored; weigh boss and mine boss make their own."[86]

A report from Culverhouse to Sloss president Thomas Seddon later that year illuminated the company's management strategy. Of central concern was the racial and ethnic composition of the workforce. African Americans made up 48 percent of the employees at Cardiff, Brazil, and Brookside; of the 52 percent who were white, most were foreign-born (Scotch or Slavic). The ideal mix, Culverhouse suggested, would be 70 percent black and 30 percent native white. Although black miners were notoriously transient—"as soon as they feel a touch of hunger they scatter to the farms"—in other ways they were especially suited to the company's needs. They exhibited greater dependence on the company in their daily living. African Americans at Sloss plowed 65 to 70 percent of their earnings back into the company store; "the balance they blow in on pay-days in Birmingham saloons, which is exactly what we want them to do as long as we can furnish them regular work." Black miners were also more likely to live in company housing. Whites, in contrast, spent only around 20 percent of their earnings at the company store (here Culverhouse distinguished between natives of southern Alabama, who were "good customers and conservative," and "our mountain natives" and the "foreign element," who "are down on the company store") and tended to avoid company housing. Drawing on the operators' ingrained racial assumptions, the superintendent referred to African Americans' intrinsic docility, which made them particularly preferable to "the foreign element." (The latter he found especially hard to handle as day laborers:

"all want to claim miners' wages as per Tennessee contract.") As for the specter of unionism, Culverhouse endorsed a proposal evidently circulating among the operators to form an "alliance with our negroes with the view of separating them and getting them to form their separate organization of labor unions." Such an endeavor might counter demagogic forces currently afoot; namely, the UMW, and "the infernal politician"—probably a reference to the Populists— "who will preach anarchy or anything to gain their votes." In the face of these ominous stirrings, "this movement with the negroes may break the matter." TCI's disturbing shift to a more accommodating labor philosophy made that company an unreliable ally. "On this negro movement keep your eye on the Tennessee company," Culverhouse warned. "I have very little faith in the officials of that company carrying out the spirit of that movement." Should the "movement" not materialize, "the southern farm style of controlling the negro is the proper way." Culverhouse felt no need to elaborate: presumably this meant expansive employer control over the black worker, spanning the realms of labor process and discipline, housing arrangements, consumer relations, social affairs, and so forth.[87]

However divergent the operators' labor strategies, certain grim realities enveloped the district, indeed, the country. As the depression extended through 1896 and into 1897, employment at the mines remained low and irregular. Weak demand for coal and iron presented operators with both the pressure and the opportunity to keep wages down. The Pratt scale was lowered in January 1896, raised back up in July, and continued to oscillate with the price of iron through the mid-1890s.[88] None of the major companies felt compelled to recognize the UMW, and the union retained no more than a skeletal presence. By 1896, reports from the coal fields had once again all but disappeared from the labor press.

That year brought the last serious Populist campaign. But after the defeats of the past five years, pessimism pervaded the coal fields. The miners' political disillusionment was captured in an exchange in the *United Mine Workers Journal* between William Mailly of Adger and "Rapparee" of Pratt Mines. Mailly, a young socialist from England who had served as the union's secretary-treasurer during the 1894 strike, had called for a labor party: "Karl Marx said 'Labor has nothing to lose but its chains,'" Mailly recalled, "and capital, through the two old parties, have [*sic*] fastened and bound those chains." "So Bro. Mailly . . . wants a third party," Rapparee sardonically replied. "Why not try a second party, Billy? There's only one party in Alabama yet."[89] When the coordinated movement of the People's and Republican parties met with decisive defeat that fall, miners' organization once again ebbed.[90] Over 1896 and into 1897 the UMW periodically resurfaced, at times to organize the miners as the expiration .

of contracts approached, at times simply to restore union membership and activity around the district. But the fruits of these efforts remained limited and transitory, the result of a depressed economy, territorial factionalism among union leaders, and the manifest power of the employers. A full-fledged revival of the Alabama UMW would have to await economic recovery and a resurgence of the American labor movement in 1898.

THE 1894 STRIKE had made race more potent than ever in shaping the perspectives and the fortunes of miners' unionism. The mass importation of black strikebreakers had been a critical, perhaps pivotal, part of the operators' arsenal. It had also altered the demography of the coal fields. As many African American "blacklegs" remained at the mines and numerous white strikers left the district, the black proportion of the labor force rose; before long, it was the largest. In some places, such as Johns, Sumter, and Dolomite, African Americans made up the overwhelming majority, if not all, of the miners.

As the presence of black miners grew, so did their militancy. An episode at TCI's Blue Creek mines illustrated the trend. In June 1895 black miners at Sumter and Johns held several meetings to consider action on such issues as dockage, firing, and the imposition upon the miners of a schoolteacher and a doctor not of their choosing. According to "A Colored Miner," General Manager George McCormack received the miners' bank committee cordially, adjusted the grievances to their satisfaction, and expressed his pleasure at seeing "his colored miners at Belle Sumter awakening to the sense of their duty in coming before him asking for justice in a legal manner." In the following month black miners at Johns tested TCI's cordiality more boldly. On the morning of July 12, "A Colored Friend" reported, one of the miners at Slope No. 4 complained to the blacksmith that his pick had not been properly sharpened. After "speaking very roughly," the blacksmith's helper raised an iron bar to strike the miner, who defended himself with his pick. At this point the bank boss ordered the disgruntled miner home, instructing him to remain there until sent for. A group of nearby miners "did not like the way their fellow-miner was treated," and called a meeting for that evening. Those attending appointed a committee to call on the superintendent to replace both the blacksmith and his helper. The next morning, the miners met at the slope to await a reply. When word came that their request had been denied, they refused to enter the mines. The superintendent ordered the recalcitrant miners to vacate the premises. Instead, they entered into a lengthy exchange with the superintendent, who finally offered a "favorable reply"—presumably a promise either to discharge or discipline the blacksmithing team. He further agreed to recognize a standing bank committee, asking only that the miners arrange future meetings at times when the mines were not running.[91] If the episode revealed TCI's comparatively benign management style, it also showed the degree of militancy it felt inclined, or compelled, to allow, and the way in which black miners took advantage of that opening.

At mines where management was less benevolent, African Americans found other outlets for resistance. A violent confrontation on July 30, 1895, between white law officers and black miners at Brookside provides a telling example. That morning at the stable, according to the Birmingham *News*, a young black driver named Charles Jenkins had approached Gus Harris, the mine boss, to request a light for his cigarette. Harris refused, prompting Jenkins to use some "improper language," which in turn riled the boss. There ensued "a little difficulty" between the two, in which stones were reportedly thrown. The black driver spent the remainder of the day doing "some loud talking." At six that evening, the two had a second encounter: Harris produced a pistol and a large whip and gave Jenkins "a sound thrashing." Returning home, Jenkins rounded up some black neighbors, who issued various threats and pronounced their readiness to "die and go to hell." Learning of the trouble, Superintendent Culverhouse determined that the malcontents must be expelled from the area. By nightfall, Deputy Sheriff Woods and two other special deputies were headed toward the black quarters to serve warrants for their removal. As they passed through a ditch, several black men emerged from the tall weeds, cried "halt," and began firing on the officers. The startled deputies returned fire, and in the battle that followed an estimated 500 shots were exchanged and "the whole camp was thrown into excitement." The volley left Woods dead and one of the deputies wounded in the leg. As the black combatants scattered to the woods and into houses, Culverhouse led a posse of white miners in armed pursuit. Two African American men were arrested: Jenkins, discovered under a bed, and George Hill, found running half a mile from Brookside, a Winchester in hand. While "sub-posses" combed the black community for other participants, Jenkins and Hill were brought to the company office, where ropes were secured around their necks, their arms were pinioned, "and the word to 'pull, boys' alone remained." The imminent lynchings were averted by Culverhouse, who ordered all hands away, and the prisoners were left under heavy guard, to be joined in coming days by nearly a dozen others. Little work was done at the Brookside mines the following day: "the whites are standing around in groups, as also are the blacks, discussing last night's riot." Although what they said is not known, there remained, in the opaque language of the *News*, "considerable feeling about Brookside" as rumors of another black "riot" at nearby Brazil swept the town. Mine production resumed the next day, but tensions persisted; "[I]n many places rifles and pistols can be seen," the paper noted, adding, "Negroes are not displaying any weapons."[92]

The growing numbers and militancy of African American miners shaped their relation to white miners and their role in the union. During the 1890–91 strike, a contributor to the *Alabama Sentinel* had warned that, should blacks ever become the majority of the miners, "you may rest assured that the continually-reducing wages that they receive, will find its way to the commissaries . . . and the negro dives . . . will grow and fester in our community."[93] The 1894 strikers had grappled with the implications of race for their prospects. The mixed ways in which the white strikers expressed their commitment to interracialism—ranging from defiant egalitarianism to bemused condescension to creative appeals to white supremacy—have been explored above. These varied approaches to the "race question" would all extend into the poststrike malaise, each taking on added intensity as the operators' use of race grew bolder and more ambitious and as African Americans became the majority around the coal fields.

During its years in the wilderness following 1894, the union retained its essentially interracial character. Most mining towns continued to have both black and white councils. African Americans still had a presence as convention delegates and districtwide officials. Black miners took the initiative in mobilizing at places where union structures had lapsed. Letters appeared regularly in the labor papers from "A Colored Friend" or "A Colored Miner." Where black miners outnumbered whites, they might predominate in the leadership; thus, when the miners of the Woodward Iron Company at Dolomite designated a committee in 1895 to request higher wages, six of the nine members chosen were African American.[94]

Interracial cooperation remained evident in localized strikes, even amid the use of black strikebreakers. During an 1895 Warrior strike, for example, the alliance between black and white miners held. Mailly praised the black miners for their "noble stand"; not one, he claimed, had faltered. The black mining towns of Johns and Sumter contributed handsomely to the Warrior relief fund. Equally revealing was the absence of racial division among the Brookside strikers that same year. The interracial spirit was especially noteworthy in this case, since the strike began in the immediate aftermath of the violent conflicts of July 30 between blacks and whites. "It was a race riot," the *News* had announced, "white men on one side and negroes on the other." But the meanings of race in the coal fields could never be so neatly cast; they were inevitably refracted through the volatile "labor question." Significantly, the *Labor Advocate* placed a different gloss on the episode. The Brookside "riot"—it did not take exception to the term—was the outcome not of an abstract strain between blacks and whites but rather of the company's gross mistreatment of its black workers. Those who had been seduced by deceptive labor agents had been "treated like very dogs. Their lives are miserable every day they live. They are brow-beaten, cursed, and driven around until their feelings are uncontrollable. Incidents like that of Tuesday night are the natural result."[95] The "race question" at the Sloss coal towns was quickly displaced, or absorbed, by the "labor question," as a strike for the right to checkweighmen and the reinstatement of the fired miners' committee began. The "race riot" of days earlier had no visible effect on the strikers' unity, arising in their public statements only as an illustration of the company's repressive ways. "It is not safe for a colored man to ask for a match to light his cigarette at Brookside in no strike," "Catch-On" wrote, "so, you may judge what it will be in time of a strike under the Sloss Iron and Steel Company's officials at Brookside."[96]

The union continued publicly to endorse interracialism in the years after 1894. When the Birmingham Trades Council voted in 1895 to admit unions composed of African Americans, the *Labor Advocate* applauded the move. For too long, it suggested, white workers had "assisted their enemy in its work by allowing a race prejudice to discolor their judgment." Evidence of good relations between black and white miners was avidly promoted as an example for all. "I am proud of the white and colored men," wrote a *Labor Advocate* agent after touring of the coal areas "I . . . find a very fraternal feeling between them, which will prove beneficial to both races, if they will only keep getting to know . . . 'that an injury to one is the concern of all.' "[97]

Inevitably, such commentary sounded a sharp dissent from the doctrine of white supremacy. But assailing that credo head-on was not the driving force behind them. Rather, union attacks on the color line were inspired and shaped by the purposes toward which they saw it being drawn; that is, by the divide-and-rule tactics of the operators. Whether in the arrangement of work or in the labor movement, the color line did not promote the supremacy of white miners so much as that of capital. UMW activists saw the racial dimension of labor relations being engineered in alarming ways. First there was the deliberate overcrowding of the coal fields with imported black labor. Then there was differential treatment by company officials of black and white miners. This could cut both ways. In some settings, black miners became the objects of especially emphatic forms of control. Elsewhere, they were treated with more paternal indulgence. In late 1894, John Jones wrote incredulously from Corona that a white who struck a black miner was "promptly discharged . . . for his nasty trick."[98]

Thus, during the dismal years after 1894, the old mix of responses among miners to the "race question" took on new life, new intensity, at times new forms. Union leaders admonished whites to cast aside self-defeating prejudice—the desire, wrote Coalburg organizer Bergen, "either to rule or ruin"—and take the lead in forging a movement of all miners. Anything but an interracial enterprise, he argued, would be doomed: blacks were now the majority of miners, and yet white initiative was critical to organization. "[L]et the white man do his part, and the negro can be taught what is right for him to do, but as the white miners hold back from organization, the harder their lot will be, as the negroes will never organize unless the whites do." Still, where black miners did lead the way in labor battles, whites were urged not to spurn them. When, for instance, African Americans spearheaded union efforts at Dolomite, "Stuck Fast" lamented that the white miners "are not going to be led by colored miners, even if they were trying to get an advance for mining coal like other places have." Given the new demography of the coal fields, Bergen added, any notion among the whites of going it alone made less sense than ever. "If it is to be a cut-throat game, and as the negro has the majority, they will likely have the game in the end."[99]

Beyond such injunctions to white miners, the union's response to the "race question" remained shaded in many tones. Whites still tended to regard black miners as objects to be uplifted, to be taught the meaning of unionism: "Let us educate the negro up to the point where he will understand that his labor is worth as much as a white man's and there need be no fear of the result," the *Labor Advocate* advised. Leaders often scolded white miners for showing less militancy than African Americans. "Give the darkey credit for having the grit to ask for his rights," "Stuck Fast" told the reluctant whites of Dolomite, "as some of the other color has not got this amount of grit."[100]

The union and the labor press continued to disclaim any aspirations for "social equality." "There need be no question of social equality enter into the minds of either the white or colored member of a labor organization," the *Labor Advocate* piously assured. "Workingmen are organized on business principles." But a unionist's reaction to capital's manipulation of race could occasionally spill over into the social realm, usually off-limits to even the most ardent interracialists. Thus, the sarcastic aside by "Ree Verba" of Brookside shortly after the 1895 Sloss strike: "The new pool room is near completed. One end will be for the colored gents and another for white folks. That's right, Billy, keep 'em apart." The social equality question might also be delicately finessed, as with the *Labor Advocate*'s enigmatic statement that, "[i]f once we become united for material equality, social equality will take care of itself."[101]

The union trafficked in the standard racist imagery of the day. The *Labor Ad-*

vocate carried reports of crimes by "niggers" and "very bad negroes" that were indistinguishable from those carried in the general white press. It found comic relief in so-called black dialect: the appearance of a company doctor at Brazil was reportedly greeted by shouts of "Bless de good Lor for dat." In more exasperated tones, unionists complained of ignorant black miners who, in the classic Sambo mold, were all too readily satisfied. Although white and black miners at Coalburg fared alike, Bergen wrote in late 1894, the latter were more easily pleased and thus had the advantage: "[A]s many live on from 5 cents to 15 cents per day and from two to ten men will live in one small room . . . these darkies are happy as it's a great blessing to be free." As African Americans flowed into the district during and following the 1894 strike, frustration over their "ignorance" was often aired. "[W]hen will the colored man begin to learn better than to be shipped around like cattle?" demanded a writer from Cardiff.[102] In many ways, then, white union miners and the labor press continued to draw upon the premises of white supremacy in coming to terms with the increasingly charged racial dynamics they encountered after the strike. But they were loath to share that perspective with operators bent on exploiting it. When "A Colored Miner" wrote of a mining accident at Sumter, the mine foreman fired off a tart reply to the miner's "flagrant misrepresentation," bringing pointed attention to the miner's race: "The correspondent who cowardly poses under the non-de-plume [*sic*] of a colored miner knows how to 'color' statements and in connection with the above has made assertions dark as he is himself black." Union official Mailly took equally pointed exception to such race-baiting as "entirely unnecessary." Mixed in with the union's portrayals of docile blacks were descriptions of the African American striker as the new "colored man" who is "beginning to get his eyes opened and to think for himself. . . . [who is] not going to be blinded by the enemies of labor much longer."[103]

White union miners expressed these contradictory views of African Americans by delineating not so much between black and white as between "genuine" miners and "scabs." During the 1895 Corona strike, "Ripper" wrote of "a lot of dirty, low-eating negroes" who had come to work; at Adger, by contrast, the *Labor Advocate* mourned the passing of Will Cantey, "a negro and one of the best men we had during the strike last summer. . . . a true man . . . trustworthy . . . [held in] high esteem." At times the two lines—the one distinguishing white from black, the other, true miner from the "low-eating"—were drawn alongside a third, yielding a more complex hierarchy of worthiness. Recalling the drastic wage reduction that TCI had proposed on the eve of the 1894 strike, "Spartacus" reflected that had it been implemented, the company "would have only employed the plantation darkies, for no white man, *nor even a practical colored miner*, could live and support his family on such miserly wages." At times the two lines could be oddly interwoven: just as white strikebreakers had earlier been called "white niggers," so the converse could apply to black union miners. Thus the striking praise offered a black miner of militant demeanor at Corona, by Mailly: "He is a black man, but underneath his ebony skin there beats a white and honest heart; would that we had more like him; would that some white men in Alabama realized and understood the principles among men as this simple, honest-hearted darkey does."[104]

The interracial impulse among white miners had been rooted in the importance of broad labor unity, along with a genuine streak of camaraderie forged in a shared class experience. It took on a new meaning, however, as white miners came to face large-scale displacement. Noting that work at Johns and Sumter— "the best work in the state"—had been reserved for African Americans in return for "loyalty to their masters," the *United Mine Workers Journal* argued that such an arrangement "does the white miners no good and the Negro but very little." A Cardiff miner raised the extraordinary suggestion that black miners who had been "colonized" at Sumter and Johns could show "their colors by saying they will share the places in these mines with white men"; nowhere, he added, had Alabama's white miners excluded blacks. "If united we stand, the color line should not be drawn on the white man at . . . Blue Creek, and if the colored miner wishes his welfare to be advanced he must share the opportunity to earn a living with his white brother."[105] The unstated reality was that the white miners had never enjoyed the power to exclude African Americans from the mines even had they wished to. The color line was not theirs to apply; in the hands of capital, it could be—was being—drawn against them.

Another potential strain between black and white miners arose from the different structures of opportunity open to them. If African Americans now claimed certain pockets of the coal fields as their own, they were also more heavily compressed at the lower strata of the labor force and had a narrower range of work available to them in other trades around the district. An exchange at the 1896 district wage scale convention suggested how these divisions could affect race relations in the miners' union. When the gathering moved to adopt a resolution abolishing the subcontracting of labor, the black miners objected so vehemently that the system was left intact. "Rapparee" explained the schism in the *United Mine Workers Journal*: when a white man is refused work at the mines, he has other options to explore; a black man denied work, on the other hand, "remains and hunts up some smart black negro driver who wants a laborer and loads coal for him at 75 cents or $1 a day till he can do better, which is a long time coming often." Sparser opportunity thus drove African Americans into greater reliance than whites on the detested subcontracting system; their larger numbers, nonetheless, gave them the power to quash initiatives within the union to abolish it. "The white man started the laborer system and they are now trying to let go of it," Rapparee reported, "but the negro will not let them."[106]

AS THE DEPRESSION pressed on into 1896 and 1897, then, collective action in the coal fields continued to flicker. It persisted in the face of sobering obstacles. The 1894 defeat, the dismal markets for coal and iron, the ongoing use of convicts, the deliberate overrecruitment of labor, the feeble state of organized labor nationwide—all conspired to weaken unionism in the mining district. But these circumstances did not fully snuff out the militancy and organization that had found root over the past decade. This tradition maintained a palpable if rickety presence in the form of District 20 (at times reaffiliating with the national UMW, at other times reemerging as the unaffiliated UMW of Alabama) and through the voices of the *Labor Advocate* and the *United Mine Workers Journal*. Collective action was kept alive through the mid-1890s by local victories here and there, together with a persisting, even growing sense of exploitation.

This fragile unionism remained mired in racial ambiguities now familiar in the coal fields: a desire to deflect the disruptive charge of "social equality"; an instinct to avoid mutually debilitating division between black and white; an inability among most whites to view African Americans as other than objects to be "elevated"; and a culture of egalitarianism deeply instilled in miners' unionism. These competing sentiments all took on greater intensity, and some new twists, during the mid-1890s. The large-scale importation of African Americans to break the 1894 strike made clear that the operators were more prepared than ever to play the race card. Nor did such recruitment cease with the strike's end: over the years that followed, the companies—above all Sloss—continued to recruit surplus labor, primarily black, not only to defeat or discourage strikes but also to prime the market for company housing and the commissary. As the labor force grew increasingly black in an increasingly Jim Crow region, union miners would have to confront the race question as never before. That test would come with the revival of the United Mine Workers at the turn of the century.

5 The United Mine Workers in the Age of Segregation

I N THE SUMMER OF 1905, the Jefferson County Board of Education encountered a peculiar demand. A protracted coal strike was underway across the district, and gradually hostility between striker and strikebreaker had spilled over into the public schools. "Youngsters on [the nonunion] side yell at the others to come and get a square meal," Superintendent I. W. McAdory reported, "while those on the other side yell 'scab.'" As a new school year approached, the board began receiving calls from striker and strikebreaker alike for separate schools for their children. McAdory was not receptive. Each mining town already had two public schools—one for white children, and one for black—and petitions had come from strikers, and strikebreakers, of *each race*. To accede to their demands would mean creating separate schools in each area for union whites, union blacks, nonunion whites, and nonunion blacks. "I can't see how it will be possible to establish four schools with the funds at our command," McAdory exclaimed, at one mining camp after another.[1]

For McAdory, the unwieldy proposition was a bureaucratic absurdity. But the miners' instinct to divide the schools along competing axes of race and union loyalty captures how powerfully both race and class shaped their identities. Certainly there was no suggestion from any quarter—black or white, striker or strikebreaker—that parents' relation to the union should *replace* color in determining where children went to school. Such a concept was scarcely conceivable in turn-of-the-century Alabama. Even had it arisen, the notion would have ignited the wrath of white opinion. Yet the miners inhabited more than just a Jim Crow region; they also inhabited a coal district, where workers' efforts to realize an independent livelihood had long clashed with the operators' drive for pliant labor, high productivity, and low costs. From Virginia Mines that same year, the *Mineral Belt Gazette* evoked the usual intensity of both race and class sentiment when it stressed the conspicuous absence of each following a bloody mine explosion: for just a moment, "[t]here was no question of union or nonunion; there was no question of white or colored."[2]

From the beginning, the miners of Alabama had struggled to reconcile the ongoing mission of labor unity with the hardening mandate of white supremacy. The years surrounding the turn of the century saw each of these pressures intensified, and the tension between them sharpened. On the one hand, this period witnessed a rejuvenation of unionism in the Birmingham district. As the depression of the 1890s receded, organization flourished across a range of local trades. Nowhere was the growth more dramatic than in the coal fields, as the United Mine Workers of Alabama shook off several years of dormancy, reaffiliated with the national UMW, and brought miners into an unprecedented era of contractual recognition. As in earlier organizing campaigns, the revived UMW functioned on an interracial basis. But if the late 1890s and early 1900s brought heightened prosperity and power to black and white miners, they also marked the coming of age of Jim Crow, as black men were stripped of the vote, mob violence against African Americans mounted, and segregation permeated virtually every aspect of southern society. From 1897 to 1904, the project of interracial unionism was at once bolstered by a dynamic economy and constrained by the region's worsening racial atmosphere. Starting in 1904, as the former turned unfavorable and the latter openly hostile, the union was forced into retreat, ending in total defeat in 1908. Yet in the very reasons for its ultimate demise lie clues as to how an interracial union had managed to function as long as it did into the era of Jim Crow.

THE SUMMER OF 1897 marked a low point for the Alabama miners. The nation's worst depression yet was in its fourth year, and demand for coal and iron continued to sag. During July a strike of 3,000 miners failed to stave off wage reductions to an abysmal minimum of 37½ cents per ton at the Tennessee and Sloss companies, the most powerful operators in the coal fields.[3]

For many, the defeat served as a prod to revitalize unionism. Hopes were raised that fall, when 150,000 miners in the midwestern Central Competitive Fields won a three-month strike, resulting in the national UMW's first interstate contract.[4] The unexpected triumph galvanized the American labor movement. The UMW looked southward, recognizing that the unorganized coal fields of central and southern Appalachia subverted the interests of miners everywhere. Both locally and within the national union, the time seemed ripe to revive organization in Alabama. On September 25, delegates from the key mining towns gathered in Birmingham to launch the new campaign. The convention voted to reaffiliate with the national union and elected two members, A. H. Gentry (white) and C. M. Coker (black), to organize locals around the district.[5]

In the months that followed, the union gradually reestablished itself across the district. Prospects were brightened by the expansion of coal and iron markets following the outbreak of war with Spain in early 1898. The organizing drive culminated in May, when delegates from the eight key mining centers—chiefly those of TCI and the Sloss company—convened in Birmingham to formally reestablish District 20 of the national UMW.[6] The shout "An advance in July!" greeted speeches by District 20's English-born president, William R. Fairley; African American vice president, S. P. Cheatham; and other officials as they toured the district. The advance materialized at the end of June, when the Tennessee and Sloss companies raised the base mining wage from 37½ cents to 40 cents per ton.[7]

Most notably, the companies signed these contracts with District 20, bringing union recognition for the first time to the coal fields. For the next six years the major operators negotiated annual contracts with the UMW and dealt regularly with the miners' pit committees and officials. In doing so, they followed the lead of their counterparts to the north, who, amid the renewed prosperity, were pursuing stable relations with "conservative, responsible" unions. These years indeed saw a startling increase in union recognition throughout America;

between 1897 and 1904, American Federation of Labor membership swelled from 447,000 to over 2 million, the national UMW's from 9,700 to 251,000.[8]

Recognition both reflected and bolstered the miners' power. In the months after the first union contract, organizers continued to build locals around the district. At the time of its founding in May 1898, District 20 claimed over 1,000 members; by the end of the year it had fifteen locals with a total of 2,300 members; a year later, the figures had risen to thirty-six locals and 6,000 members; in 1900, fifty-one locals and 6,700 members; in 1901, sixty-one locals and 8,000 members; in 1902, seventy-three locals and 11,000 members; in 1903, ninety-five locals and 14,000 members. At the hub of District 20 was Pratt City's Local 664, which by 1900 numbered nearly 1,500, making it reputedly the largest UMW local in the country.[9]

This growth was part of a broader resurgence of unionism in and around Birmingham. By 1899 the district had 10,000 union members, including iron and steel workers, coke workers, iron ore miners, printers, machinists, blacksmiths, railway workers, clerks, carpenters, bricklayers, tailors, stonecutters, cigar makers, team drivers, powdermakers, and musicians, as well as coal miners. At the turn of the century, District 20 had links with a revitalized Birmingham Trades Council and two newly established statewide organizations, the Alabama State Federation of Labor and the Union Labor League. A branch of the Socialist Party was active within and beyond the coal fields, and periodic visits by Eugene Debs drew large crowds.[10]

District 20 pressed the familiar range of issues regarding working and living conditions. Even at its height, some of the union's goals—the eight-hour day, abolition of the convict lease, organization of the entire district—would prove elusive. But in many ways District 20 made impressive gains. The pay scale rose from a minimum of 37½ cents per ton in 1897 to upward of 55 cents two years later. In 1899 the union won an across-the-board 25 percent increase for the day laborers, cementing broad union loyalty from that group.[11] Under union contracts the weighing and assessing of coal was systematized and was monitored by miners' checkweighmen.[12] So too was the disciplining of miners, as the union imposed a check upon the company's power to fire or dock its labor.[13] Pressures to patronize the company store were widely curtailed.[14] Corrupt practices in the distribution of cars and rooms underground were also scaled back.[15] Union contracts ratified a custom under which miners were entitled to take temporary leave from their work. (A miner could remain away for up to three days without losing his job, or even having his room reassigned.) Upon the death of a miner at work, production was suspended from that moment until his funeral; likewise, work would cease for the funeral of a miner's wife.[16] Finally, labor subcontracting was eliminated at large operations around the district, affirming the union principle that, as Fairley stated, "every man should have an equal share of the wealth that he produces." At unionized mines where subcontracting was found to persist, a miners' committee would order it discontinued in favor of a more egalitarian "buddie" arrangement.[17] Dramatically, then, the UMW enhanced the miners' livelihood and sense of independence. One miner made the point by describing the dismal alternative. "If you work at a non-union mine, you are required to disclaim any connection with any labor organization; you must work for wages and under conditions fixed by the company. In fact, you must do anything that the 'boss' requires of you. He can dock you, he can fine you, he can lay you off, or he can fire you out, and you have no appeal. . . . It is union and liberty, or it is individualism and slavery."[18]

The years surrounding the turn of the century were, by and large, a time of accord between District 20 and the operators. "Our relations have been very pleasant," one union miner observed in 1903. "We have both learned to get along better."[19] Labor conflict, however, had not vanished from the coal fields, even at places governed by UMW contracts. Miners might still be penalized for union activities. When the Republic company resolved to rid itself of James H. Orange, an "agitator" on the local pit committee, it took to checking his coal several times per day for slate.[20] Republic's displeasure with Orange had its roots in an episode in 1901, when he rallied the miners to walk out over the company's attempt to abandon its obligation to lay the track in the slope. The miners insisted that unwritten customs were implicitly embedded in the contract; a company official countered, "[i]f it was a custom . . . we had a right to stop it." Such skirmishes punctuated the era of recognition, over a variety of demands: a wage increase for day laborers, the eradication of subcontracting, the reversal of a contested firing, the elimination of deductions for schools and doctors. In 1902, over 4,000 TCI miners walked off for the right to have members' donations for striking anthracite miners in Pennsylvania deducted by the company.[21]

In a number of ways, large and small, the local union became a part of the social atmosphere of the mining towns. "We have the I.O.O.F. [Odd Fellows], the K. of P. [Knights of Pythias], the Red Men and the Masons, and the U.M.W. of A.," reported a resident of Brookwood. At Republic, the Odd Fellows met regularly at the local UMW hall. When Cardiff won the Alabama championship football cup in 1898, District 20 president Fairley and national UMW executive board member Fred Dilcher were on hand to address the celebration. At Pratt

City, the traditional monthly payday "holiday" became the occasion for union gatherings. District 20 locals served as soliciting agents for the miners' hospital in Birmingham.[22]

AS WITH THE Greenbackers, the Knights of Labor, and the first incarnation of the UMW in Alabama, the meaning of District 20 could never be confined wholly to the miners' conditions and relations with capital. Reemerging at a time when both the spirit and the letter of Jim Crow were spreading, miners' unionism confronted the race question as never before. Here the UMW faced a familiar dilemma, merely sharpened by the growing presence of black miners and mounting hostility to racial equality.[23]

Segregationism hardly bypassed Birmingham. The thoroughness of its codification was illustrated by an 1899 city council ordinance that barred whites and blacks from playing dominoes, cards, dice, pool, or billiards together.[24] Jim Crow cast its shadow across the coal fields as well. "[W]hile white and colored miners worked in the same mines, and maybe in adjoining rooms," black UMW national organizer Richard Davis reported from the district, "they will not ride even on a work-train with their dirty mining clothes together. . . . You may even go [to] the postoffice at Pratt City, and the white man and the colored man cannot get his mail from the same window. Oh, no, the line is drawn."[25]

Given the racial composition of the miners, union adherence to black exclusion would bring certain defeat. But just as fatal would be a perception, on the part of the public or the miners (especially whites) themselves, that the union had become a vehicle for "social equality" between the races. To survive, it had to find some way to resist the divisive pressures of Jim Crow without perishing in a frontal assault on that cherished "way of life."

As a pioneering essay by Paul B. Worthman revealed, District 20 encompassed both black and white miners.[26] Scattered evidence suggests that black miners joined the union in rough proportion to their presence in coal fields.[27] The joint involvement of blacks and whites flowed naturally from their common lot as miners. But shared experience did not on its own ensure interracial empathy. The social distance and wariness between black and white miners, the superior structural position that whites on balance enjoyed at the mines, the gathering white intolerance of interracial association (along with black advancement)—all these realities tested prospects for collaboration between the races. More than ever, the forging of a union that straddled the ubiquitous color line had to be an ongoing project. From the outset, District 20 dispatched both black and white organizers to tour the coal fields. UMW gatherings often brought together black and white miners, who were typically addressed by speakers of each race. District 20's officialdom was racially mixed as well—although, as discussed below, within significant limits. By custom, the vice president was always black, and African Americans had a presence on all major committees.[28] Black miners were also visible among the delegates to district and national conventions.[29]

As in the early 1890s, union leaders stressed how racial division played into the operators' hands. At a Walker County rally in 1900, Henry C. West called on black and white miners to "shake hands over the pick and shovel or these companies will have us just where they want us and both the white man and negro will toil for their victuals and their clothes and not much of that." Such appeals were often trimmed to the respective interests of black and white miners. West advised black miners to ignore company efforts to turn them against "poor whites." "You have often heard the expression 'I'd rather be a negro than a poor white man.' Now, I'll tell you, the poor white man and the negro stands [sic] in the same column when it comes to earning bread." Union appeals to black miners were sometimes cast in the language of racial uplift. Black District 20 vice president Benjamin L. Greer argued in 1904 that by sticking with the union, blacks would enhance their self-respect, not to mention their standing in the eyes of whites. Thus he grafted the tone of Booker T. Washington onto a pro-union clarion call that would have mortified the Wizard of Tuskegee. Whites were urged to accept African Americans in the union, if only for self-protection. The coal companies, West pointed out, advised "ignorant whites" to avoid the UMW " 'because they take negroes in the order,' and in every possible way they will take advantage of the ignorant negroes and induce them to scab."[30]

District 20 went beyond rhetoric in contesting the color line. On its insistence, annual contracts decreed (in the words of the 1901 contract) that "[n]o discrimination is to be made in the distribution of work against the colored miners, but all competent colored men are to have an equal chance at all work." The UMW challenged Jim Crow practices within other unions as well. When in 1900 the Birmingham Trades Council (BTC) called on District 20 to shun establishments that failed to use union labor, black delegates to the miners' convention raised an objection: did not constituent unions of the BTC, such as the bricklayers and the carpenters, exclude African Americans, and did not the BTC itself bar them from its meetings? Black vice president Silas Brooks, the *News* reported, "made a very strong speech on this line." The district president responded by appointing a committee—composed of Brooks and two other black delegates, national organizer Fairley, and national UMW president John Mitchell (in town for the convention)—to investigate the BTC's racial policies.

Soon afterward, District 20 petitioned the BTC to "open wide their doors to colored organized labor of the district." Within a few days, the BTC repealed the color bar from its constitution.[31]

With equal effect, the UMW at times confronted Jim Crow outside the labor movement. When the merchants who owned the Birmingham hall where District 20 customarily met objected to the presence of African Americans at its 1901 convention, the UMW took umbrage. "The Negro could not be eliminated," national organizer William Kirkpatrick told the merchants. "He is a member of our organization and when we are told that we can not use the hall because of this fact then we are insulted as an organization." The convention voted to move future meetings to Bessemer and to advise locals to sever trade with Birmingham until the merchants apologized. Before long, the merchants did. To gauge how sharply the union's stand tacked against the broader regional currents, consider the actions of another, contemporaneous convention: just weeks before, the state of Alabama had ratified its revised, disfranchisement constitution.[32]

While such acts clearly flowed from the practical imperatives of miners' unity, they reflected a growing camaraderie between black and white members as well. This spirit was manifest during the 1900 joint scale convention, as union representatives killed time awaiting a delegation of operators. A white member, E. S. Scott, rose to sing "The Honest Workingman," and a number of delegates joined in the chorus:

> It's a glorious union,
> Deny it who can,
> That defends the rights
> Of a workingman.

At the conclusion of Scott's performance, Fairley called for a song from "one of the colored brethren." Charley Farley obliged, ascending the rostrum to sing "We Are Marching to Canaan." "It had the old-time camp feeling lilt," the *Labor Advocate* reported, "and the colored delegates crooned the lines and broke into the chorus":

> Who is there among us
> The true and the tried
> Who'll stand by his fellows
> Who's on the Lord's side?

White delegate Jack Orr followed with a well-received rendition of "Silver Bells of Memory." The convention then turned its sights on Fairley himself, issuing "a storm of calls" for a tune from their English-born leader. Fairley put on a show of hesitation but, egged on by the black singer Farley, finally relented and launched into "Give Me Back My Heart." After white delegate Richard Hooper followed with a "serio-comic recitation" of "The Old Pack of Cards," calls arose for a number by Silas Brooks, the black vice president. He vigorously declined, as did Ed Flynn, the white chairman of the wage scale committee. Black delegate J. T. Allen filled the void with a "powerful" rendering of "I Am a Child of the King." "Just as the chorus died away," the *Labor Advocate* concluded, "the operators entered the hall."[33]

Much is captured in this apparently frivolous moment. The very frivolity suggests an atmosphere of interracial fellowship spilling beyond the purely expedient. Shared purpose inevitably fostered shared identity. But the episode also underscores how alive racial identities remained. The delegates by no means ignored the backgrounds of those they called on to sing; indeed, they made a point of alternating between black and white. The singers themselves drew songs largely from their distinctive cultures, relying upon those of their own race to sing along. District 20 was very much the *joint* endeavor of black and white unionists, with all the commonality and difference implied in the term.

This camaraderie came through in the labor press as well. Black individuals and associations friendly to the union were described with a respect that appeared at least as much earnest as calculated. "The colored Odd Fellows paraded around our camp in uniform," a miner reported from Cardiff. "They made a splendid showing. They are a very intelligent body, progressive and ardent supporters of unions pertaining to their interests." In 1902 the *Labor Advocate* offered an admiring life-sketch of Alex Bennett, a black union miner turned saloonkeeper. Having begun life as a porter in Mobile, Bennett had come to the coal fields after he was denied a demand for higher pay. Before long he was an avid unionist, joining first the Knights of Labor and then the UMW. In the latter, he became "a prominent leader among the colored brothers," serving as vice president for the racially mixed local at Warner Mines and as a delegate to the 1901 district convention, where "he gave good results." Bennett was a "hustler from the word go in whatever he undertakes," the *Labor Advocate* observed. It encouraged readers to patronize his new venture, the Economy Saloon, where customers would find a piano, a pool table, "cool beer on tap," and a variety of whiskies, including the Union Label brand. (Interestingly, the plug for Bennett's establishment was not directed explicitly to black readers. Perhaps it was simply assumed that only African Americans would consider going there; perhaps, on the other hand, the custom of interracial fraternizing at miners' saloons, glimpsed by the Pinkertons in 1894, persisted into the new century.)[34]

For the operators, recognizing the union meant coming to terms with racial unorthodoxy. Many did so, to a degree, although some refused to meet with black officials. Where operators and local authorities opposed the UMW, the union's interracial practices courted staunch resistance. White and black organizers traveled together at their peril. In 1900, white national organizer Kirkpatrick and black vice president Brooks were greeted at Newcastle by what the *Labor Advocate* called a "rough reception," including "stones hurled at them, pistols fired promiscuously, and other indignities offered."[35] When Joe Hallier, a white national organizer, and Benjamin Greer, the district's black vice president, ventured into Walker County in 1903, they were accosted at the Horse Creek depot by a group of white men. After beating the two, the men forced Hallier to kiss his black associate, and then, the *United Mine Workers Journal* conveyed with dramatic opaqueness, "commanded him to perform other indignities upon the person of Greer that nothing but the most degraded brute could conceive, and can not be laid before the public."[36]

However striking its racial policies may have been, District 20 was in its own ways constrained, and influenced, by the wider racial order. The conflicting imperatives of labor unity and white supremacy yielded an interracialism that was highly qualified, often cramped, at times sharply circumscribed. Never fully submerging race, the union forged what might best be called a collaboration between the races, at once accommodating and testing the boundaries of segregation.

The union's structure reflected its ambiguous approach to race. While several locals were racially mixed, the typical mining camp had one for each race. How the latter arrangement became the norm is difficult to trace. It cannot be presumed that it represented an exclusionary impulse on the part of white members, imposed upon their black associates; indeed, at times the initiative for separate locals evidently lay with African Americans. In Altoona, where 400 members were equally divided between black and white, black miners elected in 1902 to withdraw from the mixed local and establish their own. From Warner Mines came word in 1901 that since it had withdrawn from the mixed local, "the colored local U.M.W. of A. here is getting along 'tip-top.'" The biracial arrangement inspired virtually no visible debate among miners of either race, so naturally did it reproduce the separation that marked all other areas of their social world (see chapter 2). One exception to this silence appeared in a 1901 letter to the *United Mine Workers Journal* from "Joshua," of Warner Mines, lamenting the division of a mixed local into racially separate ones: "[B]y separating the locals in that way the union in Alabama will lose ground." Joshua's race is not evident, nor precisely is the reason he felt that a biracial structure would undermine the union. Perhaps he feared that it would encourage company manipulation of the color line or that blacks would suffer for a lack of "guidance" from whites or that a proliferation of smaller locals, regardless of race, would weaken the union. What remains most notable, however, is the dearth of reservations on record about the biracial system. Far more attention—generally from opponents of unionism—flowed toward the joint presence of black and white miners in a common organization on *any* terms.[37]

However cordial in tone, however much rooted in the broader patterns of the miners' lives, the color line remained clearly drawn in their union. Black and white members adjourned from common district conventions to attend separate banquets; they marched in the same Labor Day parade but retired to separate parks. Nor, as noted above, did the union repudiate racial hierarchy in its own leadership. District 20 positions, filled annually by election, were distributed between the races according to a carefully constructed formula. The presidency was reserved for white members; the vice presidency, for blacks; the secretary-treasurer's position, for whites. The various committees usually comprised a modest majority of whites. District convention delegations were likewise apportioned by racial quotas.[38]

If the union's interracialism was motivated largely by pragmatic concerns, so too was the determination to contain it within a clear racial hierarchy. When in 1904 a convention delegate called for the popular election of officers (rather than indirect election by delegates), the district president quashed the notion. Direct vote, he cautioned, might lead to the election of a black president, which could only mean the destruction of the union. A black delegate rose to concur, pledging that members of his race would oppose the popular vote, not despite but *because* African Americans would likely capture all the offices in District 20 and thus doom the union in a Jim Crow environment.[39]

The limits of interracialism remained evident in the union press as well. Sprinkled amid frequent praise of black union miners were the derogatory images of African Americans all too familiar in the Jim Crow South. Even sympathetic depictions might be couched in the language of white supremacy. A white member from Brookwood reported the "amusing zeal" with which the black miners responded to the union. The self-styled "Blue Kid" assured readers of the *Labor Advocate* that the predominantly black miners of Blue Creek were "white inwardly" when it came to union loyalty.[40]

At other points white superiority found expression in less benign forms, untempered by even patronizing camaraderie. Borrowing the style of the mainstream press, the labor papers ran hysterical news stories about "brutal" black criminals, reported on minstrel shows at the mining camps, and depicted blacks as eminently manipulable ("colored dupes," as a writer to the *United Mine Workers Journal* described black miners willing to bribe the mine boss for access to coal cars). Union papers at times adopted the degrading lingo of Jim Crow, breezily referring to "coons," "darkies," and "niggers."[41] As in earlier times, such depictions were generally reserved for African Americans who were hostile, or indifferent, to unionism; they were seldom applied to union blacks. Race was not the only lens through which white miners regarded black miners. It was, however, always there.

In one area—political rights—miners' interracialism at the turn of the century was more narrowly constructed than before. However gingerly, previous miners' campaigns had engaged the issue of white supremacy in the political arena, characterizing the Democrats' racially charged positions as part of an effort to distract the public and divide labor. As recently as 1894, the miners had regarded a gubernatorial election as vital to their cause. But by the end of the decade, even as unionism made unprecedented strides, the UMW had disengaged substantially from politics. The withdrawal was not total. District 20 continued to endorse candidates for state mine inspector, press for improvement and better enforcement of mine safety legislation, and advocate laws requiring semimonthly pay and abolishing convict labor and the commissary system.[42] But in the activities of its organizers, the focus of its newspapers, and the content of official statements and published correspondence from miners, local and state politics had largely receded from view.

This retreat was all the more notable given the debates convulsing Alabama politics at the turn of the century and culminating in the revised constitution in 1901. The new constitution, which effectively disfranchised African Americans and many whites, sealed the defeat of the biracial Populist coalition, in which the miners had played no small role. Moreover, what modest attention the proposed revisions did receive in labor circles was scarcely characterized by alarm. Despite scattered expressions of support for black voting rights, the *Labor Advocate* ultimately urged readers to support the new constitution and sought to disabuse them of the "bugaboo that white men will be disfranchised." At times the labor press even carried syndicated pieces endorsing black disfranchisement.[43] Understanding the absence of alarm or even public discussion among the miners in the face of this conservative coup requires some conjecture. It may have reflected, in varying measures, demoralization following the defeats of the 1890s, a feeling that improved conditions made access to politics less vital, and a sense that the disfranchisement of blacks was unstoppable and, in the eyes of some, wholly reasonable. Whatever the explanation, District 20's resistance to racial exclusion in hiring and in its own organizational structure evaporated in the realm of politics, where the color line appeared either less threatening or less offensive or more unassailable—or some blend of the three, a blend that surely differed for black and white miners.

On another level, the UMW drew the boundaries of interracialism with resolute clarity, that is, in its ritual denials of a taste for "social equality." Here was a specter, at once vague and volatile, on which no respectable southerner could differ—even established black leaders took pains to distance themselves from that unthinkable outcome. District 20 bristled at any insinuation that it promoted social equality, and sought strenuously, even reflexively, to deflect the charge. Recognizing its devastating potential, black officials joined in the disclaimers: "The negro," Vice President Brooks assured an audience in 1901, "is not asking for social equality. We do not ask for recognition in your homes. We only ask for a chance to work and earn an honest dollar."[44] There is no evidence that any vision of social equality, whatever that might have meant, animated unionists of either race. To have advanced such a position would have placed them vastly beyond the boundaries of legitimate discourse in turn-of-the-century Alabama. Their frequent efforts to disavow the poisonous idea testify in part to their own real feelings and in part to the narrow breathing room available to any interracial movement in the Jim Crow era.

Indeed, as with earlier miners' campaigns, an indissoluble mix of self-preservation and genuine solidarity gave shape to every dimension of District 20's racial practices—from the remarkable spirit of interracialism at one end to the self-imposed limits upon that spirit at the other. Both deliberately and instinctively, District 20 negotiated these cross-currents in ambiguous fashions.

Both the extent and the limits of interracialism came to light at Blocton, where the meeting hall of the white local burned down in 1901. To assess the situation, the white local and the black local held a joint meeting in the latter's hall. There, one participant reported, the black miners offered "enthusiastic" support for the financing of a new hall for the white local: "Brother Starr, a Hercules in size and oratorical power, said that the colored miners would be as proud of an appropriate hall for the white miners as they would for themselves."[45] A magnanimous gesture of solidarity across the color line in support of a racially defined local—here was an approach that defies sharp dichotomies of race and class consciousness.

The instinct at once to confront and to finesse the matter of race was captured in a conversation between a union miner (A) and a nonunion miner (B), carried in the *Labor Advocate* under the heading "An Adger Correspondent

Reports a Talk on the Negro Question." "What are your objections to the United Mine Workers?" A asked B. At the top of B's list was the specter of racial equality, and proximity, at the mines:

> B: Well, one objection is that it allows the negro an equal showing with the white man. For instance a union man might go to the mine some morning and have no "buddy" to work with him. "You will have to work with Jim Smith (negro) today," and he would have to go.

Significantly, A responded that B's fears were unwarranted—not because the working together of black and white miners as "buddies" would be *acceptable*, but rather because it was *unlikely*, and thus nothing to worry about:

> A: You never knew of a case like that, did you?

The nonunion miner had only a lame reply:

> B: No, but it might happen.

Expanding his point, A noted that nonunion mines did not protect whites from having to work on equal terms with blacks; on the contrary, they routinely worked together as laborers:

> A: Did you ever work in a room with a negro?
> B: Oh, yes, many a time.
> A: How was that?
> B: Well, you see it was at an unorganized mine and the negro was my laborer and sometimes I would have two laborers—a negro and a white man.
> A: And these laborers were upon a perfect equality so far as their work was concerned, I suppose?
> B: Oh, yes, I paid them the same wages and expected the same work from them.
> A: Did you ever work as laborer yourself?
> B: Yes, I used to.
> A: Ever work as a laborer in a room or heading with a negro?
> B: Yes, sir.
> A: How much per day did you get?
> B: I got a dollar an' quarter.
> A: And how much did the negro get?
> B: He got the same as I did.

Having extracted a concession that blacks and whites toiled together on equal terms in nonunion mines, the union miner moved to clinch his argument:

A: Well, then, what advantage has the non-union man over the union man on the negro question?

Miner A may have meant the question rhetorically, but B had a reply:

B: Oh, you see, I am a contractor now, and if I work with a negro, he is my laborer and I am his boss.

So the subcontracting system that prevailed at nonunion mines granted white miners power over black laborers; *that* was the advantage to the "non-union man." In response, the union miner did not challenge B's aspiration to white supremacy, preferring merely to remind him that the racial hierarchy that B cherished was not really so pervasive at nonunion mines:

A: But are all white men contractors?
B: No, the majority of them are laborers.
A: And are all the negroes laborers?
B: No, a good many of them are contractors.
A: Do any of the negro contractors have white men for laborers?
B: No, it hasn't come to that yet, but God knows how soon it's going to.[46]

Whether real or contrived, what stands out about the exchange is its appearance in the labor press. In this case, the union chose to contest not the moral premise of white supremacy, but rather the notion that white miners could advance that vision by rejecting the union.

The self-consciousness, and real ambivalence, with which many white UMW members discussed black miners surfaced with revealing subtlety in the testimony of District 20 member Adam Pow before an arbitration board in 1903. Walker Percy, a lawyer for the operators, was seeking to establish that black miners were less reliable, more prone to absenteeism—a familiar theme among southern employers. "Is it not a fact," he asked, "that a large number of negro miners and less thrifty miners are always looking for days to stay out?" Pow squirmed, apparently uneasy about the racial wedge implied in the question. "A small percentage," he allowed, hastening to add, "I would not say a large percentage." "A noticeable percentage?" Percy pressed. "Yes, sir." "A larger percentage than if they were all Scotch miners?" Pow decided he had conceded enough: "I could not tell that." Percy, confident he was merely stating the obvious, persisted: "When you have a considerable percentage of negro miners . . . have you not got to provide a surplus of labor to cover . . . the negroes who will take a considerable number of days off?" Pow, still reluctant to be drawn onto this racially divisive terrain: "I could not say whether they do or not." He then tried to deflect the question with something of a non sequitur: "I suppose if there were not negroes there would be white men, and if there were not white men there would be negroes there." Percy tried again, inviting Pow to confirm that "[w]here you have negroes you must make some provision for negroes laying off." Pow worked another dodge: "I suppose if I was in their place I would too." But in the end Pow capitulated, acknowledging the reality that black miners were often more transient (see chapter 2). When the exasperated Percy asked, "Is it your experience that the negro miner and the negro laborer is less disposed to work with money in his pocket [than] white miners?," Pow grudgingly allowed, "You don't need to ask me that. That is a fact."[47]

The Blocton meeting hall episode, the dialogue between miner A and miner B, and the tortured exchange between Pow and Percy encapsulate the tensions, strategic and ideological, residing within the union's racial atmosphere. Ultimately, District 20 brought black and white miners into common association, but never contemplated wiping out the color line entirely; it asserted their shared concerns, but avoided an all-encompassing equality; it advanced black material conditions, but deemphasized black political rights; it honored black unionists, but remained open to the notion of black inferiority.[48] Within these constraints, and for a time, an interracial union could achieve a tenuous stability in the Birmingham mineral district. Notably, operators in these years seldom raised the inflammatory race issue, the point on which the union was most vulnerable.

BY 1903, however, warning signs for this relative era of good feelings were visible. For the first time in five years of recognition, annual bargaining between District 20 and the operators broke down. Only the intervention of a board of arbitration, grudgingly agreed to by the operators, averted a districtwide strike.[49] The year that followed revealed that the settlement, rather than projecting labor stability into the future, had merely deferred its collapse. Renewed efforts to organize locals at the Walker County towns of Cordova and Horse Creek met with the raw resistance—including mass firings and armed guards—long associated with that area. Blue Creek miners striking through the fall and winter against the reintroduction of subcontracting were compelled by court injunctions to conduct daily meetings in the guise of religious services at a nearby church.[50] Relations between miners and operators grew chillier that spring. Heavy-handed measures against unionists—firings, arrests, violence, suppression of meetings, importation of cheap immigrant labor—provoked bitter strikes at the Pratt Coal Company mines at Mineral Springs and Arcadia, the Little Warrior Coal Company mines at Littleton, and the TCI mines at

Blocton. When a popular union miner was shot dead by a mine boss during the Blocton strike, between 1,500 and 2,000 people attended his funeral.[51]

Conflicts such as these were not unknown during the previous five years, but new developments portended larger struggles ahead. At the end of 1902 the economy had entered a slump that would last for the better part of two years. The slowdown coincided with mounting hostility among employers nationally to the labor movement. In 1903 the National Association of Manufacturers (NAM) launched an initiative to reverse the gains unions had made in recent years. The open-shop drive was spearheaded with fire-and-brimstone passion by NAM president David M. Parry, who declared an "open and square fight to the finish with the unions." Local Parryite outfits such as "citizens alliances" and "anti-boycott associations" cropped up across the country. Supported by a flurry of bills and injunctions undercutting the closed shop and restricting labor's rights to picket, boycott, or conduct sympathy strikes, the open-shop drive placed unions drastically on the defensive.[52] In September 1903, the Alabama legislature passed a draconian antiboycott bill. Potential repercussions for the coal fields became evident soon thereafter, when a Blue Creek striker was arrested for violating the law. "We are now up against civic law as well as the law of supply and demand and the law of necessity," the *Labor Advocate* reflected. In November, unionists from around the state—the majority from the UMW—met in Birmingham to establish the Union Labor League, envisioned as a political vehicle to resist the reduction of Alabama workers to "peon labor."[53]

By summer 1904, against this backdrop of economic downturn and mounting antiunionism within and beyond the coal fields, many regarded the coming contract negotiations uneasily. Their apprehension proved warranted. From the start it was clear that the major operators had cooled to collective bargaining. While the smaller commercial companies were willing to renew the union contract, the large furnace operators such as TCI, Sloss, Republic, and Alabama Consolidated signaled a preference to restore the system of contracts with individual miners. Although they agreed in late June to return to joint wage scale negotiating, their terms—a 7½-cent reduction in the minimum wage, a restoration of monthly pay (from semimonthly), an extension of the workday to ten hours for outside labor—ensured a deadlock. No one, the *Labor Advocate* commented, "is willing to guess what the intentions of these gentlemen are." But a statement by TCI chairman Don H. Bacon to his stockholders offered a glimpse. The nominal sticking points, he explained, were relatively inconsequential. It came down to who would run the mines. "The authority of your representatives over the property in their charge, as to the manner in which the work should be done, as to what should constitute a fair day's work, and as to who should be employed, had to be restored and maintained." Concluding that a return to the open shop was not only possible but imperative, Bacon pronounced the cost of a long strike "a necessary investment" if TCI were once again "to fully control its own operations, untrammeled by union restrictions." By the end of June the major operators had succeeded in provoking a strike; while the commercial companies signed extensions of the union contract, 9,000 miners for the furnace companies walked out. Shortly afterward, Sloss company president Maben bluntly reconfirmed the operators' intent to end the closed shop, declaring that there was "no possible point of contact" between the UMW and the operators.[54]

Anticipating a shutdown at the mines, the iron and steel companies had accumulated large stockpiles of coal and coke. Once the miners walked off, they procured alternative labor in the usual fashion. Convict miners were shuffled around to maximize production. Strikebreakers were soon assembled as well. Two weeks into the suspension, TCI made the first move to resume production.[55] By mid-August trainloads of strikebreakers, mostly black, were arriving regularly from Tennessee, Virginia, West Virginia, and Kentucky. Beginning at Blue Creek and spreading through such mining towns as Dolomite, Coalburg, Sayreton, Blocton, Brookside, and Blossburg, production gradually picked up. Meanwhile, the operators moved to evict strikers from company housing, to make room for their replacements. Evicted miners found themselves arrested by deputy sheriffs for trespassing on company property. As the courts brushed aside legal challenges to the evictions, tent colonies arose to shelter the families of strikers.[56]

Disillusioned miners streamed out of the district. Those who remained pursued a variety of activities, from hunting and fishing to baseball and pitching dollars, to passing time at the railway stations where they might some way or another convert arriving "blacklegs." Union leaders toiled to keep the strike alive, distributing relief and mobilizing local rallies.[57] But by the end of August, pessimism was mounting, and an air of violence descended on the district. Clashes among union miners, strikebreakers, and deputies erupted at Dolomite, Graves Mines, Brookside, and Adamsville. The most serious incidents occurred at Dolomite. On the night of the 16th, the town was treated to a barrage of gunfire and dynamite, in an apparent effort to intimidate the "blacklegs." Intimidate them they did. Many attempted to depart the following day, only to find that the company, in barring all not working at the mines from the grounds, had effectively hemmed them in. Another explosion the following week blew up a strikebreaker's house, and the trial of the union miner subsequently arrested—"a black burly negro," in the fevered language of the *Age-Herald*—became a cause célèbre among the strikers.[58]

The miners' efforts to head off strikebreakers received a harsh blow from Thomas G. Jones, who as governor had done so much to thwart the miners' strike of a decade earlier. Now a federal judge, he enjoined union officials "from interfering with the employees [or] with the operations of the mines and works" of those furnace operators whose headquarters lay outside the state. Elaborate state injunctions shortly followed. Deputies assigned to enforce them monitored union meetings and arrested strikers who confronted strikebreakers. In mid-September those working for TCI were required to sign "yellow-dog" contracts, pledging that they were not union members. Soon the company was reopening mines in its Pratt division, the District 20 stronghold. By winter, union membership had sunk to barely 3,000, down from over 10,500 a year earlier. Those who remained settled in for a prolonged fight to the finish. Weekly provisions of food and supplies kept strikers afloat.[59]

And so it went for another year and a half. The union fought on with grim tenacity. A small number of strikers returned to work at nonunion mines. The tedium of protracted idleness was alleviated by social gatherings and animated mass meetings. In July 1905 over 6,000 people attended a union barbecue at Blocton, featuring ball games, dancing, a parade, music, and speeches. The national UMW, deeply interested in keeping southern Appalachia in the fold, poured over half a million dollars into the strike, while calling on strikers to remain firm.[60]

Time, however, was on the operators' side. A steady flow of strikebreakers, primarily black although increasingly of southern and eastern European extraction as well, arrived from around the country and abroad. Union leaders were routinely arrested, although seldom convicted, for violating the antiboycott law. As the usual spiral of violence escalated—beatings and shootings of strike officials, dynamiting of strikebreakers' homes, etc.—public sympathy for the union waned. The next year brought little relief. In August 1906, District 20 voted by a margin of over two to one to discontinue the 26-month-old strike.[61]

THROUGHOUT THE STRIKE, the miners' unity had been tested anew by the divisive potential of race. Considering the spread of Jim Crow in these years, the atmosphere surrounding the strike was remarkably free of racial hysteria. There were few of the overheated cries of "social equality" and "black domination" that greeted other examples of interracialism or black advancement during this period. If race figured little in the rhetoric of the antiunion forces, however, it played a prominent role in their strategies. Evictions and strikebreakers were first seen at predominantly black mining centers, such as Blue Creek and Dolomite. Viewing black miners as the Achilles heel of the strike, the operators tapped African Americans as their primary source of strikebreakers, suggesting a renewed effort to inject corrosive racial tensions into the miners' cause. A further effort to pry blacks from the union came from local black ministers, who toured the district urging miners of their race to return to work. "It is to a certain extent the cloth against the union," the *News* observed.[62]

But while the operators attracted many African Americans from outside the district to replace union miners, there is little evidence that they succeeded in dividing black and white strikers. District 20 officials sought strenuously to retain the loyalty of black miners, appealing not only to their common interest with white miners, but also to their own distinctive interests. Speaking before an audience of 3,000 at the Pratt City opera house, Fairley warned black miners that a defeat would return many of them to the status of laborers in a revived system of subcontracting, performing work from which others drew pay. The UMW sought as well to dash the notion that blacks were leaving the fold. According to one "Bessemerite," those African Americans who were brought to Blue Creek as strikebreakers in 1894 were now "the men who are principally making the fight" there. The *Age-Herald*, no friend to unionism, confirmed in August that not a single miner at Blue Creek had returned to work on the company's terms. The union could even show a sort of defensive pride in the diversity of its ranks. "As is well known," the *Labor Advocate* asserted, "the miners' organization is composed of the most diversified types of mankind of any of the trades organizations, from our native Alabama negro to the delver from foreign shores. . . . Among all this heterogeneous membership it would seem but human nature to find some who were vicious and lawless, but none so far have been proven so."[63]

However much District 20 weathered the racial tests in 1904, it continued to function within a Jim Crow order that it could neither embrace unmodified nor reject outright. Efforts to foster bonds between black and white miners within the prevailing racial culture found expression in the speeches of Vice President Greer. "I want to speak to the black man, the colored man, the negro," he began at a Pratt City meeting. "Let us in this crisis prove to the white man and the world that we are fit for a union. We have been given . . . suffrage, it has been justly said, too soon. It was taken away from us because some of us sold our votes for a dollar. Let us not throw away our union birthright." In one passage, Greer thus paid homage to the dominant assumption that African Americans had been implicated in their own disfranchisement, and then called on black miners to rechannel their civic aspirations through the UMW. The union was their birthright, and yet it was not an inalienable one, any more than was the vote; they would have to prove their worthiness to their white brethren through staunch loyalty to the strike.[64]

Greer's oratory captured the blend of boldness and pragmatism that marked the relation of black miners to the union. His formulations suggest that the philosophies (or at least rhetorical strategies) of the black middle class and black unionists in the New South were not always so incompatible as is often supposed. To be sure, the union's tendencies toward interracial mutuality, collective action, and resistance to the dominance of management—all components of a class agenda—ran counter to Booker T. Washington's advocacy of accommodation, individual self-improvement, and cooperation with the employer—all ingredients of a conservative, racially defined project. Yet there is much in Greer's message that bore a striking resemblance to that of Washington: his willingness to affirm the popular white belief that blacks had been corrupted by a premature right to vote; his emphasis on social betterment through education and stability; his insistence that African Americans must prove their worth to whites in order to advance. In the end, it is not self-evident where the conflict between the two approaches ended and the overlap began. Rather than reject outright the gospel of the southern black bourgeoisie, black unionists like Greer drew upon that gospel and adapted it to the contours of unionism. In a world that offered the miners compelling cause both to cooperate across racial lines and to affirm identities of race, the lines between Washingtonianism and unionism, between accommodation and boldness, between race consciousness and class consciousness, could become elusive indeed. Ultimately, it is clear that most miners on union terrain, black and white, found ways to reconcile the claims of race with those of the labor movement.

IN THE AFTERMATH of the union's defeat, the miners' situation looked grim. Many who had struck were now blacklisted. Thousands sought work in other trades or left the district altogether. The mines were now worked largely by newcomers, many freshly arrived from other lands, who had come as strikebreakers. The economic panic of 1907 sent iron and steel prices plummeting. The acquisition that year of TCI by United States Steel introduced a corporate presence as staunchly antiunion as it was powerful.[65] District 20 nonetheless retained a foothold in the coal fields. From 1905 through 1907, membership hovered between 3,500 and 4,000. Many union miners worked for those companies that continued to negotiate union contracts—the commercial operators or the Alabama Consolidated.[66]

District 20's last stand began with a campaign to defend these contracts. As

1908 approached, the Alabama Consolidated announced a 17.5 percent cut in miners' wages, effective January 1, well before the contract's expiration. Perceiving a clear attempt to shake off one of the few remaining bastions of union recognition, the miners walked off. In June, as the strike entered its sixth month, District 20 began its annual negotiations with the commercial operators. The latter proposed a reduction in the maximum mining wage from 57½ cents to 47½ cents per ton—the same 17.5 percent cut that had sparked the Alabama Consolidated strike. The commercial companies also served notice that they would no longer sign with the union should it fail to accept their terms.[67]

The national organization decided the time had come to call all Alabama miners out on a do-or-die strike for recognition. On July 6, approximately 4,000 miners, primarily at the commercial mines, answered the call. By the second day, the strike had spread to the major furnace companies. Unionists scrambled across the district, reviving dormant locals, addressing mining communities from street corners and rallies, mobilizing picket lines, and setting up union commissaries. By the second week the union claimed over 8,000 new members.[68]

The district settled in for another long strike. Once again rail lines bore car after car of strikebreakers, largely black, from all over the nation. Again, thousands of miners and their families were evicted to make room for their replacements. Tent colonies funded by the union were erected on land leased from sympathetic farmers. Several hundred deputy sheriffs were dispatched to the mining district at the behest of sheriffs E. L. Higdon of Jefferson County and J. O. Long of Walker County. Higdon posted 2,000 announcements citing laws against labor boycotting, intimidation, and the like.[69]

The use of deputy sheriffs and company guards stirred predictable resentment in the mining communities. Brandishing pistols and rifles, they broke up meetings, closed churches, arrested miners and their families, evicted them, prevented them from moving about freely, subjected them to abusive language, and forced them to return to work. The sheriffs and guards were also prepared to use their weapons, and over the weeks a number of strikers were killed or wounded. For good measure, the operators requested gatling guns from the state to protect their property and strikebreakers. Strikers did what they could—at times surrounding the mines—to impede the flow of "blacklegs." Women turned out to jeer the new arrivals. Trains carrying new strikebreakers became the targets of armed attack. These actions inspired heated denunciations of labor violence, obligatory disclaimers from the union, and ultimately a proclamation by the governor announcing the deployment of state troops to quell "disorder and lawlessness."[70]

A charged atmosphere greeted the first arrival of troops. "The train was the most unusual that ever steamed out of Birmingham," the *News* observed. "The gondola car at the head of the train presented a very formidable appearance, with the fringe of gun barrels bordering the side." Large crowds of blacks and whites, "much wrought up over the passage of the 'war' train," watched the troops embark for the coal fields. As the train pulled past a spot near Pratt City where deputies had earlier shot several African Americans, "the track was lined with negro miners whose remarks were anything but complimentary to the 'men who had come to make us work for nothing.' " In late July, over 500 troops, mostly young white men from the Black Belt, patrolled the district. "Everywhere," the *United Mine Workers Journal* reported, "one sees guns and hears the tramp of the soldiers." Strikers initially made a show of embracing the troops as their protectors. But soon the troops themselves were arresting strikers in large numbers. "[I]t seems," one striker concluded, "that they are just as bad as the company guard." "We haven't any governor!" another cried at a mass meeting as union leader William Fairley recounted the excesses of the troops. "I rather differ with my friend," Fairley replied, "I think we have too much governor!" Meanwhile, the New Orleans *Daily Picayune* reported, Birmingham was abuzz with "countless rumors of battles with slaughter in the outlying districts."[71]

By August it appeared that the coal belt was in for another protracted strike. Union membership had grown to 18,000, embracing the vast majority of miners. Affirming the importance it attached to the battle, the national union dispatched Vice President John P. White to assume command, along with over two dozen national organizers; it would ultimately pour over $400,000 into the strike. The operators moved to strengthen their hand by resurrecting the Alabama Coal Operators' Association (ACOA).[72]

In ways both familiar and new, the conflict tested the cooperation of black and white miners. Respectable figures of black Birmingham denounced the strike. Most black miners, however, embraced the union. Once again the sight of black strikebreakers failed to drive a wedge between black and white strikers. Miners of each race attended the same meetings, spoke from the same platforms, and inhabited the same tent colonies (although these were divided into black and white sections). Black strikers at Adamsville confronted black strikebreakers; white strikers at Republic rallied behind the prosecution of a white deputy sheriff who had killed a black striker. If the union spent little time trumpeting its interracial stand, neither did the stresses of battle generate discernible racial tensions. Race in fact figured little in public discussion during the first month of the strike.[73]

But by any measure—systematic passage of segregation and disfranchisement codes, lynchings and riots against blacks, shrill expressions of white supremacy and injunctions against social equality—the South in the early years of the century had become an increasingly foreboding arena for interracial unionism. During August, District 20's adversaries injected an incendiary theme into the public discourse, one that would abruptly tilt the balance of popular opinion against the miners. As never before, they played the race card.

Starting in mid-August, a series of antiunion voices—newspapers, political figures, and operators themselves—raised a drumbeat of allegations that the union was encouraging social equality between the races. The ACOA spoke of bands of armed strikers, black and white, roving the district. An *Age-Herald* headline blared, "The Social Equality Horror." Governor Comer announced that legislators were "outraged at the attempts to establish 'social equality' between white and black miners." Opponents of the union took pains to accentuate the sexual dimensions of social equality. The *News*, generally less strident than rival dailies, added that the strikers' tent colonies (their biracial layout notwithstanding) had engendered "the too intimate association of whites and blacks in the camps, developing by degrees racial familiarity, if not equality, [and inflaming] the public mind." "[T]oday in this district," former Birmingham mayor Frank Evans intoned with an odd but ominous imagery, "we behold pluralism in the sex of devils, embodied in male and female. . . . these devils have sought to teach the negro that he should affiliate with white men and with white women." Intimacy between blacks and whites recalled the "terrible days of Reconstruction." As he witnessed the "co-mingling of white and black . . . in the very presence of gentle white women and innocent little children, I thought to myself: has it again come to this?"[74] Union detractors also raised the hot-button issue of interracial rape. Evans reported that white women had been subjected to "criminal assault by Negroes" in the strike district. Comer warned that the presence of five to ten thousand blacks in a state of vagrancy at the tent colonies was a circumstance "too dangerous to contemplate," a threat "to the integrity of our civilization."[75]

To maximize the sexual implications of interracialism, opponents broadened the meanings of the union to include women. Evans highlighted efforts "to organize the women of both races . . . into female unions known as the 'Woman's Auxiliary.'" Perhaps the most inflammatory language came from the pen of society columnist Dolly Dalrymple, who took time from her usual cheery fare to voice her outrage: "White women and black women meeting on the basis of 'Social equality' indeed! White men holding umbrellas over black speakers! Black men addressing white men as 'brother!'" Good taste prevented her from describing relations between blacks and whites across the gender line; only vague allusions would do. Social equality in the coal fields was "an unspeakable crime," the very thought of which "caused the women to shudder. . . . It is monstrous!" Never before had interracial unionism in the district encountered such fevered race-baiting.[76]

The union strained to refute the charges. UMW advocate Duncan McDonald accused opponents of devoting "their entire time in lying about 'social equality,' 'women's auxiliaries,' and other base falsehoods in order to cover up the real issue." The operators fell back on the "ghastly spectre of race hatred and social equality" in order to avoid defeat, argued Vice President White. Finding themselves beaten on "the industrial field," they turned to "drastic methods"; by injecting the "race question," they played "with fiendish skill upon the minds of the people." The UMW, he continued, "did not go to Dixie Land to preach the doctrine of social equality; neither did we practice it." The *Labor Advocate* added its own blunt reassurance: "There is not at present and never was and can never be 'social equality' between whites and blacks in this state."[77]

In their determination to shake the charge, the UMW leaned over backwards to cast their movement in narrowly economic terms. The miners' cause, White bristled, was merely "an industrial one." However real the irritation and obligatory the disclaimer, such defenses rang disingenuous. The UMW was never, after all, a purely bread-and-butter or workplace-centered association. No less than their husbands, Priscilla Long has noted, the women in American coal towns experienced "the all pervasive atmosphere of the coal company and its superintendent." The range of concerns the union addressed—from mine safety to company stores, wages to housing—inevitably engaged the wider community. So too did the UMW's almost evangelical promotion of mutualistic values—a sweeping moral vision that rendered it, in the phrase of Herbert Gutman, a "secular church." Never did the union emerge more fully as a community institution than during strikes. The mass evictions of miners' families, the provocative intrusion of troops and company guards, the vital roles of both women and men in confronting strikebreakers, supplementing livelihood, and sustaining morale—these developments turned labor battles into broad-based, social struggles. The UMW's expansive presence in the mining community bolstered the strikers, but it also left outsiders more receptive to the insinuation that its racial practices amounted to "social equality."[78]

At times UMW leaders went further in their disclaimers, answering race-baiting with some race-baiting of their own. In a public statement Fairley noted the ACOA's frenzied warnings about "large mobs of armed negroes" in the union. "There is nothing said," he countered, "about negro deputies being appointed by Sheriff Higdon on the recommendation of the coal companies, one of whom shot indiscriminatingly, endangering life of innocent women and children at Pratt City." When Higdon denied the charge, insisting that African Americans served only as spotters for deputies, Fairley was unappeased. "This method is worse than regularly commissioning a negro. . . . If the sheriff has armed negroes running around loose without commissions, the citizens of the community, and especially the striking miners, are living in grave peril." Fairley reminded a mass meeting that, for all the hysteria about "mobs of negroes" unleashed by the union, it was in fact the operators, not the miners, who had brought blacks to the coal fields to act as strikebreakers. Nor was he averse to stressing the race of white victims of repression in order to heighten the affront. He called attention to "the outrages perpetrated on John Burros, a white man, who was dragged from his bed at Wylam . . . and hung by his neck until his tongue and eyes protruded," and "the handcuffing of a white woman to a man in Walker county . . . on the trivial charge of trespass." When Comer invoked the horrors of black idleness and social equality as a pretext for suppressing the strike, UMW leaders called his bluff by proposing to send every black striker out of the state and "make it a white man's strike." (As they presumably anticipated, the governor declined to call the union's bluff, finding other reasons for the impending clampdown.) To what extent did Fairley's attempts to pin the presence of black deputies, black strikebreakers, even black strikers on the operators represent (or play upon) genuine indignation? To what extent was it outrage over the operators' efforts to defeat the miners? To what extent a desire to highlight the hypocrisy of antiunion race-baiting? To what extent, that is, did such rhetorical gambits reflect true conviction and to what extent an instinct to survive in a Jim Crow atmosphere? At some point the motivations behind the union's racial strategies become impossible to unravel.[79]

The intentions behind such disclaimers may have been tangled, but their effectiveness—or rather ineffectiveness—was all too clear: ultimately, they brought the strikers little protection. Within the volatile racial climate of 1908, the social equality charge readily inflamed public opinion. "The scene has been shifted," a committee of union officials informed national vice president White. "You are no longer dealing with the industrial question, but with the racial problem." Comer seized the moment to deal the strikers a series of crippling blows. In late August he prohibited the union from holding public meetings. He then ordered that the tent colonies, by now swelled to 70,000 people, be cut down by the militia. Soon, in a manner both somber and dramatic, the job was done. At Republic, an observer reported, troops marched upon the makeshift settlement, and "with their swords and bayonets cut the ropes, slashed the canvas and reduced all to a complete wreck." Comer cited unsanitary conditions—a charge the union vehemently denied—but amid mounting hysteria over the collaboration of black and white miners (and their families), avowed concerns about hygiene were not easily separable from the assault on interracialism. The governor himself offered a less lofty explanation in private to District 20 leader J. R. Kennamer: "You know what it means to have eight or nine thousand niggers idle in the State of Alabama, and I am not going to stand for it."[80]

Whatever the pretext, the effect was devastating. "The militia is . . . tearing down our tents at Republic and other places, and piling them up with the people's furniture in heaps," one striker wrote; "Boys, it is hard." While the troops went about destroying their camp, strikers conducted a prayer meeting along the nearby railroad tracks. As thousands scrambled to recover their worldly possessions, Comer summoned the union leaders and issued an ultimatum: call off the strike or he would convene a special session of the legislature to enact a vagrancy law under which every striker could be arrested; in essence, they would be returned to the mines as convicts. More quietly, union officials received warning that if the strike persisted, "blood would flow in the streets of Birmingham" and eight of the most prominent leaders "would swing." On August 31, the UMW announced the strike over, effective the following day.[81]

For the second time in three years, a districtwide strike had come up empty. The 1908 defeat was far more conclusive, and the miners knew it. The coal fields were now entirely open shop. The miners, brooded the *Labor Advocate*, were "the same as serfs. . . . [They] will have to dig coal at whatever price and under any condition." The bleak prediction was soon borne out, as working and living conditions deteriorated markedly. Word came from the mining camps of stringent new work rules conferring greater power to the companies, a return to extensive dockage of pay, increased pressures to patronize the company store, draconian restriction on local movement, checkweighmen now chosen by the company, and so forth. The coal fields were reportedly riddled with detectives, and miners with union ties were discharged en masse. Hundreds of miners, demoralized and out of work, departed the district. "Goodbye," one wrote to the *United Mine Workers Journal*, "may we meet above, where we won't be bothered with governors."[82]

The mass departure of miners, and the distress of those who remained, nearly wiped out the union. Within a few months membership had plummeted to 762.[83] In early 1909, William Leach of Cordova recalled wistfully how, for but a few weeks during the strike, local miners had defied the rough regime of Walker County and joined the union. "The Empire local met every day, 351 strong—open air exercises, with prayer—and that good old song was sung at every meeting, 'We Will Overcome Some Day.' " Leach recalled the scene for inspiration. "Dear Brothers, let us not drop that old song, but still sing it. If we stick together we will overcome some day, for that old tyrant governor soon will step down and out."[84] But the union would languish well beyond Comer's departure in 1911. District 20 retained a slight presence in the years ahead, but would not revive until the wartime mobilization nearly a decade later.

FOR YEARS AFTERWARD union miners remembered the strike of 1908 as a would-be victory cruelly denied. They recounted the fierce repression by the antiunion forces, and then the injection of the toxic race issue.[85] There is no way to determine how pivotal repression and race-baiting actually were to the final outcome; given the weakness of the economy, the availability of nonunion labor (free and convict), and the formidable resources of the operators, the strike might well have failed even had Alabama been an oasis of civil liberty and racial justice. However devastating the race card may have been, it was but one weapon in the daunting arsenal of the operators and their allies.

What remained remarkable, though, was the lack of any visible erosion in relations between black and white miners. Had Comer not quashed the strike when he did, such erosion might in time have begun. Still, two months—and in a larger sense, four years—of struggle had brought out little of the racial scape-goating that had plagued the Populist movement in its death throes.[86] To be sure, in its ritual denials of social equality, and its periodic efforts to throw race-baiting back upon its adversaries, the union paid rhetorical ransom to the claims of white supremacy. Nor should we assume that such statements were purely pragmatic; in some measure they conveyed genuine sentiment as well, and thus revealed the internal limits of the union's interracialism. But these disclaimers and countercharges were also meant to provide a shield for collaboration across the color line. There is no evidence that they spilled over into recriminations against the black members, who played a significant role in the union's ranks and leadership; to the contrary, even in the strike's bleak aftermath, white officials took pains to stress the contributions of black miners. "There are no better strikers in the history of the United Mine Workers in any district than the colored men of Alabama," Kennamer declared before the national convention of 1909. "They struck, and struck hard, they fought for their rights and fought manfully, and when their things were thrown out of their homes they took them out in the woods and put them under trees and said as long as we would feed them they would stay on strike."[87]

The resilience of interracial unionism in the Alabama coal fields extended into the new century a tradition dating back to the Greenbackers of the late 1870s, the Knights of Labor of the 1880s, and the early UMW of the 1890s. For most white miners, the growing presence of blacks only heightened the practical importance of that tradition. But whatever its continuities, the nature of labor interracialism was not static over time. Consider how circumstances had evolved from the great strike of 1894 to that of 1908.

In two ways, the climate for interracialism shifted markedly. The first occurred in the sphere of politics. The 1894 strike took place in the Populist era, a time when miners of each race avidly followed politics, attended campaign rallies, and voted in large numbers; a time when the competing parties debated matters of significance to mining communities; a time when a battle for the statehouse was seen as pivotal to the outcome of a coal strike. The political context for the 1908 strike was far less open or fluid. The Populist challenge had by then long fizzled, many miners had been disfranchised, and the Alabama variant of Progressivism embodied in New South industrialist Braxton B. Comer—had little to offer the people of the coal fields.[88]

Second, the racial atmosphere of Alabama, never hospitable to the joint movements of black and white workers, had by 1908 become decidedly hostile. It was not always so. For a time, labor interracialism in the coal fields had found room to survive even amid the expansion of Jim Crow. During the era of union recognition, it caused comparatively little stir. One of the more notable moments of reaction—the 1901 effort by Birmingham merchants to prohibit an interracial meeting—was an exception that proved the rule: when the union protested, they backed down. And yet that rule was itself an exception to the broader entrenchment of white supremacy around the region. Why had the spectacle of blacks and whites attending the same rallies, speaking from the same platforms, sharing leadership positions, and articulating a language of common purpose and solidarity not provoked a more intolerant reaction throughout this period? How, in short, had an interracial union been allowed to survive and even thrive so conspicuously into the era of segregation?

The rise and fall of District 20 suggests that, while there was an inevitable tension between interracial unionism and Jim Crow, the two were not necessarily incompatible. Under two circumstances they could coexist: (1) when the union kept its racial practices squarely within certain bounds, deferring above all to the visceral sanctions against "social equality"; and (2) when influential employers felt no urge to rid themselves of the closed shop (or indeed welcomed its stabilizing effects). Such was the state of affairs between 1898 and 1904—an equilibrium that enabled an interracial union to survive into what W. E. B. Du Bois baptized as the century of "the color line."

But the equilibrium was a delicate one, and it would not last. By 1903 the large operators had lost patience with the union shop. A year later, as the economy sagged and open-shop fervor swept the nation, they had determined to end it. Now the miners faced an adversary that had not only daunting resources at its command but also a motivation to play the dreaded race card against their union. In the do-or-die battle of 1908, the operators and their allies did just that. Although the UMW remained vocally committed to a limited, non-"social" brand of interracialism, its enemies smeared it as a threat to southern civilization. Race-baiting never found a more combustible setting: the coarsening racial temper of the region provided the atmosphere; the conspicuous role of white women in an interracial strike, the tinder; the "social equality" charge, the match. The two circumstances that had enabled District 20 to exist in a Jim Crow context—the readiness of the operators to recognize it and the perception that its racial practices were, if anomalous, relatively innocuous—had both evaporated by 1908. The defeat of the union was not the result of some timeless or irresolvable clash between its racial inclusionism and the broader culture of white supremacy. Rather, the tension between the two gave the union a vulnerability that the operators and their allies might exploit with great effect under a particular set of conditions: chiefly, a growing inclination to restore the open shop and a volatile racial atmosphere upon which to draw.

Ironically, the devastating impact of the social equality charge in 1908 provides clues as to how an interracial union like District 20 managed to function as long it did in a Jim Crow society. Although the specter of social equality had for decades loomed as a boundary for miners' interracialism not to cross, before 1908 it had remained more latent than active in public reactions to the Greenbackers, the Knights of Labor, and the United Mine Workers. Indeed, the collaboration of black and white miners in organized labor long enjoyed greater breathing room than would have interracial association in any other part of their world. Why the anomaly? Here the gender composition of the mine labor force provides a key.[89]

In contrast to most areas of the mining community, the mines themselves were an exclusively male environment. The absence of women at the mines and their marginal place in unions served quietly but significantly to open up space for interracial unionism in the constraining landscape of Jim Crow. The drive toward segregation was, after all, steeped in notions of gender. Lurking behind frenzied condemnations of "social equality" were the provocative images of miscegenation, amalgamation, the despoiling of white womanhood. "Nowhere were the ethics of living Jim Crow more subtle and treacherous," Jacquelyn Dowd Hall observes, "than when they touched on the proper conduct of black men toward white women. . . . Any transgression of the caste system was a step toward 'social equality'; and social equality, with its connotations of personal intimacy, could end only in interracial sex." Though the specter of "social equality" cannot be reduced entirely to sexual associations, such imagery lent the term its deepest emotional power. "Sex was the whip," Nell Irvin Painter writes, "that white supremacists used to reinforce white solidarity." The singular power of the social equality charge flowed from its capacity to link black advancement or interracial activity in far-flung endeavors— schooling, worship, recreation, politics, social movements—to the lurid imagery of miscegenation. While this fear cannot on its own explain the spread of Jim Crow, it did serve to reinforce the segregation impulse, to embarrass any tendency toward racial mixing in the social world, to render it intolerable, even unimaginable.[90]

"[T]he more closely linked to sexuality," Edward Ayers has insightfully observed, "the more likely was a place to be segregated." The critical link between racial separation and the sanctity of white womanhood lost much of its resonance in the world of mine labor, from which women, black and white, were wholly absent. And so the spectacle of black and white miners collaborating in labor unions, while disquieting to many, could not so readily quicken Jim Crow feeling as, for example, blacks and whites attending the same churches or schools. This is not of course to suggest that the masculine aspect of miners' work singlehandedly fostered interracial unionism. White women were out of the picture in many trades around the New South where interracialism never materialized. But where black and white men already felt the impetus to mobilize, the absence of white women reduced the power of race-baiting demagoguery to discredit their endeavor. In short, as bold as it was for any association to operate across the color line on terms approaching racial equality, it was greatly more daunting for such an enterprise to extend at once across the color line *and* the gender line. It is perhaps no coincidence that the two other settings in the New South where interracial unionism flourished as conspicuously as it did in the coal fields—the timberlands and the waterfront—also shared with mining a distinctively "male" aura; both types of work were markedly remote from the presence of women, and both were commonly perceived as quintessentially "masculine."[91]

As the 1908 strike showed, the union was hardly immune to the effects of racial hysteria. Amid a poisonous Jim Crow atmosphere, the gender-exclusive character of mining could not exempt interracialism from the sexually charged implications of social equality. But for many years the all-male composition of the labor force allowed black and white men more room for joint activity in the union movement than in other areas of their social world. The relative absence of women from the sphere of miners' work and organization had never guaranteed interracialism a safe haven in the mineral district. It had simply muted—at least for a while—one of the shrillest cries against integration of any sort. Although this effect does not alone explain the reemergence of miners' interracialism amid the rise of Jim Crow, it does help us to understand what made such a remarkable phenomenon possible.

6 The United Mine Workers in the World War I Era

I

F THE DeBardeleben Coal Company boasted the Birmingham district's quintessential corporate welfare program, then Robert Wesley Taylor was its quintessential functionary. Educated at Tuskegee (for which he later worked as a financial agent), Taylor served as principal of DeBardeleben's black school at the model town of Sipsey, where he oversaw various "colored" social activities as well. In these capacities he aspired to be a role model for black miners and their families. Diligent, articulate, upright, Taylor blended an austere bearing toward his charges with an impeccable deference toward his patrons. Routinely he sent upbeat reports to company officials about the achievements and goodwill resulting from the paternalistic programs at Sipsey. To illustrate the message, Taylor had students at the school, or perhaps the town's black brass band, prepare special performances in honor of President Henry T. DeBardeleben, Vice President Milton H. Fies, and their families. The company reciprocated in classic fashion: in 1916, DeBardeleben presented Taylor with an inscribed gold watch. "He is a good negro, smart & *knows his place*," Fies wrote DeBardeleben.[1]

It was presumably with a raised eyebrow, then, that in late 1916 Fies received a letter from Taylor of uncommon candor, marked "personal and confidential." Based on comments he had heard at Sipsey and elsewhere over the years, Taylor had concluded that "the Negro labor of the Alabama mining district is an incoherent mass, united in nothing save the conviction that Capital has the 'cards marked and stacked' against it." He traced this troubling sentiment to "the economic heresy that Capital and Labor are antagonistic rather than Co-operative." Such thinking, Taylor warned, "does not make for a high degree of efficiency on the part of Labor." To rectify the situation, he proposed the launching of a biweekly periodical, "with some such name as THE NEGRO MINER," to be distributed throughout the district with the discrete financial backing of the operators. As editor, he would hire a correspondent in each camp to report the local happenings, including the "names of newcomers and whence they came and the names of those who go away and where they go." Now and then the paper would feature a particular camp in depth, highlighting the good fortunes of its leading miners by running pictures of their homes and of their "sons or daughters educated." And the paper would have a didactic function: "I would encourage the planting of gardens and the raising of chickens and pigs to reduce the high cost of living; I would preach the gospel of thrift, of regular work and of reliability; I would emphasize the importance of sanitation, correct diet, sufficient sleep, etc., as the foundation of physical fitness, and point out how intellectual power and moral excellence have their basis in things physical." Such an enterprise, Taylor suggested, would "increase the efficiency of labor and at the same time bring it into more sympathetic and helpful relations with Capital."[2]

Fies was receptive to the scheme, noting that it "might come in handy." Nothing, however, came of it; it is likely that the operators considered the idea superfluous. With the emergence that year of the "labor crisis" in the coal fields, Birmingham's middle-class black establishment had mobilized a spirited campaign to impress upon the black miners just those lessons Taylor had in mind.[3] Still, Taylor's letter illuminates a moment of transition between the eight quiet years ushered in with the resounding defeat of the United Mine Workers in 1908 and the turbulent era of World War I then germinating. His message pointed toward the advent of elaborate welfare programs in portions of the coal fields in the years following the 1908 strike. It also suggested the growing centrality of African Americans to the mine labor force and to the operators' management strategies. Through a blend of paternalistic initiatives and often smothering repression, the operators had made headway in diminishing labor transiency and had succeeded almost entirely in preventing a revival of unionism. But Taylor's letter also hinted at the limits and fragility of the operators' achievement: company officials must have been particularly unsettled to learn that class discontent infected even those employees they had always assumed to be the most docile—African Americans.

If such ominous sentiments were not new, it is telling that Taylor chose the fall of 1916 to disclose them. A new spirit was afoot in the district, indeed throughout industrial America. As national preparedness boosted industrial markets, local workers began to exploit widened opportunity. That spring had seen the start of the Great Migration, which in a few years would bring approximately half a million black southerners to the industrial North. The black population of the mineral district became a significant tributary into this exodus. Those who remained at the mines were growing more openly assertive, and talk of a fresh union drive spread for the first time since 1908. Surveying the local "labor situation" that fall, DeBardeleben informed a customer with a touch of hyperbole that "we have been confronted with conditions probably without parallel since the Civil War."[4] American entry into the Great War the following spring would only intensify the operators' worries. Wartime industrial expansion afforded miners options both within and beyond the district that they could scarcely have imagined in the lean years following 1908. Suddenly they enjoyed a strong hand on the labor market, with an ability to play operators against each other for better wages, conditions, and treatment, and to mobilize collectively. By the summer of 1917, a dramatic organizing drive had rallied the large majority of miners, black and white, back to the banner of UMW District 20.

As during the years surrounding the turn of the century, the wartime revival of District 20 was propelled by a robust economy. But another continuity from earlier times worked against the union's prospects: a persisting regional hostility to interracial organization. For union miners, the old tension between two competing instincts—to submerge the color line and to shield their movement against the explosive charge of "social equality"—emerged afresh in the wartime era.

If these familiar circumstances both encouraged and constrained the revival of District 20, new forces shaped it as well. First, the union's quest for an interracial following was challenged more vigorously than ever by a concerted campaign on the part of black middle-class leaders to keep black miners loyal to their employers. Although such figures had sought influence in the coal fields before, they had never achieved it to any substantial degree before the 1910s. No less significant was a second new player, the national government. Federal agencies created to mediate wartime labor relations played a dual role in the coal fields, functioning at once to enhance and to delimit the miners' power. After the war, prosperity ebbed, and the protections of federal labor policy receded. A showdown strike in 1920–21 brought the union again to conclusive defeat. This chapter extends the story of interracial miners' unionism into this extraordinary period.

THE ALABAMA COAL FIELDS were a very different world at the end of the new century's first decade than at the beginning. The crushing of the 1908 strike had returned the mineral region conclusively to the era of open shop. Prospects for a revival of unionism were dampened over the next several years by a weak economy. The UMW did not vanish entirely from the district. Around the coal fields a handful of activists carried on. In scattered localities union contracts remained intact. Even where there was no union contract, miners here and there still mobilized effectively.[5]

Overall, however, for nearly a decade District 20 retained at best a feeble presence. Demoralization following 1908 was compounded by sharp repression of union activity. At most mines, "yellow-dog" contracts forbade union membership; miners found to have joined the union, or even to read the *United Mine Workers Journal*, were subject to discharge, eviction, and blacklisting. The coal district was riddled with "spotters and spies," a District 20 official declared in 1913. The effects of defeat and fear showed in the numbers: during 1909, 46 locals, including the once formidable Pratt City Local 644, ceased to exist; the following year, another eight were abandoned. No new locals were organized during this time. In December 1911, District 20 reported a mere 27 members, and seldom in these years did the figure exceed a few hundred. A 1913 letter from longtime UMW stalwart R. A. Statham conveyed how weak the union had become: "We have no local here any more. There were three of us until a year ago. One died, one got a bossing job, and I was left alone." A committee sent that year by the national union reported that "fear bordering on terror" rendered prospects for District 20 altogether dim.[6]

Regular communications to the *United Mine Workers Journal* (and, less commonly, the Birmingham *Labor Advocate* and the *American Federationist*) conveyed a litany of grievances that would be all too familiar to veterans of the mid-1890s: low pay and myriad deductions, poor ventilation and frequent mine explosions, cheating on weights and relentless dockage, overrecruitment and underemployment, decrepit housing and coercive commissaries, arbitrary mine bosses and repression of unionism. At some camps conditions approached what the union described as "peonage." The Alabama Fuel and Iron Company, for example, required its miners to obtain written permission before leaving company premises; black miners there, UMW national organizer William R. Fairley alleged, were "whipped by the mine deputies when they desire to lay off a day or two, and forced to work."[7]

HOW THE MINERS ranked their grievances shifted somewhat from earlier periods. The leasing of convicts lessened in the years following 1908. The Tennessee company set the trend. Sensitive to its image amid Progressive Era abhorrence of the convict lease—a mood inflamed by the 1911 Banner Mine explosion, which claimed the lives of 128 convicts—TCI terminated the use of convicts in 1912. The other major lessee of convicts did not rush to follow suit; the Sloss company, more markedly "southern" in its management style, continued to use them into the 1920s. Following TCI's initiative, however, the practice no longer stood as the miners' paramount issue.[8]

Subcontracting, meanwhile, inspired growing indignation. Largely eradicated during the union's heyday, the employment by miners of laborers was reintroduced extensively after its defeat. This trend had actually originated during the strikes of 1904 and 1908, when white farmers hired by the operators would bring their black farmhands to labor for them in the mines. Once the open shop was restored, subcontracting continued to spread, particularly to the smaller and more isolated mining camps. From Townley, Walker County, a miner described in 1916 how the whole mine was contracted out to a few men, who "pay us any old price." "We miners," he added, "haven't been making enough to live on and support our families. The children have to go hungry and without clothes. . . . but the contractor can have everything he wants to live on and dress his family and go everywhere."[9]

The defeat of unionism and the deteriorating condition of miners were accompanied by a vital demographic trend: African Americans made up a growing proportion of the mine labor force, approaching and perhaps at times passing three-fifths during the 1910s.[10] In part this shift was a legacy of 1908. Earlier defeats of major strikes each left the ratio of black to white miners marginally higher than before: whites tended to enjoy a broader range of opportunities in other trades and regions than blacks, while many of the predominantly black strikebreakers remained after the strike. But the rising presence of African Americans at the mines also reflected the operators' abiding dogma that such workers were the most pliable, and hence the least threatening to the open shop. When the Alabama Coal Operators' Association (ACOA) established a Central Labor Bureau following the 1908 strike, the first place TCI president George G. Crawford suggested it seek labor was the "boll weevil district" of Mississippi.[11]

Life in the mineral district was transformed by one further development: the rise of corporate welfare programs. Leading the trend was TCI, whose acquisition in 1907 by United States Steel brought it onto the cutting edge of early-twentieth-century employer paternalism. TCI's welfare approach actually had its genesis before the arrival of U.S. Steel. Shortly after taking the helm in 1902, President Don Bacon launched the construction of more attractive company housing and initiated medical services and professionally staffed health facilities at the mines and furnaces. Like most corporate welfare plans during the Progressive Era, TCI's had a variety of purposes. The company's move into steelmaking at the turn of the century heightened its appetite for skilled and dependable labor. Widespread turnover, absenteeism, and overall alienation—exasperating enough to managers at the coal mines and blast furnaces—were even less tolerable at the steel mills, where round-the-clock production required reliable labor. Welfare programs also promoted health and safety measures, to diminish the disruption and embarrassment of high accident and disease rates. Finally, Bacon's program coincided with the company's move to restore the open shop; in fact, the first improved miners' dwellings built under Bacon's stewardship were initially occupied in 1904 by strikebreakers.[12]

But it was with TCI's absorption into the gargantuan U.S. Steel Corporation that its welfare program really took off. By the early twentieth century U.S. Steel had emerged as the leading force, if not the embodiment, of corporate paternalism in America. Under Elbert Gary, it sought to win employee loyalty through measures such as profit sharing, wage incentives, medical services, pension benefits, and safety programs. The welfare programs of U.S. Steel were designed to create a labor force socialized in the values of efficiency, diligence, rootedness, obedience, and aversion to unionism. Upon assuming the presidency of TCI under the new regime, George Crawford, a young southern engineer, set about developing a "modern" scheme of labor relations at his mines and furnaces.[13]

Beginning soon after his arrival in 1907, Crawford assembled an army of professional health and social workers to develop educational, medical, and recreational facilities for TCI's mine, furnace, and mill workers. In the coming years there followed conveniences such as bathhouses; instructional programs in domestic skills, hygiene, and other elements of "proper" living (conducted by both black and white professionals for employees of their respective races);

and a range of cultural activities, to provide training and entertainment in music, theater, literature, and so forth. Starting in 1909, TCI launched four model communities: Fairfield, on the western edge of Birmingham, designed primarily for steelworkers; and Docena, Bayview, and Edgewater for miners in the coal fields to the west of Birmingham and Pratt City. Emulating such fabled showpieces of industrial paternalism as Pullman, Illinois, and Gary, Indiana, these new towns boasted broad, tree-lined streets, parks, libraries, YMCAs, new churches, schools, lodge buildings, and handsome commissaries. The company sponsored recreational and cultural activities, including dances, free movies, study groups, and pageants. New housing featured indoor plumbing, central heating, and hot and cold running water. The company developed an ambitious health program, designed to root out widespread diseases like malaria and hookworm, improve sewage disposal, monitor the quality of food sold through the commissaries, and enhance industrial safety. The centerpiece was a centralized apparatus for medical and dental treatment, which took modern form in 1919 with the opening of a hospital in Fairfield.[14]

During the 1910s several other operators followed suit. Perturbed by chronic labor shortage—a problem that ACOA president George McCormack attributed largely to the "cheapness of human life in this section"[15]—and determined to prevent union "agitators" from regaining influence, the ACOA took measures to expand TCI's example around the district. Toward this end, it recruited sociological experts, held conferences, and circulated pamphlets on such themes as sanitation, housing, living habits, truck gardening, commissary practices, and industrial safety.[16]

The operators' self-congratulatory assessments of these "modern" initiatives implicitly conceded that the miners' traditional grievances had been valid all along. "The old idea that the mere provision of shelter was all that was required, is banished," McCormack claimed in 1913; companies were learning "to take an interest in sanitation, cleanlines [sic] and the prevention of disease." That year ACOA secretary-treasurer James L. Davidson reported with satisfaction that some company stores had abolished the check-discount system (see chapter 2), while conceding that at other commissaries prices were "seemingly high." The following year he singled out for approval those mines where a foreman could no longer "arbitrarily or for spite, discharge a man and throw him out into the cold world without a good cause." Still, few operators—not even TCI's nearest rival, the Sloss company—showed much interest in pursuing the welfare approach.[17]

Advocates of corporate welfare measures described their progress in glowing terms. The results, state chief mine inspector C. H. Nesbitt asserted in 1912, "have been marked . . . by greater stability of labor, phenomenal freedom from sickness of employees, more working days per man, greater efficiency and consequently greater output of coal per man per working day, and greater than all, a wonderful reduction in the number of accidents." A recent study of the TCI welfare program substantially confirms his report. In 1916 Davidson could look back on the past eight years as a golden era, especially when contrasted with the years that went before—a dark era marked by regular "lock-outs, strikes, bloodshed, destruction of property, [and] the expenditure of hundreds of thousands of dollars before a settlement could be reached."[18]

The extent to which the miners embraced these paternalistic strategies is difficult to gauge. District 20 remained dormant for a number of years, but the 1908 defeat and subsequent squelching of union activity readily explained that outcome. The succession of cheerless reports in the labor press suggests a further explanation, one that goes beyond bland contentment. Unionists mocked the welfare programs as cynical efforts to mask the harsh realities of life in the coal fields. In 1913 the *United Mine Workers Journal* carried a withering assessment of operator paternalism by Ethel Armes (author just three years earlier of a boosteristic history of the Birmingham district). The educational, health, and social facilities celebrated by the ACOA and its allies, she argued, were found at only a few of the district's 276 coal companies, and even there the quality of the facilities was uneven.[19] Little evidence remains of how the miners themselves regarded these programs. Only as the coming of war spurred bottomless demand for labor, and breathed fresh life into the union movement, would their loyalties be genuinely tested.

THE YEARS surrounding World War I were an epoch of unprecedented working-class ferment. The wartime preparation of 1916 opened dramatic new opportunity for workers in all sectors. "The abundance of employment between early 1916 and the summer of 1920," David Montgomery writes, "gave millions of workers the confidence to quit jobs and search for better ones and to go on strike on a scale that dwarfed all previously recorded turnover and strike activity." Union membership soared, nearly doubling from 2.6 million in 1915 to 5.1 million in 1920. Workers were emboldened not only by expanded material options, but also by the war's ideological thrust: large parts of the rejuvenated labor movement drew upon the imperative of defending democracy abroad in calling for "industrial democracy" at home.[20] These trends engulfed Birmingham. From the eve of America's involvement in the Great War through the aftermath, Birmingham workers in many trades organized toward such goals as an eight-hour day, higher wages, and union recognition.[21]

This reawakening swept the coal fields as well. In mid-1916, the operators began brooding about their "labor situation." They focused most heavily upon their black workers—and not without reason. It was in 1916 that hundreds of thousands of African Americans began pouring out of the South, in response to industrial opportunity generated by war preparedness and curbs on immigration.[22] The Great Migration has traditionally conjured the image of black sharecroppers seeking refuge from the bleakness of southern tenant life in the promised land of the urban North. The depiction of an overarching transition—from countryside to city, from farm to factory, and from South to North—does capture a broad strand of the black exodus.[23] But the Great Migration did not bypass the Birmingham district.

Birmingham occupied a complex position in the flow of black migrants, serving at once as a starting point for those heading north, a gathering point for those en route to the North, and a new home for those leaving the countryside farther south. Although black departure from the district began long before 1916, it swelled dramatically that year. Each day northbound trains carrying scores of African Americans steamed out of Birmingham and nearby Bessemer.[24]

The exodus encompassed all strands of the district's black community, including blast furnace laborers and skilled artisans, ore miners and domestic workers, steelworkers and middle-class professionals.[25] But, as a U.S. Department of Labor investigation concluded, "[t]he Negroes most sought after in the Birmingham district have been coal miners."[26] And the sporadic availability of work—during 1916 the mines were plagued by railroad car shortages—gave black miners an added incentive to head north.[27] Operators were not slow to appreciate how the dearth of cars imperiled their labor supply.[28] The Labor Department reckoned that "fully half" of those who had left did so because the mines were short of cars.[29] "The exodus is seriously felt at the coal mines in Alabama," the Birmingham *Age-Herald* noted in early 1917. It would persist more or less unabated into the 1920s.[30]

Migration accounted for most of the mine labor shortage, but in 1916 local movement by miners from employer to employer was back on the rise as well. Operators sought to stabilize their labor supply by competing for each other's workers. "Come here at once you can get a track job with Bill Goodwin $3.30 per day," scrawled J. F. McCurdy, a night foreman at TCI's Docena mines, to a DeBardeleben employee, in a typical effort. While vowing to confront TCI over McCurdy's intrusion, DeBardeleben doubted that he could reverse what was now a fact of life. "They are short of men . . . and their foremen are going to make desperate efforts to recuperate their forces."[31]

Along with the black exodus and rising labor turnover, a resurgent unionism rumbled through the district in 1916. Mirroring a national trend, unions took root that summer and fall in numerous trades, embracing streetcar operators, bakers, printers, chauffeurs, office employees, meat cutters, and cooks and waiters—all demanding, and some obtaining, such commonly sought goals as the eight-hour day, higher wages, and recognition. Before long, organization spread to skilled metal workers in the city's vital iron and steel industry.[32]

Meanwhile, the coal fields were abuzz with talk of a revived District 20. After two years of silence, letters from Alabama began to reappear in the *United Mine Workers Journal*. "We long for organization," wrote a miner from Margaret; he described squalid conditions, infrequent pay, and the much detested "shack rouster," or "company devil," who "rides around and sees that you do not lay off unless you are sick." "We want you all to help us out here," another wrote from Townley, "for we've got to do something or starve." There were signs that African Americans, who now comprised the majority of Alabama miners, were particularly anxious to organize. "[A]lmost every negro miner in the district has sworn never to give up the struggle until we have planted the banner of union-ism in every camp in the B'ham district," a black miner reported from Acton. The impression he conveyed of widespread alienation among black miners and their openness to unionism is arrestingly confirmed in the letter from Robert Taylor cited at the outset of this chapter. Certainly the operators and the black middle class did not take the loyalty of black miners for granted. Starting in 1916 and continuing through the World War I era, Birmingham's black professional leadership mobilized a crusade, in coordination with the operators, to encourage black miners to remain true to the companies and repudiate unionism. So concerted and energetic was their campaign, that its themes and effects deserve close examination.[33]

The history of Birmingham's black middle class remains largely untold. In its outlook, its activities, and its institutional foundations, it resembled its counterparts around the urban South. Ideologically it conformed, with apparent unity, to the worldview of Booker T. Washington, an approach defined by accommodationism, gradualism, and self-help, tempered by polite but vocal opposition to the coarser excesses of the racial order, such as lynching, inflammatory stereotypes, and denial of voting rights.[34] The professional stratum of black Birmingham asserted its influence through a network of churches, fraternal lodges, newspapers, financial and business enterprises, schools, and civic organizations.

How intermeshed these institutions were is suggested by the careers of some of Birmingham's leading black figures. Oscar Adams served as editor of the Birmingham *Reporter*, principal of an elite black school (the Tuggle Institute), secretary of a fraternal burial fund, president of the Consolidated Amusement

Company, and a director of the Prudential Savings Bank. Reverend Dr. P. Colfax Rameau, editor of the *Workmen's Chronicle*, was also president of the Southern Afro-American Sociological Congress. Reverend John W. Goodgame of the Sixth Avenue Baptist Church was also a director of the Alabama Penny Savings Bank. Reverend Dr. J. H. Eason, pastor of the Colored Jackson Street Baptist Church and president of the Negro State Baptist Association of Alabama, also ran Birmingham Baptist College. Physician U. G. Mason was a director of the Prudential Savings Bank and had in earlier days served as local contact for Booker T. Washington. A number of these figures were active in Republican Party politics as well.[35] Mirroring black communities throughout the United States, the church served as the social core of black Birmingham. Of the scores of black churches that had proliferated around the young city, several—above all the Sixteenth Street Baptist Church and the Sixth Avenue Baptist Church—enjoyed extraordinary influence. Their functions ranged from Emancipation Day and Fourth of July celebrations to a series of social "uplift" activities that extended well beyond their immediate flocks.[36]

Efforts by Birmingham's African American establishment to assert moral leadership in the coal fields had had little effect before the 1910s. But the defeat of the UMW, the influx of new black labor at the mines, and the rise of corporate welfare schemes offered the black middle class their best opportunity yet to project influence. Some black professionals, such as Robert Taylor, signed on with one of the larger operators as full-time teachers or ministers. Others sought to uplift black miners through centralized structures. One such effort, launched in the aftermath of the 1908 strike, surfaced in a series of epistles in the black weekly *Baptist Leader*, under the heading, "The Improvement of the Miners' Camps." Written by W. R. Pettiford, a leading black Birmingham minister since the 1880s and president of the Alabama Penny Savings Bank, these pieces instructed black miners in respectable behavior. Pettiford urged his readers to be "constant, regular, and continuous" in their labor (and in particular to cease abandoning work on payday); to avoid "careless" labor; to cultivate harmonious ties with employers and appreciate their efforts "in renovating the quarters . . . building garden fences, as well as [providing] schools and churches"; and to otherwise improve themselves by shunning alcohol, attending church, joining reading clubs, watching their health, tending gardens, raising livestock, and, above all, saving their earnings.[37]

In stressing these points, the *Baptist Leader* was not acting alone; it was in fact a voice for an emerging network of black organizations and individuals (settlement workers, literary clubs, "Bible Bands," etc.) devoted to "improving the miners' camps." Local clergy were the hub of this network. Especially with the recent wave of segregation and disfranchisement, the *Baptist Leader* stressed, the black ministers had a role in the coal fields that extended beyond the spiritual: "Recognizing our situation as a people in the body politic in which we develop no statesmen or lawmakers and only a few business men, it leaves largely to the ministers the work of leadership not only in [their] ministerial capacity, but as leaders giving shape to public sentiment and advising along all lines among negro people." To facilitate this mission, local black Baptist leaders established a Ministerial and Economical Institute in 1908 to prepare ministers to counsel miners in matters of economics, family life, health, education, religion, and conduct in the community and at work.[38]

Thus, efforts by the black bourgeoisie to "elevate" African American miners preceded the World War I era. Their moral concerns were reinforced by material considerations. For black teachers, ministers, editors, and doctors, employee welfare schemes yielded financial support and direct employment. Prominent figures such as Adams, Rameau, and Eason showered the major operators with requests for funding and for forums to address the black miners, and more often than not such requests were favorably received. The ties between coal companies and black professionals had a social dimension as well. Leading operators were usually among the influential whites invited to ceremonial occasions in black Birmingham, such as Emancipation Day exercises or the numerous festivals and concerts that punctuated the year.[39]

Beyond the material benefits, the efforts of black professionals to remold the black miners were driven by an abiding belief, rooted both in doctrine and in experience, that it was with capital that the most promising alliances for African Americans could be struck.[40] Having seen the cracks in the edifice of interracial unionism, having witnessed the capacity of the operators and the state to demolish that edifice, having watched the creation of model towns that offered black miners conditions tangibly superior to those prevailing in the Black Belt or in earlier mining camps, and having tasted the influence and financial support that they could derive from these paternal enterprises, the black middle class found little cause to resist and every reason to embrace the prudent accommodationism of Booker T. Washington.

As turnover, migration north, and talk of unionism swept the coal fields, black middle-class leaders intensified their efforts to sway miners to conservative principles. Those who served full-time at the mining towns as teachers, pastors, etc., enjoyed an ongoing platform for their message. Newspapers such as the Birmingham *Reporter*, the *Workmen's Chronicle*, and the *Baptist Leader* offered another conduit. The major operators were more than happy to arrange for their circulation among the miners. With American entry into the war, leading black figures crisscrossed the mining district, hammering home the message of the hour. That message was as multifaceted as the Tuskegee-based philosophy it drew upon and the "labor problem" it aimed to overcome. Its themes were distilled in a speech by Dr. U. G. Mason to black miners at Sipsey: "I want to appeal to you to live clean lives, to be temperate in all of your habits, to put in each day a full day's work, to save your money, and in all things be steady-going, trustworthy men. If you have any grievances, take them to the boss and lay them before him like men. Don't be deceitful; don't shirk; don't cringe; stand up like a man." And the union was no place to stand up like a man. Here too, the late Booker T. Washington was alive and well. "Don't bite the hand that fed you," the Birmingham *Reporter* admonished. Only an all-black union, one that functioned primarily as a fraternal or self-improvement society, would do. "If the Negro must organize," Reverend A. C. Williams of the Sixteenth Street Baptist Church preached, "let the organization be a purely Negro one, offered by Negroes, and working only to promote the Negroes' efficiency and welfare."[41]

WHAT DID BLACK MINERS make of this gospel, and of those who imparted it? A protracted exchange in the pages of the Birmingham *Labor Advocate* and the *United Mine Workers Journal* during 1916 allows us to glimpse their reaction. Neither the black middle class nor white union officials could have been entirely pleased. Black correspondents decried the poverty and oppression of the mining camps and insisted that black miners eagerly awaited a revival of the UMW. Describing the iniquitous subcontracting system and the infernal "shack rouster," a miner from Acton claimed that blacks were treated as "no more than slaves" and were "praying for deliverance" through the return of organized labor. A miner from Republic detailed coercive practices at the commissary, the poor quality of black schools, the "inhuman" shack rousters, and declared that black miners "almost to a man recognize . . . that they should be organized and that the union would mean much to them." A white miner writing from Flat Creek confirmed that black miners were growing increasingly receptive to unionism. "There was a time when the Colored miners were very prejudiced against the union," he noted, but "[now] they are ready and waiting for some kind of move to be made."[42]

Such commentary reveals only so much about the relative appeal of conservative black and union campaigns to the black miners. Given the source, they indicate more about the tone of black unionism than the extent of it. More striking were the comments in these same letters concerning black middle-class figures, above all the charismatic editor of the *Workmen's Chronicle*, P. Colfax Rameau.

Pro-union miners granted that Rameau enjoyed tremendous popularity among the black miners. One white miner referred to "a preacher and social worker, Dr. P. C. Raymeau [sic], who the negro miners think all the world of, and anything that he might tell them is law and gospel." Black union miners confirmed the impression: a letter from the Bessie and Palos camps signed by "Colored Miners" remarked that "the Negro miners are simply wild about this Raymeau [sic]." They went so far as to suggest that unless he were converted to the cause, "there is not any way on earth for the union to make any progress among the Negro miners." As would be expected, the *Workmen's Chronicle* spoke often of Rameau's popularity among black miners; that the labor press concurred lent much credibility to the claim.[43]

Pro-union blacks, moreover, joined in the admiration. The "Colored Miners" from Bessie and Palos described a sermon by Rameau at Bessie Mines delivered to a packed house, including "nearly every man and woman of his race." "To give the Negro preacher justice," the writers reported, "he was a wonderful speaker and for more than one hour he held that vast crowd . . . perfectly spellbound with his wonderful argument on the 'Economic moral and industrial uplift of any race or people,' and his sermon has been the talk of the mines for the past two days." That Rameau admonished black miners not to look to the white man for their "industrial salvation"—a clear repudiation of organized labor—did not diminish the writers' praise; it only redoubled their determination somehow to convert the dynamic editor to the labor movement.[44]

The admiration of black union miners for this avowed foe of unionism confounded the one assumption shared by middle-class blacks and union leaders alike—that their respective endeavors were irreconcilable. That these miners could endorse aspects of each outlook (or at least each messenger) caught white union leaders off balance, and they responded in varying ways. J. B. Wood, vice president of the Alabama State Federation of Labor, praised Rameau's character—"Dr. Rameau is a fine type of intellectual negro. . . . I admire [his] spirit"—but faulted his vision, juxtaposing Rameau's promotion of individualism to the labor movement's ethic of mutuality. "The only hope that the Negro Miner has for real industrial freedom," he argued, "is to become a part of the great Miners Organization and put a stop to his own exploitation, and that of his fellow White Miner by those who attempt to play color against color, and wages against wages. Collective action will remedy these conditions; individual action can not." Other white unionists responded in less measured tones. R. A. Statham stridently denounced Rameau as a paid lackey for the operators, who had won over black miners—even those in the union—through his "charm." Statham's solution to the "Rameau problem" involved nothing so gentle (or unlikely) as converting him to unionism. He preferred to puncture the leader's mystique. Let Rameau's black supporters organize separately under his banner, he suggested sarcastically. "Send him to the company's headquarters and demand that schack [sic] rouster be abolished. . . . Just try him on this one issue and see if the Negro can work out his own salvation. If he fails to remove the schack rouster, turn him aside, join the United Mine Workers, obey its laws, and if the schack rouster is abolished[,] I will donate $50 to any charitable institution Dr. Rameau may choose." Elsewhere, Statham bluntly proposed that the miners drive the likes of Rameau from the district.[45]

Significantly, black unionists showed no inclination to join in the attack. At most, they acknowledged that Rameau's intentions outstripped his power to do good. "We, the negro miners of the B'ham district," one wrote, "have nothing to allege against the Dr. P. C. Raymeau [sic], as he is a great and good man, with a big, honest heart . . . who has always tried to make better conditions, but he is like we are, he is handicapped for the lack of funds to push forward the work." (The UMW, he added, had greater resources to improve the miners' condition.) From Republic, another black correspondent was less apologetic: "There is no use for Mr. Statham railing down on the Dr. P. C. Rameau, notwithstanding the fact that we want to see the United Mine Workers organized in the coal fields of Alabama, for he is an honorable man and truly a leader of the race and loyal to the rank and file of the Negro miners in this district. . . . We of Alabama are proud of Dr. P. C. Rameau." For this writer, Rameau's appeal lay largely in his persuasiveness. All who had heard the editor's recent address at Republic, the writer observed, "thank God that the race has such a sane leader." But his high regard for Rameau lay also in the tangible benefits he could deliver: "Dr. Rameau was out there in the interest of helping people in their effort to raise money to build a school and at the conclusion of his wonderful address he came down and with the assistance of the teacher raised a neat little sum of $94.40. . . . I am only making mention of this to Messers Statham and Wood, what Dr. Rameau is doing for the miners and their children." And yet Rameau himself would have been less than gratified by the writer's next point: that black miners were "not at all satisfied with conditions as they exist today at the different camps" and that they would benefit from a revival of unionism.[46]

These commentaries—lasting for only a matter of months, and discussing but one representative of the black middle class—can only begin to show how far its accommodationist gospel resonated among black miners or affected their relationship to work, company paternalism, and a renascent unionism. However, the exchange indicates much about the miners' sense of identity. It reconfirms the limits of models that counterpose class and race consciousness, as if they made up two ends of a seesaw. The letters indicate that black miners could find ways to appreciate the tangible benefits of corporate paternalism *and* of unionism; to admire the stature of established black figures *and* to embrace or discard their ideas with a selective eye; to applaud the power of such figures to enhance the welfare services of the mining camps *and* to use their time, spend their earnings, and express their loyalties in ways that defied the moral urgings of their benefactors. Solidarities of race and class thus mingled subtly in the miners' sentiments. White unionists and black middle-class leaders may both have found these dual solidarities exasperating, and those now inclined to view race and class consciousness as mutually exclusive could only find them perplexing. For the African Americans in the coal fields, the blending of these solidarities posed no insuperable contradictions. On the contrary, it represented a rational way of extracting whatever benefits—from sustenance to pride, education to empowerment—could be derived, from whatever source, in a world where conditions were hard, livelihood precarious, mobility limited, and alliances, of any kind, unreliable.

IF BLACK MINERS' regard for the black middle class was not wholly at odds with an inclination to move about, head north, and unionize, the tension among these impulses was undeniable. They were able to reconcile them only because circumstances had not yet forced a resolution. That would change, however, with America's entry into the European conflict in spring 1917.

The resulting surge in industrial production opened employment opportunities for all the district's workers more widely than ever. Labor turnover now exploded. Operators took to calling it the "labor situation," presuming the meaning to be self-evident. "The labor situation is extremely acute," Fies wrote DeBardeleben that spring. "Our labor situation continues precarious," DeBardeleben wrote a customer awaiting an overdue order in July. Miners came and left with alarming frequency. Fies estimated that the company retained only one in five men it recruited for any length of time. Of thirty-six men recruited by one of its agents that September, only six remained by mid-October. Compounding the rise in turnover were high levels of absenteeism. DeBardeleben fretted that barely 50 percent of his miners were showing up even half the time. A U.S. Fuel Administration official likewise lamented the miners' habit of "working only part time and very irregularly."[47]

If sporadic demand for labor spurred much of the prewar rise in turnover, an insatiable need for labor fueled it during the war. The Galloway Coal Company president illustrated how this competition worked: "A neighbor operator . . . raised the labor scale 2½ cents per ton. I am informed that another operator is paying 5 cents advance. I thought it was bad faith to go along in this manner, so I instructed our Superintendent . . . to raise the wage scale 20%. I thought maybe that would bring about some conditions more favorable to ourselves." The bidding war for labor spilled over into other branches of the iron and steel industry. Coal producers found that to retain their labor they had to offer wages commensurate with those at the blast furnaces, coke ovens, and steel mills.[48]

Operators competed for labor not only through rates of pay but also through its frequency. Although federally imposed regulations dictated that paydays at the mines be held no more than twice per month, many operators paid their miners weekly, some even daily. Having frequent paydays, the Alabama fuel commissioner noted, was essentially no different from paying bonuses and premiums and as such "causes a constant shifting of men from the semi-monthly pay-day mines to those that pay daily and weekly." Operators also began actively raiding each other's labor. "You can readily understand that this sort of practice will completely demoralize the labor in this County," wrote Fies to an operator he suspected of recruiting employees away from DeBardeleben. Through such efforts the operators aggravated the very problem they were seeking to address.[49]

The urgency of unflagging production during the war triggered a series of mine wage increases—some the outcome of market competition, others decreed by the U.S. Fuel Administration. If the increases were meant to stabilize labor, they also, in the operators' eyes, exacerbated transiency. Unencumbered by ethics of hard work or acquisitiveness, they argued, miners would stay on the job only long enough to indulge their short-term fancies. An Empire Coal Company official made the point in August 1917: "Since the first of March the miners in this district have had four 10% increases in wages, and with the usual result with that class of labor, they are not now working on an average of more than one-half time, being able to make enough in those 3 days to run the balance of the week, which seems to be about all they care for. In other words, they are getting too much money." Although in one sense "that class of labor" referred to miners generally, the depiction was applied most liberally to African Americans. "Fully 75% of the miners and mine workers of this district," DeBardeleben wrote the state fuel administrator, "are negroes whose whole thoughts are, as a rule, to satisfy their immediate wants and hence make irregular workers." "As I see it," he added in a letter to Rev. Williams of the Sixteenth Street Baptist Church, mine wages had been raised "until it now has placed the earning capacity of a man beyond his capacity of want; or, in other words, his earning capacity has increased more greatly than his education. At the present rates, mine labor can supply their weekly wants with one-half effort, therefore, we do not get full time at the mines." And heightening the operators' troubles, large numbers of black miners continued to depart for coal fields and urban centers to the north.[50]

With wider opportunity for miners came greater independence, as they took more time away from work, bid employers against each other, and brushed aside managerial directives that a year or two earlier they would likely have accepted as law. From the operators' perspective, alternative sources of livelihood had made their employees, even supposedly "docile" African Americans, harder to control. "The negroes of our section," Fies complained, "do not realize that they have any responsibility or obligations with reference to the part that all men should play in our country's affairs at this time. I have argued, preached, and fired with very little effect." Although never one to shrink from a challenge, here Fies could only exclaim in frustration, "[f]rankly, the problem is too hard for me to solve. Discipline has no effect because if you discharge a man he knows he can go to Bessemer and get a ticket to Cincinnati or some other place up North." "Something should be done," he declared in a speech to the ACOA, "to prevent a miner, who has been justly docked or fined for loading rock in his coal, from leaving his work and going to a neighboring mine and pursuing the same tactics." For the wartime operators, "labor shortage"— always a social concept measuring power relations no less than supply and demand—was closely connected with labor productivity and subservience. "We have been greatly disappointed in the increase of our output," DeBardeleben confided to a client. "We find that the increases given labor, especially the miners, has decreased their efficiency. . . . We have so completely lost discipline in the camp and in the mines that it is a hard matter to know one day what the output will be the next." Elsewhere he captured the link between the miners' availability and manageability with a timeless reflection: "It is an easy matter to handle labor when there are two men for every job, but it is most difficult to handle labor when there are two jobs for every man, and the latter condition exists at present."[51]

DURING SPRING 1917, circumstances seemed more auspicious for the Alabama UMW than at any time since 1908. The high demand for labor in the coal fields and the reawakening of unionism elsewhere around the district emboldened miners to organize. So too did another signal development: the unprecedented extension of federal power into labor relations around the country. Responding to a surge in worker militancy—the first nine months of 1917 would witness nearly 3,000 strikes nationally—and to the imperative of wartime production, Secretary of Labor William B. Wilson oversaw the development of a coordinated labor policy. It was a two-pronged approach. One was to snuff out radical dissent, a strategy unleashed with particular ferocity against the Industrial Workers of the World. The other was to actively stabilize labor relations through a cluster of agencies, such as the Fuel Administration, the Railway Administration, the Shipbuilding Labor Adjustment Board, the President's Mediation Commission, the United States Employment Service, the Division of Negro Economics, and, most comprehensively, the National War Labor Board (NWLB). Comprised of equal numbers of businessmen and AFL leaders and co-chaired (along with former president Taft) by the left-leaning advocate of "industrial democracy" Frank P. Walsh, the NWLB was empowered to resolve labor disputes in accordance with Progressive Era principles of collective bargaining, the eight-hour day, and the right to organize. Although bound by its mandate to consider "prevailing conditions" in particular regions—a sop especially to southern business interests—the NWLB inevitably involved an intrusion into southern class and race relations on a scale not seen since Reconstruction. Certainly the expanded federal presence afforded the Alabama UMW extraordinary room for maneuver.[52]

On June 3, 1917, the UMW held a rally at Brookside that attracted 3,000 people. So began District 20's first organizing drive in nine years. In the weeks that followed, the union, bolstered by an infusion of funds and five organizers from the national organization, held mass meetings in scores of towns across the coal fields. By mid-June, District 20 had thirty new locals, claiming a total of 9,310 members, and a speech by national UMW president John P. White drew 5,000 miners.[53]

Antiunion voices denounced District 20 (along with other unions in the district) as subversive to the war effort: an agent of the Kaiser, the communists, or both. ACOA vice president Charles F. DeBardeleben lumped the UMW together with the Industrial Workers of the World "and other extreme labor unions," saying the miners' union had "imbibed the German Kultur." "A mine district run by the grasping officials of the miners' unions is an autocracy," Walter Moore, president of the Empire and Eldorado Coal Companies, declared in June. "If the miners are wise," the Age-Herald added, "they will not listen to those who would put a stumbling block in the nation's path."[54]

If the question of patriotism put District 20 on the defensive, no less did the irrepressible race question. Black middle-class leaders pressed their campaign to keep black miners out of the union. In countless speeches at the mining camps and articles in the Birmingham Reporter and the Workmen's Chronicle, they joined leading white figures in denouncing labor organizers as traitors, bent on hampering the war effort and inducing black miners to rebuff their benefactors. If they did that, the *Reporter* warned, "the operators will look around at what they have done to give you better living conditions, better schools for your children. . . . hospital accommodations . . . [better] houses and sanitary conditions. . . . and they will wash their hands of you." AME Pastor W. H. Mixon invoked the biblical words of Isaiah: "If ye be willing and obedient ye shall eat the good of the land; but if ye refuse and rebel, ye shall be devoured with a sword, for the mouth of the Lord has spoken for it."[55]

Black professionals also stressed that the UMW was a "white man's union" with no real commitment to the welfare of black miners. They pointed out that, although African Americans now made up a clear majority of miners, District 20 and its constituent locals were still led primarily by whites. They highlighted efforts by white union leaders to bar black members from attaining a voice commensurate with their numbers. At a District 20 convention, the Birmingham *Reporter* observed, President Kennamer had refused to allow a black delegate to introduce a resolution encouraging "the principle in all local organizations that the race in majority shall be expected to hold chief offices, whether the majority be white or colored." Black members might vote for it, Kennamer explained, "but it will cause such a discussion and debate that it may destroy the purpose of the union, and of course I will have to rule it out of order." The *Reporter* saw the ruling as evidence that "the colored miner in the union is a tool to the white miner"; under a fair arrangement, it argued, black miners in many locals "should be given all the offices." Black middle-class leaders made their case against the union to federal officials as well. The Alabama UMW is "not the leader of my people and [has] no right whatever to represent them," Rameau wrote President Wilson, adding that the black miners would "gladly work if let alone by these labor agitators."[56]

Thus, District 20 was thrown on the defensive both by the old controversy of race and by the newer one of wartime patriotism.[57] The Alabama UMW found it easier to respond to the latter than to the former. Patriotism during the "war for democracy" was a double-edged sword. Echoing the rhetorical strategy of the AFL leadership, the union presented its crusade as consistent with, indeed the embodiment of, the nation's fight against autocracy. "The non-union miners of Alabama are as much an industrial autocracy as Russia was a political autocracy or Germany is a military autocracy," wrote one contributor to the *Labor Advocate*. "The men working in and around the mines," he added, "HAVE NO MORE TO SAY AS TO THE CONDITION UNDER WHICH THEY MUST WORK AND LIVE THAN HAD THE PEOPLE OF RUSSIA." The *United Mine Workers Journal* described the unorganized camps of Alabama as "feudal principalities." "Don't be a slacker," R. A. Statham intoned, using the wartime epithet for those who shirked their patriotic duty. "Your fellow miners in other states are calling for volunteers to help them protect the price of labor and conditions of employment. Your most patriotic duty now, fellow miners, is to join the union." Thus, like unionists around the country, District 20 deflected challenges to its wartime loyalty by casting its mission in patriotic terms.[58]

To the extent that "labor patriotism" gave the union rhetorical high ground, it also deflected the predictable volley of race-baiting.[59] But only indirectly: experience had taught miners' unions that public debate over the meanings of interracialism seldom augured well for their cause. On occasion the UMW indulged the temptation to crow that its enemies were failing to prevent black miners from joining. "It is reported that [the large operators] have made contracts with nearly all the negro preachers for them to use all their power to religiously influence the negroes not to join the ranks of the United Mine Workers," the *Labor Advocate* noted in early July, "but . . . it seems that this idea . . . does not have much effect."[60] Such claims were not hollow. Before the end of July the union boasted 123 locals, encompassing 23,000 out of 25,000 miners in the state. At this extraordinary moment of opportunity, in the face of spirited opposition from both the operators and the black professional class, the vast majority of miners of each race had again rallied to the promise of unionism.[61]

Not only did the operators have their own rhetorical arrows to shoot at the union, they also enjoyed collective resolution and formidable muscle. From the outset the ACOA declined to deal with District 20. Instead, it resorted to the time-tested methods of intimidation. A number of operators deployed gunmen in the vicinity of union meetings, and by mid-July hundreds of miners had been fired and blacklisted.[62]

In late July the union held its first district convention in nine years, to formulate a wage scale and list of demands. Two hundred sixty-six delegates representing 22,689 miners attended the gathering at Birmingham's Capitol Park. There were "very few old faces" among the delegates, the *Age-Herald* observed; the lean years since 1908 had given rise to a new generation of unionists. The demands they adopted, however, would have been familiar to their predecessors: union recognition, reinstatement of all miners fired for unionism, an eight-hour day, a wage of seventy-five cents per ton, the right to checkweighmen, the abolition of subcontracting, and the restoration of semimonthly pay. The one grievance that veterans of earlier struggles would find conspicuously absent was the convict lease, whose presence at the mines had diminished substantially over the past decade.[63]

Participants in earlier campaigns would also recognize the racial dynamics of the convention. While no figures exist on their relative numbers, blacks and whites were both strongly represented. The union once again conformed visibly to Jim Crow codes: black delegates sat on one side of the hall, whites on the other. And yet such separation continued to occur within a unified endeavor. If separate was not equal—the balance of district offices tilted once again toward whites—neither was it wholly a mechanism for racial subordination. African Americans played an active role. Black delegates "[are] recognized and freely express their views on each subject, just as the white members," the *Age-Herald* observed. At times they took independent positions, as when they led the fight (ultimately unsuccessful) to close the convention to outsiders. Also familiar was the manner in which District 20 cast its racial approach before a wary citizenry. Public statements once again bracketed the union's interracialism with a clear disavowal of "social equality." "If I thought the negroes wanted social equality," district president Kennamer told the convention, "I would not be here today. All the negro wants is an opportunity to work under decent conditions."[64]

Despite the breadth of miners' support for the UMW, the operators declined to negotiate. In response, the union set an August 18 strike deadline. Explaining their position to dismayed federal officials, the operators stressed the perils of union recognition where African Americans predominated. In such a setting, Pratt Consolidated Coal Company vice president Erskine Ramsey warned the chairman of the U.S. Committee on Coal Production, "a democratic closed shop would be under local management of negroes which would be fatal to efficiency." The mass firing of miners continued. U.S. Attorney Robert Bills reported to the attorney general that each side had expressed an interest in federal intervention—the operators to curtail "agitation," the miners to protect against violence by company guards. The situation, he concluded, was "very tense." A replay of the raw clashes of earlier decades appeared imminent.[65]

The circumstances surrounding this showdown, however, differed in two key ways. First, the economic situation was far more auspicious for the miners than those greeting the strikes of 1894, 1904, and 1908; this time, demand for labor was virtually inexhaustible. Second, the appearance of a federal bureaucracy committed to maintaining stable labor relations offered the miners the promise, and to some degree the reality, of external protection scarcely imaginable in earlier times. It gave unions like District 20 a powerful entity, staffed largely by current or former labor officials, through which they could pursue recourse before matters came to a strike. In 1917, the operators were no longer the preeminent power in the Alabama coal fields.

The federal government was not, of course, omnipotent, nor was it unequivocally committed to the cause of labor. But as a strike loomed in mid-August, Secretary of Labor William B. Wilson (himself a former UMW official) stepped in, calling on the miners to postpone their deadline while he worked to hammer out an agreement with each side. District 20 complied, although the operators remained unwilling to meet with UMW representatives. On August 25, Secretary Wilson proposed a "Memorandum of Agreement" under which miners discharged for unionism would be reinstated, while most other issues— the eight-hour day, subcontracting, wages, and recognition—would be deferred. The proposal, reluctantly endorsed by union negotiators Kennamer and John L. Lewis, provoked a wave of "indignation meetings" and local walkouts across the Alabama coal fields. After three days of contentious executive sessions, the district convention rejected Wilson's proposal. It delayed the strike, however, pending further attempts from Washington to broker an agreement.[66]

Efforts to resolve the deadlock fell next to the U.S. Fuel Administration (USFA), established in August by President Wilson under the 1917 Lever Act. Headed by Harry Garfield, the USFA was granted broad powers to impose order on the critical but notoriously inefficient coal industry. Labor relations fell under the new agency's purview. After the unwieldy intervention by the secretary of labor proved fruitless, Garfield summoned representatives of District 20 and the ACOA to each meet with USFA officials. Under those less than voluntary auspices, negotiations in Washington proceeded through the fall among delegations for the Alabama miners, the ACOA, and the USFA. Back in Alabama the miners were "very restless," former UMW official and now U.S. commissioner of conciliation William Fairley reported in late October, due to "the uncertainty of when their demands will be disposed of." Finally, a proposal was completed in mid-December. Known as the "Garfield Agreement," it went far toward addressing the miners' demands: among its provisions were the right to join any union with impunity, reinstatement of all miners fired during the organizing drive, the right of miners to assemble freely, the appointment of an umpire to adjudicate disputes, the miners' right to checkweighmen of their own choice, semimonthly pay, and, for the first time ever in the Alabama coal fields, an eight-hour day. (The latter clause was added at the last minute by Garfield, after the union threatened to reject the entire proposal and launch a strike without it.) The union did not get all it asked. A key omission was recognition; coal companies were bound to meet with mine committees as grievances arose, but not explicitly with the union as such. The miners were constrained from holding down output or interfering in any way with the operation of the mine. The question of wage rates was deferred for half a year. Still, the proposal offered the miners their best contractual terms since 1903, and in some regards the best ever. A District 20 convention the following week approved the proposal overwhelmingly.[67]

The operators waited several more weeks before accepting these terms, and that acceptance came only grudgingly. Nor did they adhere consistently to the agreement. In February 1918 nearly 7,000 miners at the Tennessee and Republic companies and a scattering of smaller mines walked off over the operators' failure to rehire blacklisted miners. They returned only upon extracting a commitment from the national UMW and the USFA to investigate and adjust any violations of the Garfield Agreement.[68]

However rocky the agreement's enforcement, the wartime actions of the federal government yielded gains to the miners that would have appeared fantastic only a year earlier. In February 1918 the eight-hour day took effect. During the spring Judge H. C. Selheimer, appointed by Garfield to monitor enforcement of USFA policy in Alabama, resolved a series of disputes in favor of the miners. Most notably, he ordered that all those discharged for union affiliation be reinstated, that the operators cooperate in installing checkweighmen selected by the miners, and that the operators continue to provide labor to push cars to and from the coal face (and thus stop shifting such dead work to the miners). In May, an advance negotiated through the USFA brought wages to eighty-five cents per ton, or $3.84 when paid by the day.[69]

For many miners, the gains extracted under Garfield fell far short of the expectations raised by the UMW's revival. Some never got over the postponement of the imminent strike of August 1917. "We were organized about 100 per cent and were ready to ask for our rights," recalled District 20 delegate McGuire to the 1919 national convention, "but were not allowed to do so on account of the war." When the leadership had called on the miners to put off their strike pending government intervention, McGuire continued, "a great many of us hated to do so . . . and a good many [members] . . . got chilled on [the union]." Miners also grew restive over the failure of their wages, increases notwithstanding, to keep up either with the cost of living or with increases in other coal fields or local occupations. Miners' wages within the district, moreover, were far from uniform. This unevenness in pay, union officials informed Garfield in summer 1918, "was causing a great deal of unrest and discontent among the mine workers" who were receiving "far below a living wage." Surveying the situation, one national UMW officer concluded that the Garfield regime had given the Alabama miners nothing more than a "skeleton agreement." Others, though, emphasized the benefits reaped through federal intervention. "The miners of Alabama," wrote J. W. Brown, "are fully aware of the fact that were it not for the big, long-arm of old 'Uncle Sam' reaching down here into this benighted den of industrial autocracy, they would still be working ten hours per day for 'the wage that means a crust.' "[70]

As Brown's statement suggests, supporters and opponents of unionism competed for influence in the coal fields through patriotic stances. The operators appealed regularly to their miners to keep production at full throttle. Black middle-class leaders blitzed the mining towns as never before, often on behalf of the Red Cross, the USFA, the black Four-Minute League, War Savings Stamp campaigns, and other wartime enterprises. Through editorials and personal appearances at patriotic rallies, Adams, Rameau, Williams, Mason, Taylor, Mixon, and Goodgame urged black miners to aid the war effort in any way they could—not least, by remaining loyal to their employers. The large operators continued quietly to underwrite much of their campaign.[71]

UMW leaders raised the banner of wartime sacrifice as well, participating in patriotic rallies and imploring members to subscribe to the Red Cross fund, purchase Liberty Bonds, and otherwise support the war effort. Their appeals were effective: District 20 locals bought over $300,000 in Liberty Bonds and over $65,000 in War Savings Stamps. Union officials joined in the call for unstinting production. "Let the miners of Alabama stand by their picks in the mines as we expect our boys 'over there' to stand by their guns in the trenches," read one appeal from district president Kennamer and secretary-treasurer J. L. Clemo. The union, they assured the miners, offered a mechanism for the just resolution of local grievances. Rather than triggering suspensions of work, grievances should be taken to pit committees or district officers, who could in turn appeal to federal umpires "The Fuel Administration and Uncle Sam will see that you get a square deal in the end," the leaders promised. No less than federal intervention in the coal fields, the spirit of wartime patriotism offered black and white miners a rhetorical opening, and cover, for collective organization. From mid-1917 through 1918, the large majority of miners, black and white, continued to disregard injunctions from operators, the press, and black professional leaders that they steer clear of an interracial union.[72]

During the war, the fortunes of District 20 were linked more closely than ever with those of Birmingham's iron and steel workers. A showdown between white metal trades workers and the major steel companies would have large repercussions for miners' unionism. On February 20, 1918, between 4,000 and 6,000 skilled white steelworkers—machinists, electricians, molders, boilermakers, pattern makers, and blacksmiths—struck for an eight-hour day. Another 12,000 laborers were idled as a result. Conducted under the aegis of Birmingham's Metal Trades Council (MTC), the strike locked many of the district's most skilled workers in a protracted battle with a cluster of staunchly antiunion employers led by TCI.[73] Before long, the operators were recruiting strikebreakers, black and white, while the Birmingham City Commission took measures to protect them and to arrest the strikers as "vagrants." With mounting frequency each side painted the other as the moral equivalent (if not actual agent) of Kaiserism. The antiunion Ensley *Industrial Record*, for example, denounced the strikers as "revolutionary" and "treasonous." In turn, Birmingham Trades Council president J. B. Wood scored the steel companies as, collectively, the "Great American Kaiser."

Such rhetoric might ennoble the metal workers' cause but could hardly on its own deliver victory over their powerful adversaries. By March it was clear that the strikers' prospects for success would turn upon two players thus far only on the margins of the conflict: the laborers at the mills and the federal government. After several weeks, strike leaders concluded that they had to mobilize unskilled workers to shut down production. That meant focusing on African Americans, who predominated among the laborers. Accordingly, the MTC invited the International Mine, Mill and Smelter Workers' Union (Mine Mill), which already had a following among black ore miners, to organize production workers along biracial lines at the steel mills.

Although Birmingham's steel mills were historically less fertile terrain for interracial organization than the coal fields, the strikers' move to unionize black mill workers achieved substantial success.[74] Efforts by black and white organizers to recruit a biracial base of production workers provoked virulent attacks—both red-baiting and race-baiting, rhetorical and physical. May 6 witnessed the local rebirth of the Ku Klux Klan (KKK), as 150 members paraded through the streets of Birmingham to oppose the steel strike. Vigilante groups unleashed a series of violent attacks on Mine Mill activists. Meetings of black steelworkers were broken up. In mid-April the home of black organizer Ulysses S. Hale was bombed. Several months later Hale and Edward Crough, a white Mine Mill organizer, were kidnapped; both were beaten, and Hale tarred and feathered as well. In protest, 1,500 Mine Mill ore miners, black and white, conducted a one-day strike in nearby Russellville.

Interracial unity kept the strike alive; the brutal reaction underscored its precariousness and how much success or defeat would hinge on protection from Washington. Upon the urging of the MTC and AFL president Gompers, the newly established National War Labor Board sent staff investigator Raymond G. Swing to the district to determine whether it should step in. Here was a critical crossroads, not only in the course of the strike, but more broadly in the outlook for both race and class relations in the South. Not since the Reconstruction era Freedmen's Bureau had a federal agency intervened so boldly in the way employers handled their workers, black as well as white. Sensing the possibility of expanded federal protection, black ore miners and mill workers, hitherto intimidated by company gunmen and the revived KKK, began pouring into Mine Mill. Swing observed "a new spirit" among the black workers, and cited employer efforts to prevent their organization as the source of "most of the bitterness between labor and capital in the Birmingham district."[75]

Appalled by the operators' heavy-handed tactics, Swing recommended swift action to settle the strike. But the employer wing of the NWLB—uneasy about the interracial complexion of the strike and not inclined to overrule "local custom" by imposing an eight-hour day—vetoed intervention. The NWLB's inaction at this crucial moment, historian Joseph A. McCartin has aptly observed, "starkly illustrated the limitations of federal intervention in the South, revealing the range of political forces that were threatened by growing power in the workplace." Left to their own devices against daunting opposition, the strikers abandoned their effort. Although U.S. Steel would later establish an eight-hour day, the steel mills of Birmingham remained open shop; the great national steel strike of fall 1919 would bypass the district completely.[76]

For the miners, the outcome was a sobering reminder that federal protections and the promise of "industrial democracy" were tenuous indeed—particularly for an interracial enterprise like the UMW. Even at the union's high-water mark, the outlook had been mixed. Not only did the operators implement the Garfield Agreement at best grudgingly, but periodic wage increases failed to match wartime inflation or, for that matter, the rates of profit flowing to the coal companies. Nevertheless, in the aftermath of armistice in late 1918, union miners felt they had achieved great gains during the war, gains they now stood to lose. Before American entry into the war, recalled black international organizer George H. Edmunds in 1919, "Alabama miners worked ten hours per day for $2.25, had no checkweighmen on the tipples, no mine committees, could not belong to a labor organization, nor could they meet to discuss their grievances." Now, he noted, "there are checkweighmen at every mine where the miners want one, the eight-hour day is in vogue all over the district, wages are from $3.44 to $4.00 per day, mine committees at all of the organized mines, and an umpire at Birmingham. . . . This is some stride in two years."[77] But as the coming of peace diminished the urgency of uninterrupted production, those gains came to appear increasingly vulnerable. While the continuing exodus of African Americans for points north helped to offset the tailing off of demand for mine labor, federal wartime agencies—those that were not promptly dismantled—became less responsive to miners' grievances.[78]

The ebbing of demand for workers and of federal intervention were national trends. The tensions they generated contributed to the great postwar strike wave that swept the United States in one sector after another, including textile workers, telephone operators, railroad workers, policemen, transit workers, stockyard workers, longshoremen, coal miners, and steelworkers. The year 1919 saw 3,000 strikes around the country, involving some 4 million workers. For bituminous coal miners, the moment of reckoning arrived on the first of November 1919, when nearly 400,000 struck for a wage increase. For the first time, Alabama miners joined formally in a nationwide strike. R. A. Statham caught the miners' mood as the walkout neared: "We won the battle for world democracy, yet there are those who would take this grand victory from us. . . . If we allow [the operators] to dictate terms and conditions of labor without labor having a voice in the case, then all our sacrifice was made in vain." After a week federal judge Albert B. Anderson enjoined the strike nationally, and district officers directed the Alabama miners to return to work, pending national negotiations among the UMW, the bituminous operators, and the government. Fuel Administrator Garfield proposed a 14 percent wage increase—a grave disappointment for the miners—and appointed a Bituminous Coal Commission to consider other outstanding issues. To District 20's dismay, assurances by Attorney General A. Mitchell Palmer that the Alabama operators would implement the wage increase, cooperate with the commission, and rehire returning strikers all went unfulfilled; rather, the ACOA left wages unchanged, snubbed the commission, and maintained a districtwide blacklist numbering hundreds of miners. If the operators had calculated that Washington was losing its zest for intervention, they were right. Kennamer wrote President Wilson in April 1920 protesting the operators' failure to implement the coal commission's recommendations. The letter was passed on to Palmer, who replied that there was nothing further the Justice Department could do. An appeal to the head of the USFA, whose mandate was to expire in early spring 1920, also proved fruitless. Like the Birmingham steelworkers—indeed, like industrial workers throughout postwar America—the Alabama miners were now on their own.[79]

WHAT BEING ON THEIR OWN meant became apparent the following year. By mid-1920 the wartime labor market gave way to a dramatic shortage of work, as the economy entered a downturn. On March 31 the Garfield Agreement expired, and the USFA passed out of existence. The Bituminous Coal Commission conducted national hearings to prepare its recommendations for the terms that would govern post-Garfield agreements. District 20 representative W. L. Harrison urged the commission to consider establishing uniformity of wages between the district and other coal fields, to abolish subcontracting, to install weight scales at all tipples, to raise wages to match inflation, and, at long last, to eradicate the convict lease. The commission met District 20 a portion of the way, recommending that the terms of the 1917 agreement sponsored by the USFA be renewed, along with a 27 percent wage increase. The ACOA rejected that initiative, offering instead an employee representation plan in which miners would have committees that operators might consult, but no real bargaining power; management would determine the terms of work. District 20 rejected those terms, and another showdown was in the offing.[80]

In April, unable to negotiate with a collective body of operators, District 20 adopted a "blue book" contract through which locals could seek employer recognition on a one-by-one basis. Companies refusing to sign would be targeted for strikes. Along these lines, the union conducted an organizing drive during the spring and summer. Fifty-five operators, nearly all commercial ones, signed agreements with UMW locals; the larger operators held unwaveringly to the open shop. The summer months witnessed low-level warfare in the coal fields, as several thousand miners took part in a series of local strikes for recognition. The response of the open-shop companies was fierce: they imported strikebreakers from the Black Belt, evicted strikers' families, cut off the water supply to their communities, used company guards to suppress meetings, and dispatched shack rousters to flush out striking miners. By late August, approximately 5,000 miners remained locked in a string of bitter local fights.[81]

The UMW concluded that the time had come for one last attempt at recognition. On September 1, national president John L. Lewis issued a call for a strike of Alabama miners against all open-shop operators. The following week 12,000 of the state's 27,000 miners joined in what would become the coal field's longest general strike yet.[82] As the miners walked off, it must have seemed to many that matters had come full circle since the summer of 1908. Once again, a hobbled union embarked on a do-or-die strike across the district; once again, the strike would be contested by political and economic forces within Alabama. With the world war long over, the federal government would show scant interest. Nor did the rhetoric of "industrial democracy," articulated so eloquently by union miners during the war, carry the same resonance at a time when the nation was returning to "normalcy."

The major operators lost little time in recruiting strikebreakers, largely black farmers from farther south. Strikers' families again faced mass eviction, and tent colonies, provided for by the union, reappeared around the district; at their peak, they held 48,000 people.[83] In Walker County a federal injunction barred strikers from "picketing or loitering." Before long, the familiar cycle of violence emerged, as strikers resorted to intimidation and sometimes gunfire to deter "blacklegs," and scores of deputies and company gunmen used similar methods against the strikers and their families. Predictably, the air of violence that surrounded the strike district—and the skewed and feverish coverage it received in the daily press—turned much of the public against the miners. September 17 brought the appearance of two outside parties that would escalate the conflict significantly. The first was UMW international organizer Van A. Bittner, designated by Lewis to take command of the strike. The same day, four companies of the Alabama National Guard, called up by Governor Thomas E. Kilby, arrived in Birmingham and fanned out around the strike district. The Guardsmen imposed a severe regime, harassing local inhabitants, shooting at strikers, and monitoring, then banning, union meetings.

Race resurfaced as a hot-button issue in the conflict. This time the majority of strikers were African American. The union continued to function on a biracial basis. In June, District 20 had divided itself into three subdistricts, each of which elected two vice presidents, one black and one white; these six together comprised the executive board of the overall district.[84] Following the traditional Alabama formula, the district vice president, J. F. Sorsby, was black, as were many local presidents and vice presidents. The national union, for its part, dispatched several black organizers to Alabama, and across the district black and white strikers could be seen walking the same picket lines and cheering impassioned oratory at union meetings.

The language of patriotism and common national purpose no longer filled the air as it had two or three years earlier, and coarse attacks on the union's interracialism reemerged to fill the vacuum. In a public statement, the ACOA identified as one of the most "gruesome and sordid" features of the miners' union its practice of "associating the black man on terms of perfect equality with the white man." Elsewhere, echoing its rhetoric of 1908, the ACOA reminded the public how District 20's interracialism spilled over the gender line as well, describing meetings where "whites and blacks, men and women, assemble together; white and black jointly officer them and jointly address them." Black middle-class leaders issued spirited denunciations of their own. Most visible were two editors who had risen to prominence during the previous decade, Oscar Adams of the Birmingham *Reporter* and P. Colfax Rameau of the *Workmen's Chronicle*. Each joined in the broader attack on UMW organizers as dangerous radicals ("I.W.W. Bolsheviks" in the words of Adams, "socialists and anarchists" in the words of Rameau) and as purveyors of "social equality." Rameau was particularly vocal, writing lengthy antiunion epistles to the governor, just as he had a few years earlier to the president and his representatives.[85]

Union leaders did not take diatribes of this sort lightly. Kennamer revealed how sensitive District 20 felt on the race question when he unleashed an ill-considered attack upon their authors. "If you have any such leaders among you [as Adams and Rameau] at Sayreton," he cried in a speech to predominantly black strikers in September, "you should take them and hang them by the neck!" Opponents of the UMW seized on Kennamer's inflammatory outburst, condemning what they painted as the union's violent, even lynch-mob intentions. Kennamer toned down his impatience with black opponents, and the UMW settled into its traditional response to race-baiting, insisting that its mission was an industrial not a social one. The operators, the union charged, "have repeatedly stated that it is the purpose of the United Mine Workers of America to place the negro upon a social equality with the white workers. We brand this as a deliberate lie." It went on to brush aside the race question as a cynical distraction from the real issues of the strike. "There is only one race question now in Alabama and that is the race between the United Mine Workers of America on the one hand, standing for a living wage . . . and the Alabama Coal Operators' Association, which stands for the lowest wages and most disgraceful living conditions that exist anywhere in America."

Once again, a districtwide strike generated little appreciable discord between black and white miners. The lines of battle between the strikers and the operators were too starkly drawn and the shared conditions of black and white miners too extensive to allow racial matters to subvert the miners' solidarity. But interracial unity was nearly the only strength they enjoyed. As the strike dragged on into winter, the miners' prospects grew dim. A sharp industrial slump late in the year depressed markets for coal, while production at the hands of strikebreakers gradually rose. Violence, intimidation, and repression of union meetings incapacitated the strike. Public opinion grew increasingly hostile to the miners' plight. By early 1921 mounting numbers of strikers were drifting back to the mines. At this point Bittner, seeing no other card to play, agreed to a proposal that Governor Kilby settle the strike, along whatever terms he thought fit. On February 22, a District 20 convention called off the strike, and its resolution now lay in the hands of the governor.

The union felt little cause for optimism. Over the course of the strike Kilby had proven no more sympathetic to the miners' cause than had Governor Jones in 1894 or Governor Comer in 1908. Kilby's settlement, issued March 12, confirmed the miners' worst fears. On virtually every point the governor sided with the operators, denying union demands for recognition, higher wages, abolition of contract labor, and rehiring of strikers. The strike, he suggested, had been "illegal and immoral," and the thousands of strikers left without work or shelter were the moral responsibility of the union. For a number of weeks, the UMW continued to provide relief. But through the remainder of the year material distress haunted the miners and their families. Over the summer, operators cut wages across the district. Once again, the Alabama UMW lay in ruins.

THE RESURGENCE and decline of miners' unionism in the World War I era shared key features with the earlier fortunes of organized labor. As in the late 1890s and early 1900s, the wartime revival of District 20 materialized amid a swelling demand for coal. The issues raised by the union also had a long history; they dated back to the earliest labor protests of the Alabama coal fields, voiced through the Greenback-Labor Party and the Knights of Labor, shortly after the founding of Birmingham. The defeat of 1920–21 likewise recalled the bare-knuckle tactics of earlier times.

More striking still were continuities in the racial dynamics of miners' unionism. At all times, District 20 encompassed both black and white miners, reflecting the practical merits of interracial collaboration, together with an authentic bond of shared experience. Time and again, the union hedged its challenge to Jim Crow—in part by instinct, and in part consciously to accommodate prevailing sensibilities—both within and beyond its ranks. And in each epoch, the collaboration held. The union's demise in 1921 occurred not through an internal collapse of interracialism, any more than it had in earlier periods, but rather to the formidable edge enjoyed by its adversaries—not least their ability to smear the union as an agent of "social equality."

For all the continuities that marked interracial unionism from the emergence of the Birmingham district through World War I, it did not function in a vacuum, nor did the circumstances that shaped its meanings and fortunes remain static. During the late 1910s, a concerted effort to mobilize American society fully behind the war effort created unprecedented breathing room and even inspiration for the joint struggle of black and white miners. Federal intrusion into southern labor relations (and inescapably race relations)—the first on this scale since Reconstruction—offered interracial unionism an alternative base of power. Like the Freedmen's Bureau, wartime labor agencies were riddled with shortcomings, deferring extensively to the racial and class hierarchies of the South. Yet it is instructive to consider who, between the miners and the operators, viewed the federal presence with hope and who with irritation. And no less significant than federal intervention in the coal fields was a wartime celebration of democracy that inevitably challenged the imperative of white supremacy in the South.

In this transformative atmosphere, even the advent of ambitious corporate welfare programs could not keep most miners, black and white, from embracing the union once it reemerged. Both materially and ideologically, wartime mobilization foreshadowed the conditions under which interracial unionism might someday reassert itself in the Birmingham district. The possibility would not arise again for another generation, when the New Deal, the Congress of Industrial Organizations, and finally the Second World War injected explosive new dynamics into southern race and labor relations.[86] Few in the Alabama coal fields could have foreseen such a radically altered landscape in the gloomy spring of 1921.

Epilogue

IKE THE DEVELOPMENT of Birmingham itself, the organization of Alabama's coal miners in the late nineteenth and early twentieth centuries followed a cycle of promise and failure, as a movement crossing the color line alternately transcended and succumbed to the culture of Jim Crow. The issues it raised reverberate down to our own times; the recent growth and contentiousness of historical debate on race and labor are but one manifestation. It is perhaps appropriate, then, to conclude with some broader reflections on the meanings of race in this troubled story.

By the middle of the twentieth century, one might readily identify a union whose ranks were divided into black and white locals, whose top positions went disproportionately to whites, whose leaders disclaimed antipathy to segregation, and whose white members often subjected their black associates to unvarnished condescension, as part of the less progressive wing of American unionism. By the late twentieth century, such features would appear markedly retrograde. What stands out about the early campaigns of the Alabama miners, however, is the degree to which they surpassed the racial norms of their time and region. Each retained an internal color line—although at whose insistence, and for what reasons, cannot be presumed—but it was the rare association in the Jim Crow South that veered so far toward interracialism as to have a color line *bisecting* it, instead of defining its boundaries. Racist thinking bubbled up commonly among white unionists, but in few other groups did it emerge not as the uncontested, dominant theme, but rather as a counterweight to another, more racially inclusive ethos.

Common class experience among black and white miners provided the impetus to interracialism; the prevailing culture of white supremacy imposed the limits. Unionism in the mineral district skirted, and at times crossed, the borders of segregation, but it was never—either in purpose or in effect—a civil rights movement. Nor could it have become one and still retained any prospect of surviving in the New South; even the qualified interracialism of miners' unionism left it vulnerable, at times fatally, to the intolerance of Jim Crow. At any rate, there is scant evidence that miners of either race contemplated such a challenge. To be sure, there are limits to what this tells us. The public atmosphere of the New South hardly encouraged freewheeling discussion of the race question in any setting, let alone among black and white unionists. And to the extent that they did thrash out the issue, the surviving evidence allows only a partial view. Ultimately, there is only so much we can know about how black and white miners identified, and debated, their options.

Had the racial climate they encountered taken a different, more tolerant turn, perhaps the miners would have responded by pressing more unreservedly against the color line—and perhaps not. Or perhaps a more liberal atmosphere would have emboldened black miners to demand a broader assault on Jim Crow than most white miners could accept, ultimately rupturing the delicate collaboration between the two groups. But there is no point in speculating how the Alabama miners of this period would have responded to tests and opportunities they did not face. If our purpose is historical understanding, then we must evaluate the actions of southern workers, black and white, against the circumstances of their own world. That is what this study has sought to do.

In the decades since emancipation, working-class southerners on each side of the color line have confronted (or not confronted) the tangled "race question" in myriad ways. Organized labor has been one—and only one—arena for this response. The history of race and unionism in the New South is just now beginning to receive the careful study that its complexity, and poignancy, demands. With the proliferation of in-depth research, alive to the peculiar constraints and possibilities of each setting, sterile generalities about race and class are giving way to a more nuanced, historically credible picture.

Only collective effort can yield a full panorama of labor's encounter with the southern racial order. Inevitably that effort will continue to produce conflicting interpretation—the issues are too resonant, the historians' sensibilities too diverse, and the history itself too varied to permit anything approaching scholarly consensus. Nor is this something to lament, for vigorous debate can only help to illuminate the haunting legacies of race and class in America. All we can ask of any contribution to this charged field is a hard look at the actions workers of each race took, and at how past experience and present options shaped those actions.

In short, we can ask only that the impulse to judge—whether to applaud or castigate—not overwhelm historical empathy. Not that moral vision must succumb to empirical rigor: such towering historians as C. Vann Woodward, E. P. Thompson, and W. E. B. Du Bois have shown how compellingly the one can reinforce the other. Indeed, to the extent that historical knowledge can inform social change, studies that subordinate celebration or indictment to the pursuit of empathy will be the most effective contribution to the present-day quest for a labor movement restored to the forefront of the struggle for racial justice.

Notes

ABBREVIATIONS

ACOA   Alabama Coal Operators' Association
ADAH   Alabama Department of Archives and History, Montgomery
BPLA   Birmingham Public Library Archives, Birmingham, Ala.
DeBardeleben Papers   DeBardeleben Coal Company Papers, BPLA
Jones Papers   Governor Thomas G. Jones Papers, ADAH
NA   National Archives, Washington, D.C.
Powderly Papers   Terence V. Powderly Papers, Knights of Labor microfilm edition
                        (available in miscellaneous library collections)
RG 67   U.S. Fuel Administration, Record Group 67, NA
RG 280   U.S. Federal Mediation and Conciliation Service, Record Group
                        280, NA
Shook Papers   Alfred M. Shook Papers, BPLA
Sloss-Sheffield Records Sloss-Sheffield Steel and Iron Company Records, BPLA
UMWA   United Mine Workers of America
UMWJ   *United Mine Workers Journal*

INTRODUCTION

1. *National Labor Tribune*, August 10, 1878.

2. Such findings were pioneered by Gutman, "The Negro and the United Mine Workers of America"; Worthman, "Black Workers and Labor Unions in Birmingham, Alabama, 1897–1904"; Green, "The Brotherhood of Timber Workers, 1910–1913"; Worthman and Green, "Black Workers in the New South, 1865–1915"; and Brier, "Interracial Organizing in the West Virginia Coal Industry."

More recent work that continues to unearth patterns of interracial unionism in the New South includes Rachleff, *Black Labor in the South: Richmond, Virginia, 1865–1890*; Corbin, *Life, Work, and Rebellion in the Coal Fields: The Southern West Virginia Miners, 1880–1922*; Arnesen, *Race, Class, and Politics: Black and White Longshoremen in New Orleans, 1863–1923*, "Following the Color Line of Labor: Black Workers and the Labor Movement Before 1930," and " 'It Aint Like They Do in New Orleans': Race Relations, Labor Markets, and Waterfront Labor Movements in the American South, 1880–1923"; Trotter, *Coal, Class, and Color: Blacks in Southern West Virginia, 1915–32*; R. Lewis, *Black Coal Miners in America: Race, Class, and Community Conflict, 1780–1980*; Gould, "The Strike of 1887: Louisiana Sugar War"; McKiven, *Iron and Steel: Class, Race, and Community in Birmingham, Alabama, 1875–1920*; and Letwin, "Interracial Unionism, Gender, and 'Social Equality' in the Alabama Coalfields, 1878–1908."

In the last several years, race and southern labor during the era of the Congress of Industrial Organizations (1930s–1950s) has received mounting scholarly attention as well. See Norrell, "Caste in Steel: Jim Crow Careers in Birmingham, Alabama"; Nelson, "Class and Race in the Crescent City: The ILWU, from San Francisco to New Orleans" and "Organized Labor and the Struggle for Black Equality in Mobile during World War II"; Kelley, *Hammer and Hoe: Alabama Communists During the Great Depression*; Honey, *Southern Labor and Black Civil Rights: Organizing Memphis Workers*; Korstad and Lichtenstein, "Opportunities Found and Lost: Labor, Radicals, and the Early Civil Rights Movement"; Halpern, "Interracial Unionism in the Southwest: Fort Worth's Packinghouse Workers, 1937–1954"; Stein, "Southern Workers in National Unions: Birmingham Steelworkers, 1936–1951." For a valuable overview of the literature on race and organized labor in the twentieth-century South, see Halpern, "Organized Labor, Black Workers, and the Twentieth-Century South: The Emerging Revision."

There was, of course, a significant body of work on black workers and the labor movement that antedated Gutman's article, including Cayton and Mitchell, *Black Workers and the New Unions* (1939); Wesley, *Negro Labor in the United States, 1850–1925* (1927); Spero and Harris, *The Black Worker: The Negro and the Labor Movement* (1931); and Northrup, *Organized Labor and the Negro* (1944). Contemporary as much as historical in focus, these studies first sketched the broad national patterns of race and labor in America, from industry to industry and union to union. They did not, however, have the benefit of the types of localized historical studies noted above. The growth and sophistication of the latter will lay the groundwork for richer, more textured overviews. The two historical surveys of the black worker to appear over the past two decades—Philip S. Foner's *Organized Labor and the Black Worker, 1619–1973* and William H. Harris's *The Harder We Run: Black Workers since the Civil War*—fall short of this kind of synthesis.

3. C. V. Woodward, *The Strange Career of Jim Crow*.

4. The charge that such historians, in their zeal to locate a transcendent heritage of working-class solidarity, have ignored the sobering realities of racism among white workers has been spearheaded in recent years by Herbert Hill. Hill's critique targets the work and so-called disciples of Herbert G. Gutman. See Hill's "Myth-Making as Labor History: Herbert Gutman and the United Mine Workers of America" and "Race, Ethnicity, and Organized Labor: Opposition to Affirmative Action." Support for and elaborations upon Hill's thesis are found in the following: Shulman, "Racism and the Making of the American Working Class"; F. R. Wilson, "Black Workers' Ambivalence Toward Unions"; Bernstein, "Herbert G. Gutman as Labor Historian"; and Fried, "The Gutman School; At What Intellectual Price." Other commentators, while suggesting that Hill's assault needs qualification, accept his assertion that Gutman glossed over racism; see Painter, "The New Labor History and the Historical Moment," and Roediger, "History Making and Politics." See also Walker, *Deromanticizing Black History*, and Aronowitz, *False Promises*, xxiv–xxvi.

5. For a rigorous rejoinder to Hill's critique of the "Gutman school," see Brier, "In Defense of Gutman: The Union's Case."

6. Historical scholarship that falls within this tradition is vast. Particularly influential investigations into informal slave resistance are Stampp, *The Peculiar Institution: Slavery in the Ante-Bellum South*; Blassingame, *The Slave Community: Plantation Life in the Antebellum South*; Genovese, *Roll, Jordan, Roll: The World the Slaves Made*. On free labor, see

especially Gutman, *Work, Culture, and Society in Industrializing America*, 3–78; and Montgomery, *The Fall of the House of Labor: The Workplace, the State, and American Labor Activism, 1865–1925*.

7. Kelley, " 'We Are Not What We Seem': Re-thinking Black Working-Class Opposition in the Jim Crow South." For an elaborated version of his argument, see Kelley, *Race Rebels: Culture, Politics, and the Black Working Class*, 17–75.

8. Arnesen, " 'What's on the Black Worker's Mind?' African-American Workers and the Union Tradition," 9. Both Kelley and Arnesen write with particular reference to black workers in the South.

9. For the dichotomous approach, see especially the Hill school cited in note 4. For examples of scholars who have begun fruitfully to move beyond it, see those cited in note 2.

CHAPTER ONE

1. C. V. Harris, *Political Power in Birmingham*, 12–14; Armes, *Story of Coal and Iron*, 226–27; Dubose, *Mineral Wealth*, 65.

2. C. V. Harris, *Political Power in Birmingham*, 12–13; Armes, *Story of Coal and Iron*, xxix, 11; Phillips, *Iron Making in Alabama*, 27–98; Birmingham *Iron Age*, May 21, 28, 1874, July 15, 1875; U.S. Bureau of the Census, *Mines and Quarries: 1902*, 680; Dubose, *Mineral Wealth*, 23–27; Riley, *Alabama As It Is*, 44–46, 54, 62–63; McCalley, "On the Warrior Coal Field," 3–15, 129–36, 258–62; White, *The Birmingham District*, 5, 33; Chapman, *Iron and Steel Industries of the South*, 35–40, 54–56.

3. Armes, *Story of Coal and Iron*, 221–22; Caldwell, *History of the Elyton Land Company*, 1–6; Dubose, *Mineral Wealth*, 57–63.

4. C. V. Harris, *Political Power in Birmingham*, 13–14 (quote on 13); Armes, *Story of Coal and Iron*, 232, 234 (quote), 251–54; King, *The Great South*, 335.

5. The production of pig iron rose comparably during these years: from 11,000 tons in 1870 to 62,000 in 1880, to 700,000 in 1889, to 1,200,000 in 1900. See U.S. Bureau of the Census, *Mines and Quarries: 1902*, 167; U.S. Bureau of the Census, *Ninth Census: Statistics of Wealth and Industry*, 3: 602; U.S. Bureau of the Census, *Tenth Census: Compendium, Part 2*, 1140; U.S. Bureau of the Census, *Tenth Census: Report on the Mining Industries*, 15: 642; U.S. Treasury Department, *Statistical Abstract for 1889*, 178; U.S. Bureau of the Census, *Eleventh Census: Report on the Mineral Industries*, 14: 347; U.S. Treasury Department, *Statistical Abstract for 1900*, 349.

6. U.S. Bureau of the Census, *Eleventh Census: Population, Part 1*, 1: 7; U.S. Bureau of the Census, *Twelfth Census: Abstract*, 149.

7. For the classic portrayal of the late-nineteenth-century and early-twentieth-century South as a "colonial economy," see C. V. Woodward, *Origins of the New South*, 291–320. To this series of obstacles to the Birmingham district's industrial development, historian Jonathan Wiener would add—and indeed emphasize—a planter class that throughout the century was hostile to the advent of an industrial bourgeoisie and which enjoyed the cohesion and political power to sharply constrain the rise of an iron and steel center. For Wiener's critique of Woodward on this question, and for subsequent critiques of Wiener, see note 28 below.

8. Moss, *Building Birmingham*, 2–7; Armes, *Story of Coal and Iron*, 2–3; Hamilton, *Alabama: A History*, 8.

9. On coal and iron production in Alabama before the Civil War (for this and following paragraph), see Armes, *Story of Coal and Iron*, 1–103; J. H. Woodward, *Alabama's Blast Furnaces*; McKenzie, "Horace Ware: Alabama Iron Pioneer," 157–64; W. D. Lewis, *Sloss Furnaces*, 19–23; Starobin, *Industrial Slavery in the Old South*, 14–15; James S. Swank, "Statistics of the Iron and Steel Production of the United States," in U.S. Bureau of the Census, *Tenth Census: Report on the Manufactures*, 2: 101–2; Wiener, *Social Origins of the New South*, 137–38. On antebellum iron production in the South generally, see Cappon, "Trend of the Southern Iron Industry Under the Plantation System," and Chapman, *Iron and Steel Industries of the South*, 98–101.

10. Jonathan Wiener contends that a dominant planter class viewed with suspicion if not outright hostility the prospect of a manufacturing sector, which it deemed a threat to its own social and economic hegemony. The Alabama planters, he notes, were not averse to industrial development per se; indeed, over the 1840s there emerged among the Black Belt elite a growing interest in industrial expansion. But such growth, they felt, should proceed only along lines that would serve, and remain subordinate to, the interests of commercial agriculture. To the exasperation of those industrialists who dreamed of a major iron manufacturing center in the mineral district, Wiener argues, the planters would deploy neither their political power nor their economic resources to encourage such an outcome. The emphasis on controlled industrialization found political expression in the southern Whig alliance of large Black Belt planters and their "junior partners," the industrialists and urban commercial interests, whose pro-bank and pro-tariff stances aligned them against farmers and smaller planters. See Wiener, *Social Origins of the New South*, 138–43; see also Genovese, *Political Economy of Slavery*, 20, 23–26, 180–220.

A series of more recent scholars have argued persuasively that Wiener juxtaposed the economic alignments of the planters and industrialists in overly schematic strokes. His depiction of a monolithic planter class staunchly arrayed against large-scale industrialization, critics suggest, is belied by the active role of a cluster of Whiggish Alabama planters, known through the state's early development as the "Broad River group," in spearheading efforts to generate a substantial manufacturing base. The chief political opposition to industrial development, some have noted, flowed primarily from middling citizens who comprised the core of Alabama's Jacksonian Democrats. See Mills, *Politics and Power in a Slave Society*; Lewis, *Sloss Furnaces*, 15–18.

11. Most prominent among antebellum Alabama industrialists was Daniel Pratt, whose "model" industrial town of Prattville in Autauga County included a gin factory along with an assortment of mills, shops, and small factories. See Miller, "Daniel Pratt's Industrial Urbanism"; A. B. Moore, *History of Alabama*, 284.

12. Lyell, *Second Visit to the U.S.*, 69–70; Lyell, "Coal Fields of Tuscaloosa"; Armes, *Story of Coal and Iron*, 51, 63, 100–103; Norrell, *Promising Field*, 22–23; Dean, "Michael Tuomey and the Pursuit of a Geological Survey of Alabama," 101–11; Riley, *Makers and Romance of Alabama History*, 146–51; W. B. Jones, *Geological Surveys and Industrial Development in Alabama*, 12–16.

13. Stover, *Railroads of the South*, 5; Mills, *Politics and Power in a Slave Society*, 268–81; Dubose, *Jefferson County and Birmingham*, 130–33; Lewis, *Sloss Furnaces*, 24–33; Wiener, *Social Origins of the New South*, 140–41, 162–63; Armes, *Story of Coal and Iron*, 108–22.

14. Armes, *Story of Coal and Iron*, 124–94; McKenzie, "Reconstruction of the Alabama Iron Industry," 178–82; Stockham, "Alabama Iron for the Confederacy"; Vandiver, "The Shelby Iron Company in the Civil War"; J. Woodward, "Alabama Iron Manufacturing"; Thomas, *The Confederate Nation*, 207–11; McKenzie, "Economic Impact of Federal Operations in Alabama"; James P. Jones, *Yankee Blitzkrieg*.

15. The account contained in this and the following paragraph is based upon Armes, *Story of Coal and Iron*, 216–23, 243–51; Bond, *Negro Education in Alabama*, 41–60; A. B. Moore, "Railroad Building in Alabama during Reconstruction," 422–41; Stover, *Railroads of the South*, 88–94, 210–18; E. Foner, *Reconstruction*, 210–11; Wiener, *Social Origins of the New South*, 148–52, 163–68; Herr, "Louisville & Nashville Railroad"; Klein, *Louisville & Nashville Railroad*, 118–22; C. V. Woodward, *Origins of the New South*, 8–11.

16. Armes, *Coal and Iron in Alabama*, 196–214, 238–42, 251–58; Riley, *Makers and Romance of Alabama History*, 143–45, 333–37; J. H. Woodward, *Alabama's Blast Furnaces*; McKenzie, "Reconstruction of the Alabama Iron Industry," 183–89; J. B. Luckie, "The Cholera"; C. V. Harris, *Political Power in Birmingham*, 14; Clark, *History of Manufacturers in the U.S.*, 2: 288–90; Birmingham *Iron Age*, July 15, 22, 1875, May 18, 1876, June 20, 1877; Milner, *Alabama*, 206.

17. Armes, *Story of Coal and Iron*, 258–62, 266–72; Dubose, *Jefferson County and Birmingham*, 556–58; Klein, *Louisville & Nashville Railroad*, 132–34.

18. Armes, *Story of Coal and Iron*, 272–77; Birmingham *Iron Age*, March 12, November 12, 1879; McCalley, "On the Warrior Coal Field," 316–17; Fuller, "Henry F. DeBardeleben," 5–6; C. V. Harris, *Political Power in Birmingham*, 15.

19. Armes, *Story of Coal and Iron*, 283–308; Fuller, "Henry F. DeBardeleben," 6–8; Herr, "Louisville & Nashville Railroad," 38–39; Dubose, *Mineral Wealth*, 135–38; Stover, *Railroads of the South*, 147–49, 220–32; Klein, *Louisville & Nashville Railroad*, 264–71; Birmingham *Iron Age*, June 1, 1881 (Cincinnati *Gazette* quote), July 26, 1883; J. H. Woodward, *Alabama's Blast Furnaces*, 37–38, 160–62; W. D. Lewis, *Sloss Furnaces*, 65–67, 74–79, 97–99, 103–22, 140–41. On the industrial gospel of the New South, see C. V. Woodward, *Origins of the New South*, 142–74; Gaston, *New South Creed*.

20. Armes, *Story of Coal and Iron*, 330–412, 423–25; C. V. Woodward, *Origins of the New South*, 127–28; Perkins, *Industrial History of Ensley, Ala.*, 17–30; W. D. Lewis, *Sloss Furnaces*, 46–47, 124–25, 130–48; Fuller, "History of the Tennessee Coal, Iron, and Railroad Company," 64–67, 90–100; Fuller, "Henry F. DeBardeleben," 9–13; *Bessemer*, August 13, 1887; Norrell, *James Bowron*, xxv. For an overview of the burgeoning economic and cultural institutions of Birmingham in 1887, see Dubose, *Jefferson County and Birmingham*, 189–310.

The dual surge in urban and industrial development spread beyond the Birmingham district into the wider reaches of the mineral region. Between 1885 and 1893 two dozen boomtowns and 31 blast furnaces sprang up around central and northern Alabama, each fueled by the dream of matching, if not eclipsing, the Magic City. But few of these towns or ironworks would survive very long, and the growth of coal and iron production would remain concentrated in the Birmingham district. See Fuller, "Boom Towns and Blast Furnaces."

21. John H. Jones, "Coal-Mining Industry of Alabama in 1889," 3; *Historical and Statistical Review*, 17–19, 53; U.S. Bureau of the Census, *Eleventh Census: Report on the Mineral Industries*, 14:347; Birmingham *Chronicle*, March 27, 1890; *Historical and Statistical Review*,

16 (boosteristic tract); C. V. Woodward, *Origins of the New South*, 127 (Carnegie quote). For a comprehensive listing of industrial enterprises in Birmingham during the late 1880s, see Riley, *Alabama As It Is*, 63–64.

22. Armes, *Story of Coal and Iron*, 407–11; *Engineering and Mining Journal*, March 3, 1893; Norrell, *James Bowron*, xxvi–xxviii; Banta, "Henderson Steel & Manufacturing Company," 153; Fuller, "From Iron to Steel"; W. D. Lewis, *Sloss Furnaces*, 91–95, 179–81.

23. Norrell, *James Bowron*, xxvi; Clark, *History of Manufacturers in the U.S.*, 2: 242–43; Fuller, "History of the Tennessee Coal, Iron, and Railroad Company," 100–11. This shift to northern capital was reinforced by the failure of Henry DeBardeleben's audacious effort to attain control of TCI in 1893.

24. Armes, *Story of Coal and Iron*, 351–52, 426–27, 430–32; Chapman, *Iron and Steel Industries of the South*, 109; Fels, *American Business Cycles*, 184–219 (the 1894 strike is covered in depth in chap. 4).

25. Armes, *Story of Coal and Iron*, 431–35, 461–67; Norrell, *James Bowron*, xxvi–xxviii; *Engineering and Mining Journal*, March 18, 1893; Fuller, "History of the Tennessee Coal, Iron, and Railroad Company," 250–67; W. D. Lewis, *Sloss Furnaces*, 226–37; Lichtenstein, "Convict Labor and Southern Coal," 39.

In the delay in the coming of steelmaking to Birmingham, Gavin Wright has persuasively argued, coal and iron capital squandered the district's most auspicious moment for development on the scale of Pittsburgh. By the turn of the century, the market for steel was shifting northward to the emerging Great Lakes industrial centers such as Detroit and Chicago. Wright, *Old South, New South*, 165–68, 171–74.

26. Birmingham *News*, March 14, 1899. On corporate consolidation, see Armes, *Story of Coal and Iron*, 451–60, 473–87. On consolidation nationally, see Lamoreaux, *Great Merger Movement in American Business*; Chandler, *The Visible Hand*, 331–39. On nationwide industrial prosperity, see Sobel, *Panic on Wall Street*, 269–71. On expanding export markets, see Clark, *History of Manufactures in the U.S.*, 3: 24–25, 108–13. On Alabama coal and iron production, see U.S. Treasury Department, *Statistical Abstract of the United States for 1905*, 492–95. On union recognition, see chapter 5 below.

27. Armes, *Story of Coal and Iron*, 467–73, 516–29; Norrell, *James Bowron*, xxviii–xxxii; W. D. Lewis, *Sloss Furnaces*, 220–21, 224–26, 244–45, 247–51; Wright, *Old South, New South*, 168; C. V. Woodward, *Origins of the New South*, 300–302.

28. The two major lines of debate each turn upon the validity of themes advanced in C. Vann Woodward's classic *Origins of the New South* (see esp. 1–22, 107–41, 291–349). According to Woodward, the New South that emerged on the ashes of both slavery and Reconstruction was dominated by a nascent industrial elite. For all their Old South–style posturing, Woodward argues, the merchants, lawyers, and capitalists who spearheaded the southern "Redemption" were imbued with the bourgeois sensibilities of the Gilded Age, and determined to usher the region onto the frontier of industrial expansion. But despite pockets of commercial and industrial growth—port cities such as New Orleans; the textile areas of the Piedmont; the centers of large-scale cotton, sugar, and tobacco cultivation extending from the Mississippi Delta to the upper South; and the coal, iron, and steel district of Alabama—the New South remained perpetually mired in what Woodward considered the grim features of a "colonial economy": predominantly labor-intensive, extractive forms of production; a one-party political system slavishly responsive to the concerns of outside capital; a labor-

repressive legal structure that strove to keep in place a cheap, nonunion workforce; and an increasingly rigid and elaborate scheme of racial segregation.

In the late 1970s, new voices began to challenge the thesis that an emergent industrial bourgeoisie held the reins of the "redeemed" South. Most prominently, Jonathan Wiener has argued that postbellum Alabama had pursued a "Prussian Road" to industrial development (borrowing the concept from Barrington Moore), under the constraining hegemony of an unprogressive but still dominant planter class. The agrarian elite, Wiener holds, worked vigorously to contain the pace and direction of industrialization, lest it subvert the labor-repressive social order of plantation society. See Wiener, *Social Origins of the New South*; Wiener, "Class Structure and Economic Development in the South"; B. Moore, *Social Origins of Dictatorship and Democracy*; Billings, *Planters and the Making of the New South*.

Recent critics have effectively questioned (1) the extent to which the postbellum elite represented the same cast of characters as the antebellum, (2) the extent to which planters who did "persist" demographically into the postbellum era had the power to impose the limits on industrialization that Wiener portrays, and (3) the extent to which the planters, after or even before the war, were so monolithically antibourgeois to begin with. For good assessments of the Wiener-Woodward debate, see Cobb, "Beyond Planters and Industrialists"; Lichtenstein, "Convict Labor and Southern Coal," 6–10.

More recently, Gavin Wright has called into question Woodward's influential argument that the New South's failure to keep pace with northern development derived from its status as a "colonial economy," marked above all by a poignant and debilitating reliance on outside capital. While Wright recognizes the colonial character of the postbellum southern economy, he insists that this was more a manifestation than a cause of its lagging development; the reason lay more fundamentally in its position as a "separate regional labor market, outside the scope of national and international labor markets that were active and effective" during the age of industrialization (Wright, *Old South, New South*, 7).

29. Armes, *Story of Coal and Iron*, 48–57 (Tuomey quote on 49); Alabama Commissioner of Industrial Resources, *Report to the Governor* (1874), 8–9; Burke, *Coal Fields of Alabama*, 4–5; Dubose, *Jefferson County and Birmingham*, 123–24; U.S. Bureau of the Census, *Mines and Quarries, 1902*, 167; Moss, *Building Birmingham*, 38, 114.

30. Birmingham *Iron Age*, May 28, 1874.

31. The following overview of the process and work conditions in room-and-pillar mines in the late nineteenth and early twentieth centuries draws primarily upon Goodrich, *Miner's Freedom*, 15–55; Brophy, *A Miner's Life*, 38–50; Dix, *Work Relations in the Coal Industry*; Shifflett, *Coal Towns*, 80–112; Long, *Where the Sun Never Shines*, 24–51; Fishback, *Soft Coal, Hard Choices*, 42–59.

The richest source on the world of work in the Alabama coal fields is *Proceedings of the Joint Convention of the Alabama Coal Operators' Association and the United Mine Workers of America, 1903* (hereafter *Proceedings of the Joint Convention*). See also Flynt, *Poor But Proud*, 124–35.

32. Trapper boys opened and closed the "trap doors" along the mine passageways that regulated the circulation of fresh air underground.

33. Birmingham *Iron Age*, June 1, 1881; Chattanooga *Republican*, February 18, 1893.

Somewhat more dryly, Assistant State Geologist Henry McCalley painted a similar picture in an 1886 report on the coal fields of Jefferson County: "Among the miners, there are to be seen Americans (principally natives), Germans, Irish, Welsh, English, Swedes, French, Scotch, Austrians, Swiss, Bavarians, and Africans (principally natives)" (McCalley, "On the Warrior Coal Field," 268).

34. There exist no aggregate data on the racial breakdown of Alabama's coal mine labor force for the period before 1889. In a survey of 389 coal miners in Bibb, Jefferson, and Walker counties (the state's three major coal-producing counties) listed in the 1880 manuscript census, Paul B. Worthman found that 163, or 42 percent, were black (Worthman, "Black Workers and Labor Unions in Birmingham, Alabama," 379).

For 1889 data, see U.S. Immigration Commission, *Immigrants in Industries: Bituminous Coal Mining*, 7: 136, 142.

There exist no precise racial breakdowns for Alabama coal miners in 1900. Of all the state's 17,898 miners (of all kinds) and quarrymen, African Americans comprised 9,735, or 54.3 percent. Since the approximately 5,000 iron miners were generally described as predominately black, it can be estimated that blacks made up around one-half of the coal miners, who comprised the large majority of miners overall (U.S. Bureau of the Census, *Twelfth Census: Occupations*, 13: 98, 222; U.S. Bureau of the Census, *Mines and Quarries: 1902*.

For 1910 data, see U.S. Bureau of the Census, *Thirteenth Census: Occupational Statistics*, 4: 434; for 1920, U.S. Bureau of the Census, *Fourteenth Census: State Compendium, Alabama*, 49.

35. U.S. Immigration Commission, *Immigrants in Industries: Bituminous Coal Mining*, 7: 136, 142. Although the 1889 and 1899 figures include all Alabama miners and quarrymen, coal miners from the Birmingham district in each case comprise the vast majority.

36. The nearly 500 miners working at the DeBardeleben Coal and Iron Company's Blue Creek mines in 1889 were composed equally of blacks and whites. The same was reported for the Woodward Iron Company's Dolomite mines in 1891. Shortly after its formation in the late 1880s the Sloss company had approximately a 5:3 ratio of white to black miners; over the next decade the black presence pulled even. At Pratt Mines, the largest mining center in the district, the ratio remained roughly equal from the 1880s into the new century. On Blue Creek: *Alabama Sentinel*, December 14, 1889. On Dolomite: *UMWJ*, December 31, 1891. On Sloss: *Alabama Sentinel*, December 14, 1889; "Report of Superintendent of Coal Mines at Brookside, Brazil, and Cardiff for Fiscal Year Ending January 31, 1896," in Sloss-Sheffield Records, 4.1.105.1 SSSIC 1892–1902; *Proceedings of the Joint Convention*, 309–10. On Pratt Mines: Birmingham *Age*, February 5, 1886; *Proceedings of the Joint Convention*, 247.

37. The most telling evidence that blacks and whites both figured significantly as skilled miners lies not in direct description, but rather in common references to the presence of "white and colored miners" at this or that locality. The use of the term "miner" in this context was not generally meant to include unskilled laborers. (See Long, *Where the Sun Never Shines*, 36.) The available sources, to be sure, offer little explicit confirmation of the extensive presence of skilled black miners; more revealing, however, is the lack of evidence that the line between skilled and unskilled labor at the mines was a cleanly racial one. Given the inflammatory nature of race in the relations between miners and operators throughout this period—a central theme of the chapters to follow—we can expect that such a stark correlation between skill and race would have been the object of regular commentary.

38. An early-twentieth-century investigation into the conditions of American labor conducted by the British Board of Trade, for example, yielded this comment on the Alabama coal

miner: "Both white and coloured men are employed in the coal mines as pick miners, but practically all the labourers are coloured." British Board of Trade, *Report of an Enquiry*, 90. See also U.S. Immigration Commission, *Immigrants in Industries: Bituminous Coal Mining*, 7: 200; Birmingham *Labor Advocate*, October 14, 1905.

39. U.S. Immigration Commission, *Immigrants in Industries: Iron and Steel*, 9: 204–5.

40. Fitch, "Birmingham District: Labor Conservation," 1528; Birmingham *Age*, December 22, 1886. W. S. Lang, a Tennessee company superintendent, suggested in 1903 that native whites were more efficient miners than the average African American, but less so than the average immigrant (*Proceedings of the Joint Convention*, 550).

Such depictions of "poor whites" of the countryside were standard fare in the New South. See C. V. Woodward, *Origins of the New South*, 109–10; Flynt, *Dixie's Forgotten People*; Newby, *Plain Folk in the New South*; Jacqueline Jones, *The Dispossessed*, 48–53.

41. On northern European immigrants: U.S. Immigration Commission, *Immigrants in Industries: Iron and Steel*, 9: 204–5 (best labor available); Alabama Commissioner of Industrial Resources, *Report to the Governor* [1871], 21 ("fallacy"); see also British Board of Trade, *Report of an Enquiry*, 90–91; U.S. Immigration Commission, *Immigrants in Industries: Iron and Steel*, 9: 203.

42. On Slavs: British Board of Trade, *Report of an Enquiry*, 88; U.S. Immigration Commission, *Immigrants in Industries: Iron and Steel*, 9: 205. On other southeastern European immigrants: U.S. Immigration Commission, *Immigrants in Industries: Iron and Steel*, 9: 161–62, 203–5, 256. For prevailing images of various European ethnic groups in the United States during the late nineteenth and early twentieth centuries, see Higham, *Strangers in the Land*; Bennett, *Party of Fear*, 163–79; U.S. Immigration Commission, *Reports of United States Immigration Commission*, 43 vols. (Washington, D.C.: Government Printing Office, 1911).

43. Giles Edwards testimony in U.S. Senate, *Labor and Capital*, 4: 384–85; James W. Sloss testimony, ibid., 288, 290 (see also 283); Brainerd, "Colored Mine Labor," 78; *Proceedings of the Joint Convention*, 550, 655.

Some promoters of the mining interests consoled themselves with the belief that blacks were more suited to mining and other industrial pursuits than to farming, perhaps because they were less subject to the ongoing supervision of whites in the latter than in the former. "[O]ur negro labor," John T. Milner claimed in 1876, "does better in the coal and iron business than in farming." (Milner, *Alabama*, 206). See also Milner, *Alabama*, 146–61; Alabama Department of Agriculture and Industries, *Alabama Opportunity*, 7.

44. On brawniness: Brainerd, "Colored Mine Labor," 79; Fitch, "Birmingham District: Labor Conservation," 1527; U.S. Immigration Commission, *Immigrants in Industries: Bituminous Coal Mining*, 7: 217.

On docility: Edwards testimony in U.S. Senate, *Labor and Capital*, 4: 386 (quote); Sloss testimony, ibid., 283; Brainerd, "Colored Mine Labor," 78–79; Birmingham *Age-Herald*, August 27, 1890; U.S. Immigration Commission, *Immigrants in Industries: Bituminous Coal Mining*, 7: 217. Brainerd described black miners as "good natured and happy . . . like children. . . . It is quite common in 'open work' on ore-veins and quarries, for the colored laborers to enliven the monotony of their task by singing some melody, keeping time with their hammers, picks, and shovels to the music. One never hears as yet of strikes among them." The operators contrasted this "manageable" nature with the "difficult" character of immigrants from southern and eastern Europe. "We cannot show a district filled with Huns, Slavs, Poles, etc., armed, fierce, and intractable," Tennessee company official James Bowron announced in 1897, "but we can show one well supplied with an orderly and docile negro population." Birmingham *Age-Herald*, 1897 [date unclear], in James Bowron Scrapbook, 101.1.1.1.1, 1895–1902, 17, BPLA.

On willingness to perform work rejected by whites: Sloss and Aldrich testimony in U.S. Senate, *Labor and Capital*, 4: 290, 483; British Board of Trade, *Report of an Enquiry*, 88.

On company store: "Report of Superintendent of Coal Mines," Sloss-Sheffield Records, 4.1.105.1 SSSIC 1892–1902; Brainerd, "Colored Mine Labor," 79; *Proceedings of the Joint Convention*, 274; U.S. Immigration Commission, *Immigrants in Industries: Bituminous Coal Mining*, 7: 199; U.S. Immigration Commission, *Immigrants in Industries: Iron and Steel*, 9: 88, 153, 190; British Board of Trade, *Report of an Enquiry*, 88.

For statements on the overall virtues of black labor: testimony of Sloss, Charles J. Hazard (secretary of the Shelby Iron Works), and Truman H. Aldrich (president of the Cahaba Coal Mining Company) in U.S. Senate, *Labor and Capital*, 4: 283, 470, 483; McCalley, "On the Warrior Coal Field," 268; Birmingham *Age*, October 29, 1886.

Johns quote from Birmingham *Age*, February 5, 1886. In an interview for this issue, Johns offered an elaborate assessment of the European immigrants, native whites, and African Americans who worked under him. It was a colorful example of the social assumptions that suffused the operators' thinking about their diverse labor force. A native of Glamorganshire, Wales, Johns had drifted far and wide, often destitute, across the United States before establishing himself as a mining engineer and later superintendent in the Birmingham coal fields. For background on Johns, see Dubose, *Jefferson County and Birmingham*, 561–65.

45. On the racial stratification of work in the early Birmingham iron industry, see McKiven, *Iron and Steel*, 41–53; W. D. Lewis, *Sloss Furnaces*, 82–91. On industrial slavery, see Starobin, *Industrial Slavery*, 14–15, 22–23; R. Lewis, *Coal, Iron, and Slaves*; J. H. Woodward, *Alabama Blast Furnaces*.

46. On the high levels of mobility built into the miners' world, see Brophy, *A Miner's Life*, 51–67; Shifflett, *Coal Towns*, 76–78; R. L. Lewis, "From Peasant to Proletarian."

47. Birmingham *Iron Age*, September 23, November 4, 1875 (Jefferson County Industrial and Immigration Society); Birmingham *Age-Herald*, October 28, 1893, April 24, 1894 (Birmingham Commercial Club); Birmingham *Iron Age*, June 1, 1882, and Birmingham *Age-Herald*, March 12, 1889, April 24, 1894 (newspaper boosterism); Birmingham *Age-Herald*, January 26, February 11, March 12, 1889 (commission on immigration); Birmingham *Iron Age*, June 1, 1882, and Birmingham *Age-Herald*, March 3, July 21, August 18, 1895 (railroads); Birmingham *News*, July 7, 1898 (labor agents).

48. Alabama Department of Agriculture and Industries, *Alabama Opportunity*, 148 (statement on Centreville). For *National Labor Tribune* reports, see chapter 3. A succession of columns in the Birmingham *Labor Advocate* in the mid-1890s upbraided "jackleg" farmers who in slack times would head off to the mines where the men were on strike. "Just as spring opens so that the weather will admit," one writer complained, "he takes a few eggs off to the market and buys a supply of powder and shot"; another asked, "[w]hy in the devil don't they work at their own trade, and if they are farmers let them farm" (Birmingham *Labor Advocate*, October 8, 15, 1895). See also *UMWJ*, October 14, 1915. Southern-born whites, the U.S. Immigration Commission noted, were not "adapted to conditions. . . . They come from

agricultural districts, and the majority are willing to work only during the winter months" (U.S. Immigration Commission, *Immigrants in Industries: Iron and Steel*, 9: 151). See also Fitch, "Birmingham District: Labor Conservation," 1528; U.S. Immigration Commission, *Immigrants in Industries: Bituminous Coal Mining*, 7: 216–17.

49. Birmingham *Labor Advocate*, February 9, July 27, 1895 (labor agents); Birmingham *Age*, October 6, 1886, *Alabama Sentinel*, February 4, 1888, and Birmingham *Age-Herald*, December 7, 1898 (planter complaints); Birmingham *Age*, March 19, July 22, 23, 25, 28, 1886, and Birmingham *Age-Herald*, August 28, 1895 (railroads); Birmingham *News*, July 31, 1899 (editor).

50. U.S. Senate, *Labor and Capital*, 4: 70 (Gardner); "Report of Superintendent of Coal Mines," Sloss-Sheffield Records, 4.1.105.1 SSSIC 1892–1902 ("scatter"). See also Brainerd, "Colored Mine Labor," 80; U.S. Industrial Commission, *Agriculture and Agricultural Labor*, 10: 919; *Proceedings of the Joint Convention*, 411–13; British Board of Trade, *Report of an Enquiry*, 89; Rikard, "Experiment in Welfare Capitalism," 59–60.

51. How operators could utilize the movement of African Americans between industry and agriculture was evoked in a letter sent by Milton Fies, vice president of the DeBardeleben Coal Company, to one Roy Nolan, a central Alabama planter, at the outset of 1917 (Milton Fies to Roy Nolan, January 2, 1917, DeBardeleben Papers, Box 4, File "No-Ny"). His purpose was, in essence, to "borrow" agricultural labor for a stint at the mines. "When several of the negroes from your section left our mines to go home for Christmas," he began, "I requested them to . . . explain to you our desires [for] additional men from your section, provided this did not interfere with anything in your section." Fies acknowledged the delicate nature of his proposition. "I understand that quite a few of the negroes in your section did not make good crops last year, and that many of them are under obligations to you." But he had a solution he thought could accommodate the needs of both his company and the planter: "It occurs to me that if you could get some of those negroes to come over and work for us that we would agree to make monthly collections for you, which you would previously arrange with any negroes that would agree to come over here and work." Fies took pains to reassure Nolan that he had "no desire to disturb any labor that is needed in your section." Having assured the planter of his good intentions, Fies turned to the details: "[W]e feel that if we could get, say, 20 or 30 men until spring, when they might want to go back to the farm, that it would tide us over, and in addition would put men, who are now idle and unproductive, to work and help the situation and ourselves generally." While there are no indications as to Nolan's response, Fies's proposal illustrates the interconnectedness of labor supplies in the coal fields and the agricultural areas, not to mention how readily the plantation fashion of labor control could be extended into the mineral district.

52. Alabama Department of Agriculture and Industries, *Alabama Opportunity*, 7 ("race exodus"); U.S. Senate, *Labor and Capital*, 4: 287, 283 (Sloss on "restless, migratory class").

53. On civic-minded whites: Birmingham *Observer*, September 2, 1881; Birmingham *Iron Age*, December 15, 1881; Birmingham *Age*, October 19, 1886, August 3, 1887; *Bessemer*, August 6, 1887; Birmingham *Sunday Chronicle*, March 24, 1889; Birmingham *Chronicle*, December 7, 9, 1889, February 7, 1890; Birmingham *Age-Herald*, January 13, March 12, 1891; C. V. Harris, *Political Power in Birmingham*, 186–96. On how to control "idle negro": C. V. Harris, *Political Power in Birmingham*, 198–202. Sloss on floating labor: U.S. Senate, *Labor and Capital*, 4: 289. In 1893 the *Age-Herald* estimated that there were about 1,500 blacks

"lying about Birmingham" available for casual work (June 16, 1893). Advertisements: Birmingham *Age*, December 24, 1886; Birmingham *Labor Advocate*, October 8, 1898. That the operators were interested in preserving this reserve of "idle" labor as much as reforming it is indicated by their staunch opposition to municipal measures to suppress saloons, on the grounds that a prohibition district would fail to retain current or attract new labor (C. V. Harris, *Political Power in Birmingham*, 193–94).

Miners' transiency is discussed here as a significant circumstance confronting the operators as they sought to mobilize a labor force in the coal fields. The causes of such movement, and particularly the way it functioned as an assertion of independence and resistance among the miners, is addressed in chapter 2. On the tendency of common laborers to shift from one trade to another in industrializing America, see Montgomery, *Fall of the House of Labor*, 58–111. On the extensive mobility of (African American and white) labor, agricultural and industrial, in the South during this era, see Jacqueline Jones, *The Dispossessed*, 104–66; Cohen, *At Freedom's Edge*; Gottlieb, *Making Their Own Way*, 12–38. On the compression of African Americans in the lowest, least desirable echelons of the district's manufacturing trades, see McKiven, *Iron and Steel*, 41–53; Kulik, "Black Workers and Technological Change," 29; Worthman, "Working Class Mobility in Birmingham," 175, 178–79, 184–85.

54. This section on the origins and growth of the convict lease system in Alabama relies on Ward and Rogers, *Convicts, Coal, and the Banner Mine Tragedy*, 26–45; Lichtenstein, "Political Economy of Convict Labor," 250–340; Lichtenstein, "Convict Labor and Southern Coal," 3–42; Going, *Bourbon Democracy in Alabama*, 170–90; R. L. Lewis, *Black Coal Miners in America*, 26–35; C. V. Harris, *Political Power in Birmingham*, 202–4; *Biennial Reports of the Inspectors of Convicts to the Governor* (1884–92); *Biennial Reports of the Board of Inspectors of Convicts to the Governor* (1894–1906); *Quadrennial Reports of the Board of Inspectors of Convicts to the Governor* (1910–18); U.S. Bureau of the Census, *Tenth Census: Report on the Defective, Dependent, and Delinquent Classes*, 21: 520; U.S. Bureau of the Census, *Eleventh Census: Report on Crime, Pauperism, and Benevolence*, 5: 15, 118, 121; Tennessee Coal, Iron, and Railroad Company, *1890 Annual Report*, 13, Shook Papers, 386.4.1.9.10. For the origins and character of the convict lease in the postbellum South, see Ayers, *Vengeance and Justice*, 185–222; Shapiro, "Tennessee Coal Miners' Revolts of 1891–92"; Cohen, *At Freedom's Edge*, 201–47; C. V. Woodward, *Origins of the New South*, 212–15.

CHAPTER TWO

1. On Pratt Mines lynching: Birmingham *Age-Herald*, January 14, 17, 19, 20, 22, 1889; *Alabama Sentinel*, January 19, February 2, 1889; Warrior *Index*, January 19, 1889. On Lewisburg baptismal ceremony: *Mineral Belt Gazette*, September 16, 1905. On Cardiff baseball game: Birmingham *Age-Herald*, August 15, 1891.

2. To the south of Birmingham were the Shelby County towns of Helena (Pratt Coal and Coke, Davis & Carr) and Montevallo (T. H. Aldrich). The remaining coal towns of any note lay in Jefferson County: Warrior (J. Pierce, Ernst and Rupp), located twenty-five miles north of Birmingham; Newcastle (New Castle Coal and Iron), ten miles north; and Coketon (Pratt Coal and Coke), six miles west. Coketon soon became known as Pratt Mines and by the early 1890s was incorporated as Pratt City. See U.S. Bureau of the Census, *Tenth Census: Report on the Mining Industries*, 15: 866.

3. Much of this growth remained confined to Jefferson County. To the northeast of Birmingham, along the Georgia Pacific Railway, arose the Sloss company towns of Brookside, Coalburg, Cardiff, and Blossburg; to the southeast, along the Louisville & Nashville, the towns of Johns, Adger, and Sumter appeared in the DeBardeleben Company's Blue Creek holdings. Farther to the south, in Bibb County, ten new Cahaba company mines along the L & N gave rise to Blocton, which quickly emerged as one of the larger coal towns in the region. By the mid-1880s a number of smaller coal operations, along with the Georgia Pacific and the Kansas City, Fort Scott & Memphis railroads, had established a presence in Walker County, to the northwest of Jefferson County. Each Walker County coal firm had its own mining community, such as Carbon Hill (Kansas City Coal and Coke), Horse Creek (Thomas Price, C. E. Mallett), Lockhart (Lockhart Coal), Corona (Coal and Coke), Patton Junction (Virginia and Alabama Coal), Patton (Deer Creek Coal), and Day's Gap (Virginia and Alabama Coal). By the first decade of the new century Walker County was the area where mine production, and consequently mining communities, were growing most rapidly. Towns like Carbon Hill, Horse Creek (renamed Dora in 1906), Cordova, and Corona doubled or tripled in size during the twenty years after 1890 to reach populations of several thousand. See John H. Jones, "Coal-Mining Industry of Alabama in 1889," 5; W. D. Lewis, *Sloss Furnaces*, 167–68, 187–89; Dombhart, *History of Walker County*, 58–64; Ellison, *Bibb County*, 168–84; White, *Birmingham District*; U.S. Bureau of the Census, *Thirteenth Census: Statistics for Alabama*, 584.

4. On mining towns: the greater Blocton area, for example, had 2,709 inhabitants in 1890 and 3,823 ten years later; for Pratt City, the respective figures were 1,946 and 3,485, with several times each figure living in the area immediately beyond the town. On mining camps: Walker County's Carbon Hill, for example, had only 568 residents in 1890 and 830 in 1900; Brookwood (Tuscaloosa County), only 473 in 1890, although it mushroomed to 2,510 a decade later. See U.S. Bureau of the Census, *Thirteenth Census: Statistics for Alabama*, 574; U.S. Bureau of the Census, *Twelfth Census: Population, Part 1*, 1: 59, 63. Before 1890 the statistical federal census provides no figures for the populations of most mining centers of Alabama. Starting in 1890 the census provides figures primarily for the larger coal towns.

5. Birmingham *Labor Advocate*, July 4, 1891 (Pratt Mines); *Mineral Belt Gazette*, October 22, 1904 (Graysville). See also Birmingham *Age*, April 20, May 2, 1886; Birmingham *Age-Herald*, October 17, 1892; Birmingham *Wide-Awake*, August 17, 1899; *Mineral Belt Gazette*, May 7, June 11, 1904; Birmingham *Labor Advocate*, December 25, 1897, September 29, 1906. What transpired inside the churches, unfortunately, is elusive: the content of sermons was seldom recorded.

Press coverage of the mineral district in the late nineteenth century, by both the daily and labor press, makes regular reference to the array of social institutions and activities described in this chapter. See especially the *Alabama Sentinel*, the *Mineral Belt Gazette*, the Birmingham *Labor Advocate*, the Birmingham *News*, and the Birmingham *Age-Herald*.

6. Regular announcements and reports of community activities during the late nineteenth and early twentieth centuries appeared in such labor and mainstream newspapers as the *Alabama Sentinel*, the Birmingham *Labor Advocate*, the Birmingham *News*, and the Birmingham *Age-Herald*. Such social networks were particularly elaborate in the larger mining communities, such as Pratt City or Blocton. For examples of balls to benefit the injured, see Birmingham *Labor Advocate*, February 16, March 16, April 20, June 8, 1895.

7. One public picnic held in 1904 at Kimberly, for example, drew an estimated 20,000 people from across the district (*Mineral Belt Gazette*, July 30, 1904).

8. Birmingham *Age-Herald*, June 8, 1891 (shovel); Birmingham *Chronicle*, February 13, 1890 (debt); Birmingham *News*, June 14, 1899, August 21, 1895 (dominoes, woman); Warrior *Advance*, May 9, 1885 (Warrior); *Alabama Sentinel*, May 25, 1889, Birmingham *Age-Herald*, April 21, 1891, March 12, 1892 (crap games); Birmingham *Hot Shots*, March 25, 1905 (Wylam); Birmingham *Labor Advocate*, November 23, 1901 (Cardiff).

Paydays were often raucous occasions. A resident of Murray found the *absence* of disorder on a payday in 1899 worthy of mention: "there were none of the brawls which sometimes mark this day of distinction," he wrote (Birmingham *Labor Advocate*, January 28, 1899). For other examples of violence on paydays, see *Alabama Sentinel*, March 3, April 21, May 5, 1888, March 21, 1891; Birmingham *Chronicle*, December 9, 1889; Birmingham *Age-Herald*, October 14, 1889.

9. *Alabama Sentinel*, April 23, 1887 (Blocton), June 2, 1888 (Pratt Mines).

10. The historical literature on racial segregation in the New South is extensive. See C. V. Woodward, *Strange Career of Jim Crow*; Rabinowitz, *Race Relations in the Urban South*; Ayers, *Promise of a New South*, 132–59; Cox, "From Emancipation to Segregation," 250–53; McMillen, *Dark Journey*, 3–32.

11. *Mineral Belt Gazette*, June 11, 1904, June 24, 1905 (commencements); *Mineral Belt Gazette*, September 16, 1905 (baptismal ceremony); *Proceedings of the Joint Convention*, 424 (funeral).

12. Birmingham *Age-Herald*, August 15, 1891.

13. Birmingham *Iron Age*, March 19, 1879 (Helena); Birmingham *Age-Herald*, February 20, 1889, *Alabama Sentinel*, February 23, 1889 (Bradford).

14. Birmingham *Labor Advocate*, September 1, 1894.

15. *Jefferson Enterprise*, October 2, 1889; Birmingham *Age-Herald*, October 6, 1889; telegrams to Governor Thomas Seay from Joseph S. Smith (September 27, 1889), L. V. Clark (September 27 and 28, 1889), S. D. Makly (September 28, 1889), and Randolph Petton (September 28, 1889), Seay Papers, Correspondence, Letters, September–October 1889, ADAH; letter from Joseph S. Smith to Governor Seay, September 28, 1889, Seay Papers; Alexander, "Ten Years of Riot Duty," 2.

16. On Griffin: Birmingham *Iron Age*, August 18, 1881. On Meadows: Birmingham *Age-Herald*, January 14, 17, 19, 20, 22, 1889; *Alabama Sentinel*, January 19, February 2, 1889; Warrior *Index*, January 19, 1889. Significantly, the rape of Mrs. Kellum received more extensive attention in the local press than did the murder. The same fate befell a black man accused of rape in Blocton in 1887 (*Alabama Sentinel*, June 11, 1887). For an account of a near-lynching of a black miner accused of rape at Coalburg in 1892, see Birmingham *Age-Herald*, August 26, 1892.

The late 1880s and early 1890s saw a surge in lynchings of African Americans around the South. See Williamson, *Crucible of Race*, 115–19; Brundage, *Lynching in the New South*.

17. The following account of the Shepherd pursuit draws upon Birmingham *News*, June 22–30, 1899.

18. No evidence exists that Shepherd was ever apprehended. He was spotted twice over the next several days in Birmingham, but each time eluded capture in wild escapes—assisted by African Americans who pointed pursuers in the wrong direction—into the black neigh-

borhoods near the Alice furnace or the ore mining town of Ishkooda. For another example of militant response by black miners to extralegal violence against an alleged black rapist, see Birmingham *State-Herald*, July 7, 1897.

19. The following narrative of the Dolomite incident draws upon Birmingham *News*, March 27–30, 1899.

20. *Mineral Belt Gazette*, October 14, 1905; U.S. Immigration Commission, *Immigrants in Industries: Bituminous Coal Mining*, 7: 200.

21. U.S. Immigration Commission, *Immigrants in Industries: Bituminous Coal Mining*, 7: 200. On one of the most prominent examples of a single-race mine, the DeBardeleben company's all-black operation at Blue Creek, see chapter 4.

22. Warrior *Advance*, May 9, 1885 (Kelly Mines).

23. *National Labor Tribune*, July 2, 1887 (Coalburg). The operators' social domination in some towns was illustrated in a letter to the statehouse from Edwin Thomas, vice president and general manager of the Pioneer Mining and Manufacturing Company, requesting a justice of the peace and a constable for the company-created town of Thomas. "The community," he explained, "is too large to be without some protection of this kind and I, as representative of the Pioneer Mining and Manufacturing Company, which owns all the land and houses in said village, hereby . . . request the appointment of John W. Minor as Justice and Yancey A. Woods as constable, a selection which I am sure will meet the approval of the citizens." Edwin Thomas to Governor Thomas G. Jones, January 31, 1891, in Jones Papers, RC2:G25, Letters M–Z, January 20–March 6, 1891, ADAH.

24. Birmingham *Labor Advocate*, July 3, 1897 (Alexander), November 25, 1899 (Thompson); Birmingham *News*, November 3, 1898 (Johns). Warm feelings for Johns, however, were not universal among the miners. On September 13, 1894, "One of the Boys" wrote the *National Labor Tribune* from Birmingham, "No one can remember any kindness he ever showed to a workingman." Frequent examples of the operators' paternal involvement in the miners' social institutions and practices can be found in the local daily and labor press.

25. On Cahaba housing: *Alabama Sentinel*, April 2, 1892. On company profits from housing: Statements of Profit and Loss in the Sloss-Sheffield Records, 4.1.105, 1892–1902; W. D. Lewis, *Sloss Furnaces*, 190; annual reports of the Tennessee Coal, Iron and Railroad Company, Shook Papers, 386.4.1.9.10 (e.g., "The profit from rents is large if substantial houses are built," TCI assistant general manager George B. McCormack observed in the 1893 annual report). On accounts of company housing: Birmingham *Labor Union*, September 11, 1886; *Journal of United Labor*, November 5, 1887; *Alabama Sentinel*, March 8, 1890; Chattanooga *Republican*, February 18, 1893; Birmingham *Labor Advocate*, July 27, 1895; *Proceedings of the Joint Convention*, 799–800; British Board of Trade, *Report of an Enquiry*, 96. TCI official on Pratt Mines: 1893 Tennessee Coal, Iron and Railroad Company annual report, Shook Papers. On miners' home ownership: An estimated half the employees of the Central Coal Company, for example, were homeowners in 1902 (Birmingham *Labor Advocate*, March 15, 1902); nearly all those of the Sloss company at Cardiff owned homes in 1904 (*Mineral Belt Gazette*, June 18, 1904). Complaints about housing quality: *National Labor Tribune* (from Coalburg), July 2, 1887; *Alabama Sentinel* (from Blue Creek), September 8, 1888, March 8, 1890. On health problems: Knowles, "Water and Waste"; Rikard, "An Experiment in Welfare Capitalism," 131–43.

26. Birmingham *Labor Advocate*, July 20, 1895 (Compton); *Journal of United Labor*,

November 5, 1887 (school strike). See also *Alabama Sentinel*, March 3, 1888; Birmingham *Age-Herald*, January 27, 1889; Warrior *Index*, May 25, 1894; Birmingham *Labor Advocate*, February 16, 1895; Birmingham *State-Herald*, November 11, 1895; *Proceedings of the Joint Convention*, 503, 513–14.

27. *National Labor Tribune*, March 22, May 3, 1884, April 23, July 2, 1887; *Proceedings of the Joint Convention*, 283.

28. *Alabama Sentinel*, November 24, 1888 (Wheeling Mines); Birmingham *Labor Advocate*, February 16, 1895 (Corona). See also Birmingham *Labor Advocate*, August 3, 1895; *Proceedings of the Joint Convention*, 210–11; British Board of Trade, *Report of an Enquiry*, 9; *National Labor Tribune*, February 24, 1883; *Journal of United Labor*, June 10, 1886, November 5, 1887; Birmingham *Age-Herald*, January 27, 1889; *Alabama Sentinel*, September 6, 1890; Bessemer *Herald-Journal*, November 25, 1897.

29. *UMWJ*, December 21, 1893 (Blocton); *National Labor Tribune*, June 20, 1885 (high fees); *Proceedings of the Joint Convention* (dynamite). See also *National Labor Tribune*, April 5, 1879, January 22, 1881, June 20, 1885, and April 23, 1887; *Alabama Sentinel*, March 22, 1890; Birmingham *Labor Advocate*, March 28, 1891; *Proceedings of the Joint Convention*, 202, 206–7, 282–83.

30. The workings of the store check-cashing system were covered extensively in *Proceedings of the Joint Convention*, 204, 226–32, 251–53, 274–76, 280–82, 407, 502–7, 580–96, 635–41. See also *National Labor Tribune*, May 16, 30, June 20, 1885; *UMWJ*, September 14, 1893; British Board of Trade, *Report of an Enquiry*, 97; U.S. Immigration Commission, *Immigrants in Industries: Bituminous Coal Mining*, 7: 199. On debt trap: *Proceedings of the Joint Convention*, 623 (Warner Mines); Birmingham *Labor Advocate*, November 6, 1897 (Brookside). See also *National Labor Tribune*, May 30, 1885, March 12, 1887; *Proceedings of the Joint Convention*, 216–17.

31. On overrecruitment of labor: *Alabama Sentinel*, March 29, 1890. This approach could pay off handsomely. During the last two-thirds of 1901, for example, profits from the company store equaled 26.4 percent of what the Bessemer Land and Improvement Company made on coal itself and 17.6 percent of the company's profits overall (Bessemer company annual report in James Bowron Scrapbooks, 101.1.1.1.1, 1895–1902, 58, BPLA). During the 1890s, profits from the Sloss company commissaries at Coalburg, Cardiff, Brookside, and Blossburg regularly reached a proportion of from 25 to more than 50 percent of total coal profits (profit and loss statements, Sloss-Sheffield Records, 4.1.105, 1892–1902). See also Warrior *Index*, May 25, 1894; Birmingham *Labor Advocate*, December 1, 1894; *Proceedings of the Joint Convention*, 260.

32. *Proceedings of the Joint Convention*, 283, 615–17, 628–30, 648, 657–59 (operator defenses); *Mineral Belt Gazette*, June 18, 1904 (Adger and Kimberly). For other positive assessments by miners of company stores, see Birmingham *Labor Advocate*, November 10, 1900, June 1, 1901, April 12, 1902; *Mineral Belt Gazette*, June 11, 1904.

33. In 1880, when Alabama coal sold at an average of $1.47 per ton, the average cost of labor per ton came to $1.02, or 69 percent of the market price; in no other state was the percentage so high (U.S. Bureau of the Census, *Report on the Mining Industries*, 15: 683). In 1895, the wages of coal miners alone comprised over 60 percent of the Sloss company's overall production costs ("Report of the Superintendent of Coal Mines," Sloss-Sheffield

Records, 4.1.105.1 SSSIC 1892–1902). In 1909, the figure for the Alabama coal fields was 65 percent (U.S. Bureau of the Census, *Thirteenth Census: Statistics for Alabama*, 682).

34. This overview of mine wage levels is derived from regular coverage in the daily and labor press. For a good company-by-company survey, see *National Labor Tribune*, March 2, 1888. The 1890 census offered a comprehensive picture of the daily earnings of Alabama's coal mine employees: underground miners, who numbered 4,110, earned a daily average of $2.15; the 1,564 underground laborers, $1.33; the 123 above-ground mechanics, $2.12; the 797 above-ground laborers, $1.25. (U.S. Bureau of the Census, *Eleventh Census: Report on the Mineral Industries*, 14: 356). For wage levels in the early 1900s, see *Proceedings of the Joint Convention*, 613–14. On the origins of the sliding scale, see *Alabama Sentinel*, August 18, September 8, 1888; February 15, October 18, 1890.

35. On deductions: *Journal of United Labor*, September 25, 1884 (Warrior); *National Labor Tribune*, July 3, 1886, April 9, 1887; *Journal of United Labor*, December 3, 1887; *Alabama Sentinel*, March 1, 1890; Birmingham *Labor Advocate*, February 16, May 25, July 27, 1895; *Proceedings of the Joint Convention*, 431–35, 674–76, 784–86, 791; *UMWJ*, October 13, 1917. On underweighing: *National Labor Tribune*, August 5, 1882; Birmingham *Age*, May 13, 1886; *Journal of United Labor*, August 20, 1887; *Alabama Sentinel*, June 16, 1888; Birmingham *Labor Advocate*, July 6, 1895. On "dirty" coal: Birmingham *Age-Herald*, September 11, 1890; Birmingham *Labor Advocate*, December 15, 1894; *UMWJ*, November 12, December 10, 1896, November 4, 1897; *Proceedings of the Joint Convention*, 139–40, 666–67, 794–96, 810–12. On dead work: Birmingham *Labor Advocate*, December 1, 1894, June 1, 1895 (observer quote), July 3, 1897; Blocton *Enterprise*, July 30, 1908 (Walker County); *National Labor Tribune*, August 5, 1882; *Alabama Sentinel*, May 24, 1890; *UMWJ*, December 21, 1893, December 26, 1895, November 12, 1896; *Proceedings of the Joint Convention*, 271–73, 797–98, 813–16.

36. Goodrich, *The Miner's Freedom*.

37. *Alabama Sentinel*, August 16, 1890, September 5, 1891 (Patton Mines quote); *National Labor Tribune*, November 22, 1879, May 16, 1885; *American Manufacturer*, January 13, 1888; Birmingham *Age-Herald*, January 27, 1889; Birmingham *News*, July 27, 1895; "1894—Guarding Miners at Ensley" (second file), in Adjutant General's Office Papers, SG15148: Accounts/Financial Records of 1908 Strike, 89–90, ADAH; *Proceedings of the Joint Convention*, 834–38. On a still larger scale, an operator might subcontract an entire mine out to an individual, who would in turn pay and manage labor at the mines (*UMWJ*, November 16, December 28, 1893, July 11, 1895, August 20, 1908; Birmingham *Labor Advocate*, March 16, 1895, August 21, 1897). The most extensive history and description of the subcontracting system in the Alabama coal fields is offered by Ethel Armes in "Evils of Sub-Contract System Exposed," *UMWJ*, April 24, 1913.

38. "1894—Guarding Miners at Ensley," (second file), in Adjutant General's Office Papers, SG15148: Accounts/Financial Records of 1908 Strike, 89–90, ADAH; *Proceedings of the Joint Convention*, 851.

39. On "equal pay" principle: *Alabama Sentinel*, August 16, 1890 (quote); *Proceedings of the Joint Convention*, 259–60. On "robbing": *National Labor Tribune*, November 22, 1879; Birmingham *State-Herald*, January 5, 1896. On glutting: *UMWJ*, January 4, 1894. On preferential treatment: *Alabama Sentinel*, August 16, 1890. On abolition of subcontracting

under union recognition, see chapter 5. On persistence of subcontracting elsewhere in the district, and its general revival following the defeat of the closed shop in 1908, see chapter 6 below and British Board of Trade, *Report of an Enquiry*, 90.

40. *National Labor Tribune*, March 31, 1883 (Pratt Mines); Blocton *Enterprise*, July 30, 1908 (Blocton); *Alabama Sentinel*, February 22, 1890 (equal turn). For an example of one such strike (at Pratt Mines in 1897), see Alfred M. Shook to James T. Woodward, June 22, 1897, Shook Papers. For more on this issue, see *National Labor Tribune*, March 22, 1884, May 16, 1885; *Alabama Sentinel*, September 8, 1888, May 19, 1890; Birmingham *Labor Advocate*, June 1, 29, July 6, 1895.

41. On firing: *National Labor Tribune*, March 19, 1887; Birmingham *Age-Herald*, April 3, 1891; *Alabama Sentinel*, August 22, 1891; *UMWJ*, January 11, 1894; *Proceedings of the Joint Convention*, 59–60, 210, 249–50. On absenteeism: *National Labor Tribune*, May 26, 1888; *Alabama Sentinel*, May 3, 17, 31, June 7, 1890; *Proceedings of the Joint Convention*, 247–48, 257, 277; British Board of Trade, *Report of an Enquiry*, 91. On tasks performed: *National Labor Tribune*, July 2, 1887 (quote); *Alabama Sentinel*, May 24, 1890. On contractual limits: *Alabama Sentinel*, February 22, 1890; *Proceedings of the Joint Convention*, 77–80.

The historical literature on battles over workplace control in the United States during this period has grown extensively in recent years. Most wide-ranging and influential among this scholarship are Montgomery's *Workers' Control in America* and *Fall of the House of Labor*.

42. *Alabama Sentinel*, May 10, 1890 (Carbon Hill); Birmingham *Age*, May 13, 1886, *Alabama Sentinel*, February 15, 1890 (less tyrannical settings).

On republican ideology in late-nineteenth-century popular movements, see L. Fink, *Workingmen's Democracy*; Oestreicher, "Terence Powderly, the Knights of Labor, and Artisanal Republicanism"; S. Hahn, *Roots of Southern Populism*; Krause, *Battle for Homestead, 1880–1892*; Salvatore, *Eugene V. Debs*.

43. Innumerable, often terribly vivid reports of routine accidents can be found in the local labor and daily press of the period. On extraordinary accidents: Birmingham *Age-Herald*, January 27, 1895 (coal washer); Warrior *Index*, June 15, 1889 (trestle); Birmingham *Labor Advocate*, August 15, 1891 (mule); Birmingham *Age-Herald*, January 27, 1895 (shaft); Birmingham *Iron Age*, October 19, 1882 (elevator). For lists of individual injuries and fatalities in the mines, see State of Alabama, *First Biennial Report of the Inspector of Mines, 1893–94* (Birmingham: n.p., 1895), 27–37, 52–54; *Second Biennial Report of the Inspector of Mines, 1898* (Birmingham: Dispatch Printing Co., 1898), 52–59; *Third Biennial Report of the Inspector of Mines, 1900* (Birmingham: n.p., 1900), no page numbers; *Fourth Biennial Report of the Inspector of Mines, 1902* (Birmingham: n.p., 1902), no page numbers. On the Virginia Mines disaster: *Mineral Belt Gazette*, February 25, 1905; Birmingham *Age-Herald*, February 21–28, 1905; Bessemer *Weekly*, February 25, 1905. On other major disasters: *Mineral Belt Gazette*, March 10, 1906; Birmingham *Labor Advocate*, December 20, 1907; Ward and Rogers, *Convicts, Coal, and the Banner Mine Tragedy*. On the common incidence of mining disasters around the nation in this period, see Jackson, *The Dreadful Month*, esp. 7–70.

44. Miners blamed: *Proceedings of the Joint Convention*, 495; State of Alabama, *Fourth Biennial Report of the Inspector of Mines, 1902*, 4. Miners admonished: State of Alabama, *First Biennial Report of the Inspector of Mines, 1893–94*, 25–26; *UMWJ*, March 9, 1905. Employees sign documents: Birmingham *Labor Advocate*, July 6, 1895. Legislation: "Miners

are killed daily by falling slate and by explosion," the Knights of Labor paper noted in a typical commentary. "[T]hese deaths occur because [our people] have not insisted that laws should be passed for the protection of the miner and the mine laborer" (*Alabama Sentinel*, December 21, 1889).

45. *UMWJ*, October 5, 1893 (TCI mines); *Alabama Sentinel*, March 15, 1890 ("outrage"); Cullman *Advance and Guide*, September 19, 1885 (beleaguered farmers). See also Birmingham *Sunday Chronicle*, February 17, 1889; Birmingham *Labor Advocate*, June 1, 1895; Birmingham *Advance*, September 24, October 1, 1883; Birmingham *Semi-Weekly Review*, April 30, 1884.

Convict labor was a prominent target of the late-nineteenth-century labor movement nationwide. See, for example, Rachleff, *Black Labor in Richmond*, 81–82, 124–27, 147–49; Walkowitz, *Worker City, Company Town*, 209.

46. *Journal of United Labor*, June 10, 1886 (Pratt company); Birmingham *Labor Advocate*, January 9, 1892 (administration of justice); *Biennial Report of Inspectors of Convicts (1902)* [*Fourth Biennial Report of the Board of Inspectors of Convicts to the Governor, from September 1, 1900, to August 31, 1902* (Montgomery: Brown Printing Co., 1902)], 21 (misdemeanors); *Biennial Report of Inspectors of Convicts (1902)*, 22 (two youths); Birmingham *Free Lance*, January 4, 1904 (vagrancy law).

47. State of Alabama, *Biennial Report of the Inspectors of Convicts (1886)* [*First Biennial Report of the Inspectors of Convicts to the Governor, from October 1, 1884, to October 1, 1886* (Montgomery: Barrett & Co., 1886)], 70. At times operators and their advocates strained to put an even happier gloss on life at the coal mine stockades. "A more cheerful looking crowd could not have been seen," the Birmingham *Evening Chronicle* exclaimed after a visit to the prisoners of Pratt Mines in 1885, "and it could have been readily imagined a picnic jollity if the stripes had not told you that the ban of the law rested upon them" (September 19, 1885).

48. *Biennial Report of Inspectors of Convicts (1898)* [*Second Biennial Report of the Board of Inspectors of Convicts to the Governor, from September 1, 1896, to August 31, 1898* (Montgomery: Roemer Printing Co., 1898)], xxiv. For individual listings of convict fatalities at the mines and their causes, see *Biennial Report of Inspectors of Convicts (1886)*, 96–97, 128–33; *Biennial Report of Inspectors of Convicts (1888)* [*Second Biennial Report of the Inspectors of Convicts to the Governor, from October 1, 1886, to September 30, 1888* (Montgomery: W. D. Brown & Co., 1888)], 48–50, 92–96; *Biennial Report of Inspectors of Convicts (1892)* [*Fourth Biennial Report of the Inspectors of Convicts to the Governor, from October 1, 1890, to August 31, 1892* (Montgomery: Smith, Allred & Co., 1892)], 36–45, 59; *Biennial Report of Inspectors of Convicts (1896)* [*First Biennial Report of the Board of Inspectors of Convicts to the Governor, from September 1, 1894, to August 31, 1896* (Montgomery: Roemer Printing Co., 1896)], 18–34; *Biennial Report of Inspectors of Convicts (1898)*, 20–30, 63–64; *Biennial Report of Inspectors of Convicts (1900)* [*Third Biennial Report of the Board of Inspectors of Convicts to the Governor, from September 1, 1898, to August 31, 1900* (Montgomery: A. Roemer, 1900)], 30–39; *Biennial Report of Inspectors of Convicts (1902)*, 43–51; *Biennial Report of Inspectors of Convicts (1904)* [*Fifth Biennial Report of the Board of Inspectors of Convicts to the Governor, from September 1, 1902, to August 31, 1904* (Montgomery: Brown Printing Co., 1904)], 34–39, 52–55; *Biennial Report of Inspectors of Convicts (1906)* [*Sixth Biennial Report of the Board of Inspectors of Convicts to the Governor, from September 1, 1904, to August 31, 1906* (Montgomery: Brown Printing Co., 1906)], 24–28; *Quadrennial*

Report of Inspectors of Convicts (1910) [*Quadrennial Report of the Board of Inspectors of Convicts for the State of Alabama, to the Governor, from Sept. 1st, 1906, to August 31, 1910* (Montgomery: Brown Printing Co., 1910)], 15, 44–52, 74–80; *Quadrennial Report of Inspectors of Convicts (1914)* [*Quadrennial Report of the Board of Inspectors of Convicts for the State of Alabama, to the Governor, from Sept. 1st, 1910, to August 31, 1914* (Montgomery: Brown Printing Co., 1914)], 134–46, 186–202; *Quadrennial Report of Inspectors of Convicts (1918)* [*Quadrennial Report of the Board of Inspectors of Convicts for the State of Alabama, to the Governor, from Sept. 1st, 1914, to August 31, 1918* (Montgomery: Brown Printing Co., 1918)], 50–67, 98–101.

49. Birmingham *Semi-Weekly Review*, November 17, 1883 (Knights of Labor paper); *UMWJ*, August 12, 1915 (investigative committee); Montgomery *Daily Dispatch*, April 9, 1889, in James Bowron Scrapbook, 1877–1895, BPLA (Dawson). For further background on public controversy over the convict lease system, see Ward and Rogers, *Convicts, Coal, and the Banner Mine Tragedy*, 45–50; C. V. Harris, *Political Power in Birmingham*, 204–7.

50. Morgan, *American Slavery, American Freedom*.

51. *Proceedings of the Joint Convention*, 247, 453–55, 462, 605 (extent of absenteeism); 456, 460–61, 551–53 (most days mines idle); 478–81, 498–99, 512–13, 563–64 (hampered operators' purpose).

Instinctively, the operators attributed these "destabilizing" habits to what they saw as the congenital "restlessness" of their miners, particularly those of certain ethnic and racial backgrounds, and above all blacks. The more one paid them, company officials were wont to observe, the less they were inclined to continue working on a regular basis (see, for example, *Proceedings of the Joint Convention*, 254–55, 649–55). Although little documentation exists to assess the operators' portrayals, they could well have contained a germ of truth—African Americans may indeed have figured disproportionately among those who declined to stay at one mine, in the coal fields, or even in the district for the long term. Such mobility would have represented a natural response to the particularly dismal conditions they often encountered.

52. *Mineral Belt Gazette*, May 14, 1904 (gardens); Birmingham *News*, April 26, 1904 (picnic); Birmingham *News*, January 5, 1898; *Proceedings of the Joint Convention*, 247, 258–59, 406–7, 550–51, 660 (payday); *Proceedings of the Joint Convention*, 424, 550–51 (funeral); *Proceedings of the Joint Convention*, 247–49 (three days away).

53. On Tennessee, see Shapiro, "Tennessee Coal Miners' Revolts of 1891–92." On schools: Night schools for convicts at the mines, established by state law, covered such areas as spelling, reading, writing, arithmetic, geography, grammar, history, and "moral instruction" [*Biennial Report of Inspectors of Convicts (1898)*, 10, 57, 59]; see also *Biennial Report of Inspectors of Convicts (1888)*, 10; *Biennial Report of Inspectors of Convicts (1892)*, 19–20, 63–66). On moral and religious teachings: Birmingham *Age-Herald*, December 5, 1904; *Biennial Report of Inspectors of Convicts (1888)*, 11, 15; *Biennial Report of Inspectors of Convicts (1892)*, 19–20, 63–66; *Biennial Report of Inspectors of Convicts (1898)*, 45–47; *Biennial Report of Inspectors of Convicts (1904)*, 59–60. During Sunday school, one instructor of the convicts reported, "the men seem to forget their unfortunate position, and throw themselves heartily into the exercises, which comprise reciting scripture from memory [and] singing of the most lively character" (*Biennial Report of Inspectors of Convicts (1904)*, 59–60). On revivals: Birmingham *Labor Advocate*, June 10, 22, 1893. On dances: Birmingham *Age*, January 29, 1886. On troupe shows: Birmingham *News*, August 25, September 1, 1898.

On dinners: Birmingham *News*, April 5, July 10, 1899, July 5, 1904. On gambling: Julia S. Tutwiler to R. H. Dawson, February 9, 1888, Seay Papers (Correspondence, January–February 1888), ADAH; *UMWJ*, July 18, 1912. On pooling resources: Birmingham *Age*, February 21, 1886.

54. On sabotaging own health: *Biennial Report of Inspectors of Convicts (1886)*, 73; F. P. Lewis, M.D., to S. B. Trapp, President of Board of Convict Inspectors, October 15, 1898, in Sloss-Sheffield Records, 1899–1908. On violence among convicts: Birmingham *News*, September 19, 1899, Birmingham *Labor Advocate*, September 23, 1899 (duel); see also *Biennial Report of Inspectors of Convicts (1888)*, 7; Birmingham *Daily News*, January 22, 1891. On assistant mine boss: Birmingham *State-Herald*, February 16, 1886.

55. Birmingham *Iron Age*, December 21, 1882 (Newcastle); *Union Labor Review*, March 31, 1888 (Loveless); Birmingham *Advance*, September 17, 1883, and *Biennial Report of Inspectors of Convicts (1886)*, 41 (hunger strikes); Mobile *Register*, June 2, 1885 (Coalburg); Birmingham *Age-Herald*, May 27, 1891 (iron sprags; see also Birmingham *Evening Chronicle*, September 19, 1885); *Biennial Report of the Inspectors of Convicts (1886)*, 48, and *Biennial Report of the Inspectors of Convicts (1888)*, 6 ("mutinies"); Birmingham *Iron Age*, December 21, 1882 (Milner); Birmingham *Evening Chronicle*, September 19, 1885 (Anti-Convict League).

56. Babe Ellis to Governor Seay, September 4, 1887, Seay Papers, Correspondence, A–L, June–September, 1887, ADAH; Will Johnson to Governor Thomas G. Jones, April 5, 1891, Jones Papers, RC2:G25, Letters, March–April, 1891 (incoming), ADAH.

57. *Biennial Report of Inspectors of Convicts (1896)*, 68–69. Requests for pardons on medical grounds, however, were by no means easy to obtain. An impassioned appeal to Governor Jones from Anna E. Penny for her husband's release from the mines on the basis that he was not fit to perform the work elicited a bland denial from the state physician of convicts—written on the stationary of the Tennessee Coal, Iron, and Railroad Company! Anna E. Penny to Governor Jones, March 1, 1891, Jones Papers, RC2:G25, Letters M–Z, Jan. 20–March 6, 1891 (incoming), ADAH; R. A. Jones to Governor Jones, March 3, 1891, Jones Papers, RC2:G25: Letters, March–April, 1891 (incoming). For lists of pardons granted, see *Biennial Report of Inspectors of Convicts (1886)*, 101; *Biennial Report of Inspectors of Convicts (1888)*, 46–47.

58. Birmingham *Age-Herald*, August 27, 1897 (feigning illness); Birmingham *Age-Herald*, March 19, 1898, Birmingham *News*, October 22, 1898, July 12, 1904, Birmingham *Labor Advocate*, September 12, 1903 (tunneling or dynamiting); *Biennial Report of Inspectors of Convicts (1888)*, 6 (sawing); *UMWJ*, December 5, 1908 (fire); Birmingham *News*, June 15, 1899 (picks); Birmingham *News*, July 25, 1898 (overpowering guards); Huntsville *Gazette*, January 20, 1894, Birmingham *Age-Herald*, December 20, 1904 (hurling dynamite); Birmingham *News*, January 21, 1898 (bloodhounds). For further examples of convict escapes, see Birmingham *Iron Age*, May 1, 1878; Birmingham *Age*, April 10, 1886, August 18, 1887; Birmingham *Chronicle*, January 16, 1890; Birmingham *Age-Herald*, October 9, 1890, May 3, 1892; Birmingham *Labor Advocate*, January 20, 27, March 1, 1894; Birmingham *News*, July 1, 29, 1895. For listings of attempted and successful escapes from the mines, see *Biennial Reports of Inspectors of Convicts* and *Quadrennial Reports of Inspectors of Convicts: Biennial Report (1886)*, 98–100, 134–37; *Biennial Report (1888)*, 44–45, 90–92; *Biennial Report (1892)*, 46–50; *Biennial Report (1896)*, 34–37; *Biennial Report (1898)*, 31–33; *Biennial*

Report (1900), 39–42; Biennial Report (1902), 52–54; Biennial Report (1904), 40–43, 56–59; Biennial Report (1906), 29–33; Quadrennial Report (1910), 53–63, 81–87; Quadrennial Report (1914), 149–68, 203–10; Quadrennial Report (1918), 68–85, 102–5.

CHAPTER THREE

1. On the National Labor Union, see Montgomery, *Beyond Equality*, 340–56, 425–47, and "William H. Sylvis and the Search for Working-Class Citizenship"; E. Foner, *Reconstruction*, 475–84; Goodwyn, *Populist Moment*, 10–19. On the early roots of Greenbackism, see Unger, *Greenback Era*, 94–97; Sharkey, *Money, Class, and Party*, 188–91; Goodwyn, *Populist Moment*, 9, 13–14.

2. Kleppner, "The Greenback and Prohibition Parties," 1549–66, 1599–1601, 1609–11; French, " 'Reaping the Whirlwind' "; Krause, *Battle for Homestead, 1880–1892*, 121–24, 129–32; Ricker, *Greenback-Labor Movement in Pennsylvania*; Commons et al., *History of Labour*, 167–71, 240–51; P. S. Foner, *History of the Labor Movement*, 475–88.

3. On Greenbackism in the South, see C. V. Woodward, *Origins of the New South*, 76–85; S. Hahn, *Roots of Southern Populism*, 226–38; Salutos, *Farmer Movements in the South*, 48–54; Hyman, *Anti-Redeemers*; Rachleff, *Black Labor in the South*, 82–108; Wharton, *The Negro in Mississippi*, 201–6; Kirwan, *Revolt of the Rednecks*, 18–26; Hair, *Bourbonism and Agrarian Protest*, 60–82; Going, *Bourbon Democracy in Alabama*, 41–60; Abramowitz, "The South: Arena for Greenback Reformers," 108–10; Barjenbruck, "The Greenback Political Movement."

4. In Alabama, Greenbackism was concentrated in the Tennessee Valley. Led by William M. Lowe, who was elected to Congress in 1878 and again in 1880, the Greenbackers and allied independents and Republicans of northern Alabama were composed largely of white small farmers who made up the bulk of the area's population. In 1882, at the peak of their influence in the state, non-Democrats held 20 of the 100 seats in the Alabama House of Representatives, 14 from northern counties. See Rogers, *One-Gallused Rebellion*, 41–55; Going, *Bourbon Democracy in Alabama*, 54–60; Roberts, "William Manning Lowe and the Greenback Party in Alabama."

5. On the 1874 election and its aftermath in Alabama, see Rogers, *One-Gallused Rebellion*, 41–44; Going, *Bourbon Democracy in Alabama*, 9–26. For the Birmingham district in particular, see the Birmingham *Iron Age* for that year.

6. On numbers of clubs: *National Labor Tribune*, August 10, 1878, October 18, 1879. On the various institutional activities of the GLP in the Birmingham district, see *National Labor Tribune*, June 29, August 10, 17, 1878. The existence of the GLP in this area was first uncovered by Herbert Gutman in the mid-1960s during his research through the one key source about it, the *National Labor Tribune*, which published correspondence from Greenback organizers around the country. Gutman reprinted a sample from the scores of Alabama letters in "Black Coal Miners and the Greenback-Labor Party in Redeemer Alabama, 1878–79." Ronald L. Lewis also touches on the GLP of the Birmingham district in *Black Coal Miners in America*, 40.

7. *National Labor Tribune*, August 10, 17, 1878. For text of the Toledo platform, see P. S. Foner, *History of the Labor Movement*, 483.

8. *National Labor Tribune*, October 25, 1879 ("Olympic"); see also February 28,

April 26, June 7, July 25, November 8, 1879, July 17, 1880. On Republican Party, see June 7, 1879, July 17, 1880.

9. *National Labor Tribune*, September 27, October 18 ("Olympic"), 1879, January 3, 1880 (uniform pay scale); *National Labor Tribune*, April 26, 1879 (conditions); *National Labor Tribune*, February 28, 1880 (Moran; also January 4, August 2, 1879, January 3, May 1, 1880). For an in-depth investigation of southern coal miners' reaction to the convict lease, see Shapiro, "Tennessee Coal Miners' Revolts of 1891–92."

10. If not entirely separable, neither were the worlds of the farmer and the miner interchangeable. The gulf, and the determination of the Greenback-Labor Party to overcome it, were illustrated in a letter from Willis Thomas. Reporting on a *National Labor Tribune* subscription drive around the district, Thomas noted the objection among Greenback farmers that, "We can never see anything in it for labor except for miners, mill workers, and puddlers. If those fellows would tell us all about corn, cotton, wheat, potatoes, rice, eggs, and so on, we would take it in a minute." Thomas duly obliged them: "The fruit was all cut and dried by Mr. Jack Frost the nights of the 3d, 4th and 6th of April. Wheat is very promising on Turkey Creek, five miles east of Jefferson. The gardens are young, but look flourishing. Cotton and corn looks well. Hail and rain fell here on the 14th and 15th, accompanied with much wind. . . . A man was killed by the lightening up Turkey Creek near Smith Mills" (*National Labor Tribune*, May 31, 1879). Another Greenbacker, Warren D. Kelley, wrote in a similar vein in the next issue. Following a routine report on the state of the mines ("Work continues dull. . . . We get about three days a week"), he proceeded without transition to the farming situation: "The young corn and cotton look nice and green. Turkey Creek wheat is bully. Eggs and butter are cheap" (June 7, 1879). On the porous line between mining and farming, see chapter 1.

11. *National Labor Tribune*, January 4 (first quote), February 22 (second quote), April 26 (third quote), 1879.

12. E. Foner, *Reconstruction*, xxv, 512–601; Perman, *Road to Redemption*, 135–277.

13. *National Labor Tribune*, September 27, 1879.

14. No hard numbers survive on the racial composition of the GLP clubs of the mineral district, but impressionistic data suggest that miners of each race figured prominently in the organization. On its biracial structure, see *National Labor Tribune*, August 10, 17, 1878, May 31, 1879. White correspondents to the *National Labor Tribune* included Michael Moran, James Dye, and "Dawson"; black correspondents included Willis J. Thomas, Warren D. Kelley, and "A Close Looker."

15. McKee, *National Conventions and Platforms*, 173 75, 191–93. Interestingly, it was instead the Prohibition Reform Party whose 1876 platform affirmed the inviolability of the Reconstruction constitutional amendments (177–78).

16. Dawson's statements are from *National Labor Tribune*, April 26, 1879. "The cause of labor," he added, "is advancing in our State. The colored race can be no more hitched up to the infernal car of Juggernaut." "D.J." statement is from August 30, 1879; Moran statement, from November 22, 1879 (for similar arguments, see statements by "Olympic," September 27, 1879, and Dawson, August 16, 1879.

17. On Kelley: *National Labor Tribune*, May 31, 1879. On Thomas: June 29, July 6, August 10, 17, 1878. The depiction of Thomas as "small, spare" is from Henry Hospun to *National Labor Tribune*, August 17, 1878. As with other local Greenback leaders, little is

known of Thomas's origins—that is, where he came from, how long he had lived in the Birmingham district or been a miner, etc. There is, however, one clue that his exposure to miners' organization may not have been as new as Kelley suggested: in one of his many published letters, Thomas prefaced a pronouncement on the principle of strikes with the words, "As once members of the Miners' National Association we say. . ." (*National Labor Tribune*, February 8, 1879). Established in 1873, the Miners' National Association (MNA) had been intended as a nationwide, umbrella federation of local miners' organizations, embodying all types of miners in all types of mines. By 1876, buffeted by the depression and the defeat of the Long Strike in the anthracite fields of 1875, the MNA had collapsed. (See Wallace, *St. Clair*, 393–98; P. S. Foner, *History of the Labor Movement*, 440–41, 459.) Thomas's remark is the only indication, if that is what it is, of an MNA presence in the Alabama coal fields. It may also indicate that Thomas, and perhaps other local Green- backers, had previously worked and participated in miners' organizations in other districts. Or his statement may refer to the Greenback-Labor Party generally as the lineal descendant of the MNA, rather than to his own personal experience.

18. All quotes from *National Labor Tribune*, August 17, 1878, with the exception of last Kelley statement (June 29, 1878) and local Democrat statement (August 10, 1878).

19. *National Labor Tribune*, August 10, 1878. The Democrat's expression of consterna- tion over the interracialism of the Greenback-Labor Party was hardly an isolated phenome- non. The newspapers regularly voiced such sentiments. Referring to black involvement in the local Greenback movement, for instance, the Birmingham *Iron Age* claimed that, to "true Democrats," it looked "like *reconstruction days had come again*" (January 29, 1879; em- phasis in original). It is possible, of course, that the story of the encounter between Thomas and the group of Democrats, and the one Democrat's subsequent tirade, was contrived or embellished, either by the Greenbacker at the meeting or by the *National Labor Tribune*. If so, its circulation in Greenbackers' gatherings and their press is all the more revealing as a window on the organization's public approach to the race issue. Rather than flinching from inflammatory allusions to carpetbaggers and black domination—language which only a few years earlier (as the Democrat wistfully recalled) had been able to delegitimize, even endan- ger, activists—the party called attention to the charges, neither confirming nor denying them, apparently confident that they would turn rancid in cold print.

20. *National Labor Tribune*, August 10 (prominent white Greenbacker), 17 (white War- rior club), 1878; August 16, 1879 (Dawson).

21. *National Labor Tribune*, September 13, 1879.

22. "The concern was run by the delegates from Warrior and Jefferson Mines," the Birmingham *Independent* reported. The Birmingham *Independent* report was published in the Montgomery *Advertiser*, July 10, 1878. The *Independent*, despite its name, was a pro- Democrat newspaper. See also Birmingham *Iron Age*, July 3, 10, 1878; *National Labor Tribune*, July 6, 1878.

23. Montgomery *Advertiser*, September 3, 1878.

24. Birmingham *Iron Age*, July 3, 1878, and Montgomery *Advertiser*, September 11, 1878 (irresponsible); Birmingham *Iron Age*, August 7, 1878 ("false doctrines"); Birmingham *Iron Age*, September 4, 1878 (race card); Birmingham *Iron Age*, August 7, 1878 (Harper); "Ad- dress of the County Executive Committee [Democratic Party] to the Voters of Jefferson County," in Birmingham *Iron Age*, July 3, 1878 (on depiction of GLP as a stalking horse, see

also Birmingham *Iron Age*, October 30, 1878). Such rhetoric was typical of that used by Democrats against the Greenbackers around Alabama. See Going, *Bourbon Democracy in Alabama*, 58; Roberts, "William Manning Lowe and the Greenback Party in Alabama," 103.

25. Sharit ran as an independent but was closely aligned with the GLP. How his and Handy's support was distributed from setting to setting is difficult to estimate. Although the *Iron Age* broke the vote down by beat, these did not always correspond to localities associated with groups such as miners. There is no mention of such mining towns as Newcastle, Jefferson Mines, or Helena, although Warrior and Oxmoor are listed. The unreliability of the election results in the Redeemed South, moreover, is legendary. On a more impressionistic level, the pro-Democrat Montgomery *Advertiser* asserted that the Greenbackers in the Birmingham district enjoyed strong support among African Americans. While the Democrats had carried the ballots for state legislature in Jefferson County by 300 to 500 votes, the *Advertiser* claimed, "the negroes and white Radicals voted solidly for the Greenbackers," adding that "the few negroes up here have a great deal to learn" (August 7, 1878). But that portrayal too must be viewed with caution in a context where substantial black support could undermine the credibility of the party that received it.

26. Birmingham *Iron Age*, August 7, 1878 (results in coal fields); Going, *Bourbon Democracy in Alabama*, 57 (hill country). On the 1878 election in Alabama, see also Rogers, *One-Gallused Rebellion*, 52; Roberts, "William Manning Lowe and the Greenback Party in Alabama," 100-108.

27. On Weaver tour: *Jefferson Independent*, July 17, 1880; *Alabama True Issue*, July 10, 1880; Montgomery *Advertiser*, September 18, 1880. (All references to *Jefferson Independent* and *Alabama True Issue* in this and subsequent notes in this chapter draw upon the research notes of Paul Worthman. I am indebted to Mr. Worthman for graciously allowing me access to these materials.) On Randall tour: *Jefferson Independent*, April 3, 1880 (Randall's visit to the Birmingham district involved a day each at Birmingham, Warrior, and Oxmoor). Quote about Greenbacker platform: *Alabama True Issue*, July 3, 1880. On Republican-baiting: *Jefferson Independent*, April 10, 17, July 17, 1880; Montgomery *Advertiser*, October 7, 1880; Birmingham *Observer*, July 29, 1880. Dizzying epithet: Tuskegee *Weekly News*, July 15, 1880, quoting Birmingham *Independent*, cited in Rogers, *One-Gallused Rebellion*, 53. Disclaimers: *Alabama True Issue*, August 21, 1880 (quote); also June 26, July 17, August 28, September 4, November 27, 1880. On the white supremacist assumptions that suffused the sentiments of the hill country Greenbackers of Alabama, see Hyman, *Anti-Redeemers*, 187-88.

28. On Jefferson County results: *Jefferson Independent*, August 7, November 6, 13, 1880. On Greenbackers in coal fields: *Jefferson Independent*, July 24, 1880; Birmingham *Observer*, August 5, 1880. Speaking of the Birmingham district in general, the *Observer* lamented elsewhere that the spirit of independentism was "rampant" (April 29, 1880). On the 1880 election around Alabama: Rogers, *One-Gallused Rebellion*, 53; Roberts, "William Manning Lowe and the Greenback Party in Alabama," 109-18; Going, *Bourbon Democracy in Alabama*, 58.

29. On Democratic attacks: Birmingham *Iron Age*, July 20, 1882 (quote); also June 15, August 3, 10, 1882. On results: Birmingham *Iron Age*, August 17, 1882; Rogers, *One-Gallused Rebellion*, 53-54; Roberts, "William Manning Lowe and the Greenback Party in Alabama," 118-20; Going, *Bourbon Democracy in Alabama*, 54.

30. Rogers, *One-Gallused Rebellion*, 54; Going, *Bourbon Democracy in Alabama*, 59; Roberts, "William Manning Lowe and the Greenback Party in Alabama," 120-21.

31. *National Labor Tribune*, November 22, 1879 ("Olympic"); see also July 5, August 16, 30, 1879, January 3, 1880. Such "superstition" smacked of backwoods ignorance to some Greenback organizers, many of whom were perhaps new to the district. Of course, a clannish reserve toward outsiders traditionally has been characteristic of mining communities wherever they exist. But this wary temperament was probably rooted in recent history at least as much as in regional or occupational tradition. In light of the price many southern workers had paid for engagement in the recent struggles of Reconstruction, a guarded response to Greenback recruiters made ample sense. This was particularly true for black miners, whose reluctance to embrace the movement unreservedly, even when approached by black representatives, was well illustrated in a letter to black Greenback leader Warren Kelley from Henry Hospun of the Colored Club at Haygood's Cross Roads. Upon a visit from the unknown Willis Thomas, Hospun wrote, "most of our boys were afraid to join. . . . Some say the Democrats have hired him to do this; others say it is to get up war; so that he had better mark his tracks here, as the white folks will upset him in a minute. If you know anything about him, let us know. . . . I hope he is the right kind of man, but we have been fooled before so much that we can't trust him yet" (*National Labor Tribune*, August 17, 1878).

32. This overview of the Knights of Labor as a national movement draws primarily on L. Fink, *Workingmen's Democracy*, esp. 3–37, 219–33; Oestreicher, "Terence Powderly, the Knights of Labor, and Artisanal Republicanism"; Laurie, *Artisans into Workers*, 141–75; Voss, *The Making of American Exceptionalism*; and Kealey and Palmer, *Dreaming of What Might Be*, 1–23.

33. The program of the Knights of Labor quoted in this paragraph was carried in every issue of the *Journal of United Labor*.

34. The concept of a "movement culture" draws on Goodwyn, *Populist Moment*, esp. 20–54. See also Rachleff, *Black Labor in the South*, 124–42.

35. For general surveys of the Knights of Labor in the South, see McLaurin, *The Knights of Labor in the South*; K. Hahn, "The Knights of Labor and the Southern Black Worker"; Kessler, "Organization of the Negro in the Knights of Labor." Among the more important case studies in this area are Rachleff, *Black Labor in the South*; Fink, *Workingmen's Democracy*, 149–77; Gould, "The Strike of 1887: Louisiana Sugar War"; Hine, "Black Organized Labor in Reconstruction Charleston"; Reed, "The Augusta Textile Mills and the Strike of 1886."

36. *Journal of United Labor*, May 15, October 15, 1880; Knights of Labor, *Record of the Proceedings of the Fifth Regular Session of the General Assembly Held at Detroit, Michigan, Sept. 6–10, 1881*, 207, in Powderly Papers; Garlock, *Guide to the Local Assemblies*, 3–9, 588–680; McLaurin, *The Knights of Labor in the South*, 43–44; Abernathy, "The Knights of Labor in Alabama," 29–39; Head, "The Development of the Labor Movement in Alabama," 50–51.

37. The chief source on the presence and activities of the Knights of Labor assemblies around the Birmingham district during the late 1880s is the *Alabama Sentinel*, started in early 1887 as the official organ of the Order around the state. See also Garlock, *Guide to the Local Assemblies*, 3–9; McLaurin, *The Knights of Labor in the South*, 44–45, 56–58, 106, 173–74, 178; Abernathy, "The Knights of Labor in Alabama," 50–64; McKiven, *Iron and Steel*, 37–39; Head, "Development of the Labor Movement," 51–52, 65–67. The 7,000–9,000 parade figure and the 4,000 membership figure are from the *Alabama Sentinel*, June 4 and April 16, 1887, respectively. The proceedings of the 1887 general assembly listed 3,951

members in Alabama as of July 1, 1887 (*Proceedings of the General Assembly of the Knights of Labor of America: Eleventh Regular Session, Held at Minneapolis, Minnesota, October 4 to 19, 1887*, 1850, in Powderly Papers).

38. On assemblies in various trades: routine coverage in *Journal of United Labor*, 1885–89, and *Alabama Sentinel*, 1887–90; Garlock, *Guide to the Local Assemblies*, 3–9. In early 1887, a group of Birmingham Knights established the nearby town of Powderly, where housing lots were purchased by Knights of Labor shareholders, and manufacturing concerns—most notably a Powderly Co-operative Cigar Works—functioned on a cooperative basis. (Of the 400 shares of the cigar works, over half were purchased by miners.) See *Journal of United Labor*, April 2, May 28, November 5, 1887; *Alabama Sentinel*, April 23, 30, 1887, April 21, 1888, January 4, 1890; Birmingham *Chronicle*, November 27, 1889; McLaurin, *The Knights of Labor in the South*, 126–27; Abernathy, "The Knights of Labor in Alabama," 64–69.

39. *Negro American*, October 19, 1887, *Alabama Sentinel*, May 10, 1890 (funeral/condolences); *Alabama Sentinel*, June 4, July 16, 1887 (parades); Huntsville *Gazette*, July 11, 1885 (Warrior); *Red Mountain Magnet*, May 12, 1888 (Blocton). The *Alabama Sentinel* regularly carried information concerning the times and locations of meetings around the district.

40. On mine committees, see, for example, letter from Brierfield in *Journal of United Labor*, December 3, 1887. On published advice to avoid mines, see *Journal of United Labor*, September 25, 1884; *Alabama Sentinel*, January 19, April 20, 1889. On raising money for strikers in other districts, see *Alabama Sentinel*, October 19, 1889. On cooperative enterprises in mining towns, see *Journal of United Labor*, April 25, 1885. On presence of National Trade Assembly No. 135 in Alabama, see *Alabama Sentinel*, March 31, April 7, 1888; *Record of Proceedings of the Second Annual Session of the Miners' and Mine Laborers' Nat. Dist. Assembly, No. 135, Held at Cincinnati, Ohio, June 1–8, 1887* (Grand Rapids, Mich.: Germania Printing House, n.d.), 64, in Powderly Papers; *Record of Proceedings of the Third Annual Session of the Miners' and Mine Laborers' National Trade Assembly, No. 135, Held at Cleveland, Ohio, August 8–14, 1888* (Columbus, Ohio: Hann & Adair, 1888), 188, in Powderly Papers. For background on National Trade Assembly No. 135, see Long, *Where the Sun Never Shines*, 148. The Knights' roles in politics and strikes are covered in detail below.

41. *Alabama Sentinel*, June 4, 1887.

42. *Alabama Sentinel*, January 28, 1888. Proposed legislation pertaining to miners concerned frequency of pay, coal screening and weighing procedures, scrip or check forms of payment, convict labor, and mine inspection by the state.

43. *National Labor Tribune*, June 20, 1885 (Harris); *Journal of United Labor*, June 10, 1886 (Coalburg). On Harris's occupation, see Abernathy, "The Knights of Labor in Alabama," 45.

44. Fink, *Workingmen's Democracy*, 18–34.

45. On the political program and campaigns of the Alabama Knights, see McLaurin, *Knights of Labor in the South*, 103–6; Abernathy, "Knights of Labor in Alabama," 73–74; Head, "Development of the Labor Movement," 68–74; *Alabama Sentinel*, September 23, 1887, June 23, 30, July 7, 21, 28, 1888.

In 1885, State Master Workman Nicholas B. Stack initiated an Anti-Convict League, which for a time spearheaded that cause. On the origins of the Anti-Convict League, see Stack's handwritten memoir, entitled "N. B.'s *own* story," Stack Papers, series 10, file 14, 14–16.

46. *National Labor Tribune*, March 12, April 5, May 3, 24, 1879; Birmingham *Iron Age*, March 12, 1879; Shelby *Sentinel*, March 13, 1879; Montgomery *Advertiser*, March 11, 1879.

47. Wage levels were by far the prevailing focus of miners' strikes around the nation during this era. Of 2,060 miners' strikes recorded for the period 1882–86, 1,726 (84 percent) were held chiefly or exclusively over rates of pay (U.S. Commissioner of Labor, *Third Annual Report, 1887*, 1010–11).

Wages were a key issue at strikes or lockouts at Warrior (1880); Pratt Mines (1879 [when the location was still known as Jefferson Mines], 1882, 1883, 1884, 1888, 1889); Coalburg (1884, 1888); Walker County (1887); Helena (1887); Brookside (1888); Blue Creek (1888); Blocton (1888); and Carbon Hill (1889 and 1890). On Warrior, see Jefferson *Independent*, October 9, 1880. On Pratt Mines, see note 46 above; Birmingham *Iron Age*, May 4, 11, June 1, 1882; *National Labor Tribune*, May 13, June 3, 1882; Birmingham *Advance*, October 1, 1883; Birmingham *Iron Age*, April 3, 1884; Birmingham *Chronicle*, December 13, 14, 1889; Birmingham *Age-Herald*, December 14, 1889; Atlanta *Constitution*, December 13, 1889 (for 1888, see discussion later in this chapter). On Coalburg, see Birmingham *Iron Age*, March 20, 1884; Birmingham *Semi-Weekly Review*, March 15, 1884. On Walker County, see *National Labor Tribune*, April 9, May 14, 1887; *Alabama Sentinel*, July 2, 1887. On Helena, see *Alabama Sentinel*, November 26, December 3, 1887; *Journal of United Labor*, December 3, 1887. On Brookside and Blue Creek, see discussion of 1888 strike below. On Blocton, see *Alabama Sentinel*, October 20, 1888; Birmingham *Age-Herald*, November 22, 1888. On Carbon Hill, see *Alabama Sentinel*, July 20, 1889, January 4, 11, 1890.

48. On strikes over faulty scales: *National Labor Tribune*, February 8, 1879; *Alabama Sentinel*, October 12, 1889; A. N. Jones to Governor Seay, August 3, 1889, in Seay Papers, Correspondence, July–August, 1889, ADAH. On location of scales: Birmingham *Age*, December 5, 1887. On definition of ton: *National Labor Tribune*, April 5, 1879. On tram size: *Alabama Sentinel*, May 28, 1887. On screening of coal: *Alabama Sentinel*, November 10, 1888, March 2, April 20, May 18, July 20, 1889, May 10, 17, 31, June 14, 1890. On theft of coal by company: *National Labor Tribune*, March 22, 1884; *Alabama Sentinel*, October 12, 1889. On "dirty" coal: Birmingham *Iron Age*, October 4, 11, 1883; *Alabama Sentinel*, November 19, 26, December 10, 1887; *National Labor Tribune*, February 25, 1888; U.S. Commissioner of Labor, *Tenth Annual Report, 1894*, 34–37; Pratt Mines *Advertiser*, August 2, 1889. On unpopular bosses: *National Labor Tribune*, March 19, 1887; Birmingham *Advance*, September 17, 1883; *Alabama Sentinel*, October 12, 1889; Birmingham *Age-Herald*, August 14, 1890. On recognition of bank committees: *National Labor Tribune*, July 22, 1882, September 15, October 6, 1888; *Alabama Sentinel*, June 8, 1889. On safety issues: *National Labor Tribune*, February 8, 1879, March 22, 1884. On dead work: Birmingham *Observer*, July 22, 1880; Jefferson *Independent*, July 17, 24, 1880. On distribution of work: *National Labor Tribune*, March 22, 1884, July 3, 1886. On company store: Birmingham *Iron Age*, May 11, 1882; *National Labor Tribune*, April 9, 23, 1887; *Alabama Sentinel*, May 28, 1887, May 18, 1889.

49. Birmingham *Iron Age*, August 21 ("remarkably fine"), 28, September 4, 11, October 30, 1884, February 26, 1885; Huntsville *Gazette*, September 13, 1884; *Journal of United Labor*, December 10, 1884, April 25, 1885 ("terrible curse"); *National Labor Tribune*, October 11, 1884; Birmingham *Weekly Iron Age*, August 20, 1885; U.S. Commissioner of

Labor, *Third Annual Report, 1887*, 36–39; Knights of Labor, *Record of the Proceedings of the Eighth Regular Session of the General Assembly, Held at Philadelphia, Pa., Sept. 1–10, 1884*, 704, in Powderly Papers ("adopt some means").

50. Birmingham *Weekly Iron Age*, July 30, August 20, 1885; *John Swinton's Paper*, August 23, 1885; *National Labor Tribune*, June 20, 1885 ("Operators make agreements"); Mobile *Register*, August 17, 1885; U.S. Commissioner of Labor, *Third Annual Report, 1887*, 36–39.

51. Mobile *Register*, August 17, 18, 20, 22, 23, September 11, 1885; Birmingham *Weekly Iron Age*, August 20 (Kelsey quote), 27, 1885; Huntsville *Gazette*, August 22, September 19, 1885. On the salience of contests over workplace control in the relations of labor and employers during the late nineteenth century, see Montgomery, *Workers' Control in America*, 1–31, and *Fall of the House of Labor*, 22–44, 210–11.

52. Huntsville *Gazette*, September 13, 1884.

53. In summer 1886, the *Journal of United Labor* described Warrior Station organizer Edward Harris as having built "a great many Assemblies of colored people" (September 10, 1886). For an instance in which African Americans predominated, see description of the assembly at Blocton in *Alabama Sentinel*, January 28, 1888.

54. *Journal of United Labor*, August 15, 1880 ("We should be false . . ."); *Alabama Sentinel*, July 20, 1889 (Alabama Knights' Assembly); Nicholas Stack to Terence Powderly, August 7, 1887, Powderly Papers, Reel 23 (Campbell's organizing); *John Swinton's Paper*, July 3, 1887 (Birmingham rally); *Journal of United Labor*, December 24, 1887 ("fraternal visits"); *Negro American*, October 29, 1887 (Aldrich). The latter item made no mention of Richson's race, but the significance holds either way. Indeed, were he white, coverage of his Knights of Labor funeral in a black paper would be all the more noteworthy.

W. J. Campbell at times recruited for the Knights elsewhere in the South. Leon Fink mentions an organizing visit to Richmond during the spring of 1886. See Fink, *Workingmen's Democracy*, 171; *John Swinton's Paper*, May 9, 1886. For decades afterwards Campbell would remain an active proponent of interracial unionism. As a representative of the Kentucky district of the United Mine Workers, he rose at the union's 1911 national convention to oppose affiliation between the UMW and the seven railroad brotherhoods so long as the latter retained the color line: "I want them first to remove lines of race, color and nationality before we unite with them. When I fail to be true to my own race I will not ask the Anglo-Saxon to repose confidence in me to represent him" (*Proceedings of the Twenty-second Annual Convention of the United Mine Workers of America* [Indianapolis: Cheltenham Press, 1911], 599). On the color line in the railroad brotherhoods, see Arnesen, "'Like Banquo's Ghost, It Will Not Down': The Race Question and the American Railroad Brotherhoods, 1880–1920." In a 1913 eulogy, black UMW national organizer G. H. Edmunds spoke of Campbell's "gentleness of manner," "earnestness," and "incorruptible integrity" (*UMWJ*, July 24, 1913).

On the racial approach of the Knights of Labor nationally, see Laurie, *Artisans into Workers*, 158–63; P. S. Foner, *Organized Labor and the Black Worker*, 47–63; W. H. Harris, *The Harder We Run*, 26–27. This ethic of inclusiveness, it might be noted, did not extend to Chinese immigrants. Motivated both by a self-protective instinct regarding labor competition and ingrained racial attitudes, the Knights joined in the campaign for their exclusion from the United States.

55. *Alabama Sentinel*, July 28, 1887 ("appeal to prejudice"), February 1, 1890 ("we of the Knights"), February 18, 1888 ("the colored people"); *National Labor Tribune*, June 20, 1885 ("Should the negroes").

56. *Alabama Sentinel*, November 12, 1887; see also April 9, May 28, 1887, February 1, 1890.

57. For examples, see *Alabama Sentinel*, August 4, September 22, 1888.

58. On the formation of the mine labor force in the Birmingham district, see chapter 1.

59. *National Labor Tribune*, September 27, 1879 ("Olympic"). The Warrior miners' identification of convicts and Italians as the chief threats to their turf was not an isolated case. In 1890 the *Alabama Sentinel*, noting the arrival of seventy-five Italians at Helena, observed that, in addition to lack of work, a sliding scale, and convicts, the Alabama miners now had "Dagoes" to contend with (July 12, 1890). Nor was the singling out of Italians as intruders unique to the miners of the Birmingham district. A similar reaction arose among the coal miners of western Pennsylvania during the 1870s. See Gutman, "Citizen-Miners and the Erosion of Traditional Rights." (I thank Ira Berlin for providing me with a copy of this 1967 unpublished paper.)

60. For an influential expression of this formulation, see C. V. Woodward, *Origins of the New South*, 229.

61. *Alabama Sentinel*, November 12, 1887 ("The man—*white or colored*"; emphasis added), April 9, 1887 ("common cause"); *Proceedings of the State Assembly of the K. of L. (1887)*, 23, in Powderly Papers (resolution).

62. *Alabama Sentinel*, November 12, 1887 ("little cause for complaint"); *Negro American*, October 9, 1886 ("noble work"); Huntsville *Gazette*, April 4, 1885 (Campbell).

63. Birmingham *Iron Age*, May 11, 1882; *Alabama Sentinel*, May 28, June 4, 1887.

64. *Alabama Sentinel*, May 14, 1887 (post office); *National Labor Tribune*, April 23, 1887 (Kirkpatrick).

65. Tom O'Reilly to Powderly, March 4, 1887, in Powderly Papers, Reel 21; *Alabama Sentinel*, February 18, 1888 ("slowly learning"), November 12, 1887 ("elevating influence"). O'Reilly, it might be noted, did not confine his bemused condescension to African Americans. Perhaps because he felt they should know better, he reserved particular exasperation for southern whites: while he sympathized with their condition, he wrote, he could not refrain from adding that "their method of spelling is uncivilized. Just imagine the following at all the depots:—'Beware of *Pic-pockets*.' And they spell it 'Eppigram'! We visited a white Assembly in Birmingham Ala, and the M.W. said 'We mean no "Anty-gony-ism" to necessary capital.' That was more than I could stand and we left the city late that night."

66. *Alabama Sentinel*, August 16, 1890.

67. Two flyers in Powderly Papers, Reel 22 (Powderly celebration; there is no such evidence of racially exclusive events held by the Knights in the mining towns, however); *Alabama Sentinel*, July 28, 1888 ("white man's fight").

68. *Alabama Sentinel*, June 11, 1887. The issue of the *Negro American* mentioned in the *Sentinel* is no longer extant.

69. *Alabama Sentinel*, June 4, 1887.

70. For numerous charges by newspapers around the nation that the Knights had provoked "social equality" at their gathering in Richmond, see *Public Opinion*, October 16,

1886. See also Rachleff, *Black Labor in the South*, 169–78; Fink, *Workingmen's Democracy*, 162–63.

71. Having denied (and sought to defang) the "social equality" charge, the *Sentinel* impugned the motives of its authors. Those who were "continually stubbing their toes on the color line," it suggested, "are not very sure of their own social standing" (June 4, 1887). It accused the Order's critics of deliberately "keeping alive a prejudice that certainly has no foundation in the true aims and objects of the Knights of Labor" (June 11, 1887). The *Sentinel* blended these points in its response to a Birmingham *Age* article entitled "Social Equality Causes Trouble among the Knights of Labor in Birmingham": "[N]ot certain of its own social standing, [the *Age*] now resorts to an absurd and self-evident assumption and tries to irritate a sore that is well-nigh healed" (June 25, 1887). Finally, the *Sentinel* turned to what it saw as the real issue lurking beneath the "social equality" charge—fear that the Knights would encourage blacks to assert their political independence: "[S]ocial equality or the color line is not so much the basis of the difficulty as the probability there is of the colored man changing his base politically, and in the future casting his vote in the direction of his own personal interests instead of those of his enemies" (June 11, 1887). The Birmingham *Age* for this part of 1887 is not extant.

72. *National Labor Tribune*, May 13, June 3, July 22, 1882; Birmingham *Iron Age*, May 4, 11, 18, 25, June 1, 1882; U.S. Commissioner of Labor, *Third Annual Report, 1887*, 36–39.

73. A settlement involving a modest wage increase negotiated by Stack quickly unraveled, as neither the miners nor the company honored it. The two parties finally reached an agreement of their own entailing a higher pay scale, but relations soon deteriorated when company officials implemented a blacklist, withheld pay for arbitrary amounts of coal for dirt or slack, and fired anyone who complained. This account of the 1887 strike draws upon *National Labor Tribune*, March 12, April 9, 23, May 7, 14, June 11, July 16, 30, 1887, February 4, 1888; *Alabama Sentinel*, May 14, 21, 28, June 4, 25, September 3, 1887, February 8, 1888; *Labor Union*, February 19, March 12, 1887; U.S. Commissioner of Labor, *Tenth Annual Report, 1894*, 34–37; *Proceedings of the State Assembly of the K. of L.* (Birmingham: Herald Job and Book Rooms, 1887), in Powderly Papers, 5–7; *Proceedings of the General Assembly of the Knights of Labor of America, Eleventh Regular Session* (n.p.: General Assembly, 1887), 1737–38, in Powderly Papers.

74. This account of the 1888 strike draws upon the following sources: *Alabama Sentinel*, March 31, May 5, 26, June 9, 16, 23, July 7, 28, December 22, 1888; *National Labor Tribune*, June 9, 1888; *Bessemer*, June 2, 1888; Birmingham *Weekly Iron Age*, May 10, 24, June 27, 1888; Birmingham *Weekly Herald*, May 9, 1888; "Report of the Director of the Sloss-Sheffield Steel and Iron Company for the Year Ending February 1, 1889," Sloss-Sheffield Records, 1892–1902, 342.4.1.105.1, 5. On the aversion of national Knights leaders to strikes, see Oestreicher, "Terence Powderly, the Knights of Labor, and Artisanal Republicanism," 52–53; Laurie, *Artisans into Workers*, 166–68.

75. On the Blocton strike, see Birmingham *Age-Herald*, November 22, December 14, 1888; *Alabama Sentinel*, October 20, 27, November 3, 10, December 15, 1888; Birmingham *Weekly Iron Age*, November 1, 1888. For the Knights' analysis, see *Alabama Sentinel*, December 22, 1888, January 26, 1889.

76. At the founding convention, a proposal that the state federation affiliate with the

Knights of Labor failed by a vote of 1,335 to 310. Little information survives about the NFMML in Alabama. See *National Labor Tribune*, August 4, September 15, October 6, 1888; Birmingham *Age-Herald*, November 22, 1888; *Alabama Sentinel*, July 13, 1889. Founded in September 1885, the National Federation had for several years functioned as a bitter rival to the Knights of Labor in the northern coal fields. Although it overlapped substantially with the Knights in both membership and objectives, it adopted a more trade unionist orientation. This approach emerged most dramatically in the federation's pioneering of the Joint Wage Convention in the Central Competitive Field, under which miners' representatives and operators met to negotiate a districtwide contract (Evans, *History of the UMWA, 1860 to 1890*, 224–39; Roy, *History of the Coal Miners*, 262–74; Long, *Where the Sun Never Shines*, 149–51).

77. The MTC was promoted regularly in the pages of the *Sentinel*, which now listed itself as the "official organ of the State Assembly of the Knights of Labor, Miners' Trade Council, and Bricklayers' Association." See *Alabama Sentinel*, July 27, August 3, September 27, October 26, November 2, 9, 16, 30, December 7, 14, 21, 1889, January 11, 18, 1890; Savannah *Tribune*, August 17, 1889.

78. On the decline of the Knights nationally, see Fink, *Workingmen's Democracy*, 228–30; Laurie, *Artisans into Workers*, 172–75; Oestreicher, "Terence V. Powderly, the Knights of Labor, and Artisanal Republicanism," 50–56. On the Order's decline in the South, see McLaurin, *The Knights of Labor in the South*, 149–81.

79. This move by the Knights of Labor and the National Federation of Miners and Mine Laborers in the Birmingham district to bury the hatchet mirrored a national trend, which culminated in January 1890 with the founding of the United Mine Workers of America (Long, *Where the Sun Never Shines*, 151).

CHAPTER FOUR

1. P. S. Foner, *Organized Labor and the Black Worker*, 64–71; Foner and Lewis, *The Black Worker*, 20–22; Laslett, "Samuel Gompers and the Rise of American Business Unionism," 76–78. The first signs of the AFL leadership's softening resistance to the color line came around 1896, when President Samuel Gompers began agreeing to admit unions with Jim Crow restrictions into the federation. By the turn of the century the AFL leadership routinely refrained from challenging racially exclusive practices of affiliated unions, on the grounds that it had to respect their autonomy.

2. Among the objects enumerated were proper rates of pay and an end to payment in scrip, safe work conditions, an eight-hour day, a prohibition on child labor, a fair system of weights and measures at the mines, weekly pay, a ban on the operators' use of private detectives or guards during strikes and lockouts, and the use wherever possible of arbitration and conciliation to resolve disputes between miners and operators (Fox, *United We Stand*, 23–25). On the founding of the United Mine Workers of America, see Evans, *History of the UMWA, 1890 to 1900*, 3–29; Roy, *History of the Coal Miners*, 243–66; Fox, *United We Stand*, 22–29.

3. In becoming a part of District 20, MTC assemblies severed all remaining ties with the state assembly of the Knights of Labor, although the *Alabama Sentinel* straddled the juris-

dictional issue by serving as the official organ for both groups. See *Alabama Sentinel*, April 12, May 3, 10, 17, 24, 1890.

4. *Alabama Sentinel*, May 10, 1890 (Conley); May 24, 31, June 14, 28, 1890 (organizing locals); March 22, April 12, 26, May 10, 24, 1890 (campaigning); January 4, 25, February 8, 15, 22, May 10, 31, June 14, 1890 (Carbon Hill); May 1, 1890 (Milldale); Birmingham *Chronicle*, February 7, 1890, *Alabama Sentinel*, February 15, 1890 (Kitchen Mine); *Alabama Sentinel*, January 18, 25, February 22, May 3, 17, 31, June 7, 1890, Birmingham *Chronicle*, February 7, 8, 10, 13, 1890 (Blue Creek).

5. Central to the new scale were a wage increase from 45 to 50 cents per ton at Pratt Mines (with corresponding increases at points throughout the district) and the abolition of the sliding scale, according to which the miners' wage rates rose and fell with the market price of pig iron. See *Alabama Sentinel*, June 1, 14, 21, 28, 1890; Birmingham *Age-Herald*, July 2, 1890; Birmingham *Daily News*, July 1, 1890.

6. *Alabama Sentinel*, July 5, 1890; Birmingham *Age-Herald*, July 2, 3, 4, 7, 8, 1890; Birmingham *Daily News*, July 1, 2, 3, 1890.

7. *Alabama Sentinel*, July 12, 19, 26, August 23, September 6, 13, 20, 1890; Birmingham *Daily News*, July 4, 7, 1890.

8. The strike affected all the major companies—including Tennessee, DeBardeleben, Sloss, Cahaba, Virginia and Alabama, and Mary Lee—and engulfed such key mining centers as Pratt Mines, Blue Creek, Blocton, Coalburg, Cardiff, Brookside, Carbon Hill, Warrior, Henry Ellen, Cordova, Corona, and Brookwood. See *Alabama Sentinel*, November 8, 15, 22, 29, December 5, 1890; Birmingham *Age-Herald*, November 7, 18, 19, 24, 25, 29, 30, 1890; Chattanooga *Republican*, December 7, 1890; interview with R. A. Statham, July 2, 1936, in Lester J. Cappon Collection, Box 1, Folder 9, Archives of Labor and Urban Affairs, Wayne State University, Detroit, Mich.; A. M. Shook to George B. McCormack, November 24, 1890, Shook Papers, Outgoing Correspondence 386.1.1.1.2.

9. Day-to-day, often conflicting accounts of the strike can be found in the Birmingham *Age-Herald*, the Birmingham *Daily News*, and the *Alabama Sentinel* (which published a daily edition during much of the strike) for the months of December and January. The strike is also described in Ward and Rogers, *Labor Revolt in Alabama*, 32–33; and Head, "Development of the Labor Movement in Alabama," 89–90. On the poststrike situation at Warrior, see *Alabama Sentinel*, January 31, February 12, 21, 28, March 7, 21, 1891; Birmingham *Daily News*, January 22, 1891. Among the provisions of the Warrior contract were the right of the company to refuse employment to anyone who in their judgment would "work injury or discontent to the works"; a commitment on the part of the miners to request no wage increase over the remainder of the contract; no restriction on the times miners worked or on the hiring by miners of laborers; the right of the company to discharge miners for absenteeism; restrictions on the times miners could hold meetings; and a bar on the stoppage of work on account of a grievance (*Alabama Sentinel*, January 31, 1891).

10. DeBardeleben made contact with labor contractors as far away as New Orleans. See Birmingham *Age-Herald*, December 3, 9, 22, 1890.

11. Shook to McCormack, December 4, 16, 1890, Shook Papers, Outgoing Correspondence 386.1.1.1.2; Birmingham *Age-Herald*, December 2, 9, 15, 17, 18, 1890, January 2, 4, 6, 7, 1891.

12. Birmingham *Age-Herald*, December 1, 1890. The credibility of this account, which the labor press did not confirm, should be weighed in light of the *Age-Herald*'s manifest inclination to render union efforts either notorious or trivial in the public eye. Even if the account were inaccurate, however, it no doubt reinforced the prominence of race as a live issue as the strike drew near.

13. *Alabama Sentinel*, December 20, 1890; Birmingham *Daily News*, January 9, 1891; *Alabama Sentinel*, December 28 (daily strike edition), 1890.

14. Birmingham *Age-Herald*, December 4, 1890; see also December 17, 1890. The Atlanta *Journal* gleefully wondered what northern Republicans would make of these peaceful relations between white strikers and black strikebreakers at Blue Creek, in light of their charges of southern whites "oppressing the negroes." In reprinting the piece, the *Age-Herald* found the point "well taken. Such a change in any Northern locality would have been the cause of bloodshed, and the negroes would have been the sufferers" (January 4, 1891).

15. *Alabama Sentinel*, December 20, 1890 ("blackleg" was a traditional pejorative for strikebreaker in mining areas). See also *Sentinel* issues of December 6, 13, 19 (daily strike edition), 28 (daily strike edition), 1890, and January 10, 1891.

16. *Alabama Sentinel*, December 13, 20, 27, 28 (daily strike edition), 1890; Birmingham *Daily News*, January 7, 1891; Birmingham *Age-Herald*, December 23, 1890.

17. Birmingham *Age-Herald*, January 2, 1891 ("sensation"); Birmingham *Daily News*, January 2, 1891 ("got rampant"); Birmingham *Daily News*, January 6, 1891 (women and DeBardeleben). If the *Daily News* can be believed, a bemused DeBardeleben invited the women around to the company office, where beer was provided and "everybody drank and felt better than ever"—after which the president reminded his guests that "fun was fun and business was business and they would have to vacate the houses," which was duly done. Perhaps mindful of the implausibility of such good cheer, the article closed with the caveat that the "story is given as it reached a *News* reporter and for what it is worth."

18. Birmingham *Daily News*, January 8, 1891.

19. *Alabama Sentinel*, January 3, 31, 1891. For background on the Lodge "force bill," see C. V. Woodward, *Origins of the New South*, 254–55.

20. *Alabama Sentinel*, December 27, 28 (daily strike edition), 1890, January 3, 17, 1891; Birmingham *Age-Herald*, December 22, 1890, reprinted in *Alabama Sentinel*, December 27, 1890; *National Labor Tribune*, December 27, 1890.

21. Birmingham *Age-Herald*, January 26, June 28, 1891; Birmingham *Labor Advocate*, June 13, 20, 1891; *Alabama Sentinel*, January 31, March 7, June 27, 1891; *UMWJ*, March 28, July 2, 1891; *Engineering and Mining Journal*, January 27, 1894. In a communication at this time to McCormack, TCI general manager Shook betrayed a breezy lack of concern over the union's capacity for battle. After recommending that the current mine wages at the Pratt division be renewed, he added: "Of course, you would be able to whip any strike that might occur, or even reduce the price 5c per ton, force a strike and then perhaps whip that, but in the long run it would not pay. I had rather be excused from going to B'ham just now to make a contract with the Miners; the weather is too hot and I have too much to do here" (Shook to McCormack, June 18, 1891, Shook Papers, Outgoing Correspondence 386.1.1.1.5).

22. Birmingham *Labor Advocate*, August 20, 1892 (UMWA or Knights of Labor); *Alabama Sentinel*, October 29, November 19, 1892 (Alabama Miners' Protective Association); *Alabama Sentinel*, March 19, 1892 (Relief Society).

23. That the Alabama miners might respond to the example of their Tennessee brethren did not escape the thoughts of state authorities. See, for example, Birmingham *Age-Herald*, July 25, November 23, 1891. On the 1891–92 Tennessee coal miners' rebellion, see Shapiro, "Tennessee Coal Miners' Revolts of 1891–92"; Daniel, "Tennessee Convict War," 273–92.

24. Rogers, *One-Gallused Rebellion*, 165–66; Ward and Rogers, *Labor Revolt in Alabama*, 37–38.

25. The following background on the Southern Farmers' Alliance draws upon Goodwyn, *The Populist Moment*; C. V. Woodward, *Origins of the New South*, 175–204; Palmer, *"Man Over Money"*; McMath, *American Populism*. Black farmers organized as the Colored Farmers' National Alliance and Cooperative Union.

26. These proposals, which would be adopted as the platform of the National Alliance convention at Ocala in December, included measures for the abolition of national banks, expansion of currency, a graduated income tax, tariff reduction, direct election of senators, federal control of railroad and telegraph companies, and, a system of government-run "subtreasuries," or federal warehouses.

27. Loathe to break with the party, Kolb pledged his support to the Democratic ticket. On the 1890 Alabama elections, see Hackney, *Populism to Progressivism in Alabama*, 14–27; Rogers, *One-Gallused Rebellion*, 165–87; Ward and Rogers, *Labor Revolt in Alabama*, 39–40; Goodwyn, *Populist Moment*, 144–45; Rogers, Ward, Atkins, and Flynt, *Alabama*, 306–8.

28. In the same spirit as the *Sentinel*'s report from Pratt Mines, "Civic" of Warrior contrasted the local "Workingman's Ticket," made up of "laboring men," with that of the "Warrior Clique," comprised of "coal operators, merchants, one banker, one editor, one ex-saloon keeper, and three or four good old farmers." See *Alabama Sentinel*, March 22, April 12, 26, May 10, 24, 1890.

29. Birmingham *Age-Herald*, August 6, 8, 12, 1890; *Alabama Sentinel*, May 24, 1890. The racial breakdown of the vote is unknown.

30. Ward and Rogers, *Labor Revolt in Alabama*, 41.

31. The convict issue remained salient in the coal fields during the campaign. Both sides were on record as favoring the lease's termination, but they differed as to the means. Jones deplored the Tennessee miners' "radical" attack on convict stockades and counseled patience. Kolb, while not condoning extralegal methods, attached greater urgency to the issue. See *Alabama Sentinel*, October 24, 1891 (Jones); Birmingham *Daily News*, November 1, 1891 (Kolb); Birmingham *Labor Advocate*, January 9, 1892 (observer). On the 1892 Alabama elections, see Hackney, *Populism to Progressivism in Alabama*, 19–27; Rogers, *One-Gallused Rebellion*, 221–27; Ward and Rogers, *Labor Revolt in Alabama*, 41–45; Rogers, Ward, Atkins, and Flynt, *Alabama*, 308–11.

32. The entire political process, the *Sentinel* went on, was laughable: "We go through the form of holding popular elections, but the whole thing is a gigantic farce; it is simply, as a rule, a time for the purchase and sale of votes, and a time also for the commission of political prostitution and moral debauchery." See *Alabama Sentinel*, December 12, 1891; Birmingham *Age-Herald*, January 2, 1892 (results). The DeBardeleben company worked with characteristic zeal to bring its miners to the conservative fold. In early November, for example, the pit boss at Johns threw what the *Daily News* described as a "regular old-fashioned miners' dance and frolic, attended by leading Democratic officials—the latter serving as 'political trimmings to finish off the bill of fare'" (Birmingham *Daily News*, November 3, 1891).

33. Birmingham *Age-Herald*, August 10, 1892. As with the 1890 elections and the voting over Democratic delegates the previous December, the racial breakdown of this ballot for the mineral district is unknown. Nor do we have records of the distinctive ways in which black and white miners thought about the respective campaigns of Jones and Kolb. The contrast between the two sides on matters of race was significant, though not pure. The Bourbon Democrats stood unabashedly for white supremacy, and were ever prepared to race-bait any reform movement, including the Farmers' Alliance/Jeffersonian Democrats. The latter took a murkier stand, pledging themselves on the one hand to the protection of black rights (including, at times, the right to vote) and adamantly denying charges that they represented a threat to white supremacy on the other (Rogers, Ward, Atkins, and Flynt, *Alabama*, 306–10).

34. *Engineering and Mining Journal*, February 24, 1894; Ward and Rogers, *Labor Revolt in Alabama*, 47–49. On the panic of 1893, see chapter 1 above.

35. Following a gathering of TCI employees to plot strategy against plans to lower wages, the company reconsidered and agreed to renew the Pratt scale unchanged. See Birmingham *Age-Herald*, June 30, July 1, 1893; Birmingham *Labor Advocate*, June 22, 29, July 6, 1893; *UMWJ*, July 13, 1893. When the Sloss company declined to follow suit, its miners at Blossburg, Brookside, Brazil, and Cardiff embarked on a strike that would continue through most of July (the outcome of the strike is not known). See Birmingham *Age-Herald*, July 20, 1893.

36. *UMWJ*, October 5, November 16, December 28, 1893; Birmingham *Labor Advocate*, October 21, 28, November 11, 18, 25, December 2, 30, 1893; *National Labor Tribune*, December 7, 1893.

37. Birmingham *Labor Advocate*, November 11 (Blocton), 18, 25, December 9 (Corona), 30, 1893.

38. Birmingham *Labor Advocate*, December 9, 1893 (Corona), February 10, 1894 ("stone wall"); *UMWJ*, September 14 (Hannigan), October 5 ("capital and its paid . . ."; "preposterous in the face . . ."), 1893.

39. Birmingham *Labor Advocate*, November 11, 18, 1893 ("Excelsor"), December 9 (Corona); *UMWJ*, October 5, 1893 ("blackest hue").

40. The concessions sought from the operators included reductions in the costs of company housing, store and mining supplies, and medical expenses; the weighing of coal at the tipple before being dumped and in the presence of a checkweighman; and an end to the subcontracting of whole mines. See Ward and Rogers, *Labor Revolt in Alabama*, 59–64. The story of the 1894 strike has been told several times, at varying lengths. The most complete account is Ward and Rogers, *Labor Revolt in Alabama*. See also Head, "The Development of the Labor Movement in Alabama," 96–105; Straw, " 'This Is Not a Strike,' " 14–24; R. L. Lewis, *Black Coal Miners in America*, 41–44; Lichtenstein, "Racial Conflict and Racial Solidarity." The general narrative of this strike can be reconstructed from regular coverage in the Birmingham *Age-Herald*, the Birmingham *Daily News*, the Birmingham *Labor Advocate*, and the *UMWJ*.

41. Ward and Rogers, *Labor Revolt in Alabama*, 63, 67, 69, 71; *National Labor Tribune*, May 3, 1894. On the 1894 coal strike nationally, see Fox, *United We Stand*, 44–45; Long, *Where the Sun Never Shines*, 153–54.

42. Ward and Rogers, *Labor Revolt in Alabama*, 68; Chattanooga *Tradesman*, May 1, 15, 1894.

43. Birmingham *Age-Herald*, April 14, 1894; see also June 12, 1894.

44. Birmingham *Age-Herald*, April 20, 1894 (Eden); Ward and Rogers, *Labor Revolt in Alabama*, 68–69.

45. Birmingham *Labor Advocate*, June 23, 1894 ("Maiden's Prayer"); Warrior *Index*, May 4, 1894; Birmingham *Age-Herald*, April 12, June 27, August 4, 1894; Chattanooga *Tradesman*, May 15, August 1, 1894 (trainloads of black laborers); *West End Banner*, June 2, 1894; Ward and Rogers, *Labor Revolt in Alabama*, 73, 80–81 (Blue Creek mines); Birmingham *Age-Herald*, May 16, July 14, 1894 (Sloss company).

46. *National Labor Tribune*, June 21, 1894, emphasis in original ("slave trade"); JHF to Governor Jones, June 3, 1894 ("bitter against the negro"); Vallens to Governor Jones, May 11, 1894 ("throwing out white strikers"; see also Vallens to Governor Jones, May 14, June 4, 8, 1894, and JHF to Governor Jones, June 3, 4, 13, 1894); Vallens to Governor Jones, May 8, 1894 ("bitter against them"). All letters in Jones Papers, "Pinkerton, 1893–94" File.

47. Vallens to Governor Jones, May 21, 1894, Jones Papers, "Pinkerton, 1893–94" File; *UMWJ*, July 5, 1894 (Bergen); Birmingham *Labor Advocate*, May 12, 1894; *UMWJ*, May 10, 1894 (also Birmingham *Labor Advocate*, May 19, 1894).

48. Moral revulsion over the Pinkertons' furtive activities during strikes should not blind the investigator to the historical value of their reports. However mercenary their efforts may have been (or perhaps *because* of their essentially mercenary disposition), such detectives brought considerable professionalism to their work. Their voluminous correspondence to Governor Jones during the 1894 strike shows every sign that they called the unfolding situation as they saw it. Indeed, despite the governor's manifest eagerness to view the unionists as irresponsible agitators, Vallens, "JHF," and in the later stages "JMP" often evenly described them as cool-headed and restrained. For a good extended case study of industrial espionage in the New South, see G. M. Fink, *Espionage, Labor Conflict, and New South Industrial Relations*.

49. JHF to Governor Jones, May 27, 1894 (Fox's Saloon); JHF to Governor Jones, June 16, 1894 (camaraderie; on the mingling of black and white strikers in saloons, see also Lichtenstein, "Racial Conflict and Racial Solidarity"); JHF to Governor Jones, June 14, 1894 ("all talking together"). For other examples of interracial miners' saloons during the strike, see JHF to Governor Jones, June 6, 1894, and JMP to Governor Jones, July 22, 1894. (All letters in Jones Papers, "Pinkerton, 1893–94" File.) Whether interracial patronage of miners' saloons was common in the Alabama mineral district during this era or occurred only during major strikes is difficult to know. The sources from this era reveal little about the extent of mixed-race saloons during ordinary times.

50. Vallens to Governor Jones, May 21, 1894 (300 strikers); Vallens to Governor Jones, May 26, 1894 (2,000 miners); Vallens to Governor Jones, June 9, 1894 ("No blackleg"). All letters in Jones Papers, "Pinkerton, 1893–94" File.

51. Birmingham *Labor Advocate*, April 14, 1894.

52. Vallens to Governor Jones, May 24, 1894 ("divided"); see also Vallens to Governor Jones, June 18, 1894, Jones Papers, "Pinkerton, 1893–94" File.

53. Birmingham *Labor Advocate*, April 28, 1894 ("We, the colored"); Vallens to Governor Jones, June 3, 1894, Jones Papers, "Pinkerton, 1893–94" File ("standing firmly"; see also JHF to Governor Jones, June 5, 6, 1894); *UMWJ*, May 31, June 28, July 12, 1894, and Birmingham *Labor Advocate*, May 19, June 2, 1894 ("as firm as ever," etc.); Ward and

Rogers, *Labor Revolt in Alabama*, 67 ("Pendragon"); *UMWJ*, June 7, 1894 (Hannigan); Birmingham *Age-Herald*, June 6, 1894 (executive committee).

54. Birmingham *Labor Advocate*, May 12, 26, June 23, 1894 (DeBard's pets); *UMWJ*, July 5, 26, 1894 (Kelso); Birmingham *Labor Advocate*, May 5, 1894; Birmingham *Labor Advocate*, May 19, 1894 (Brookside).

55. Birmingham *Labor Advocate*, May 19, 1894 (Brookside); *UMWJ*, July 5, 1894 (white writer); Birmingham *Labor Advocate*, June 9, 1894 (black miner).

56. Birmingham *Labor Advocate*, May 26, 1894.

57. Birmingham *Age-Herald*, June 6, 1894 (letter to merchants and citizens); Birmingham *Labor Advocate*, June 23, 1894; Warrior *Index*, April 27, 1894.

58. Vallens to Governor Jones, May 5, 1894 (women beating pans); *West End Banner*, May 12, 1894, and Vallens to Governor Jones, April 22, May 8, 1894 (Horse Creek, evictions, rumors); JHF to Governor Jones, May 22, 1894 (Bud Hanley's; see also Vallens to Governor Jones, May 23, 1894); Ward and Rogers, *Labor Revolt in Alabama*, 76–78, 81–85 (Holman, Glover, regiment). All letters in Jones Papers, "Pinkerton, 1893–94" File. For a brief catalogue of troop movements during the strike, see Alexander, "Ten Years of Riot Duty," 3.

59. Vallens to Governor Jones, April 22, 1894, Jones Papers, "Pinkerton, 1893–94" File (convict lease); Ward and Rogers, *Labor Revolt in Alabama*, 88 (2,500 citizens). In early May a crowd of over 500 attended a performance at the Opera House by the Birmingham Female Minstrel Troupe to benefit the miners (*National Labor Tribune*, May 10, 1894).

60. JHF to Governor Jones, June 1, 1894, Jones Papers, "Pinkerton, 1893–94" File ("starvation"); Ward and Rogers, *Labor Revolt in Alabama*, 91–99; Vallens to Governor Jones, June 18, 1894, Jones Papers, "Pinkerton, 1893–94" File (Hannigan; Vallens, the resourceful Pinkerton, had graciously offered the aged organizer a ride from the rally). On the ambiguous resolution of the national strike, see Fox, *United We Stand*, 45–47. Operators in some districts, particularly the Central Competitive Field, met most of the miners' demands, while elsewhere employer intransigence, state repression, and gradually rising production brought the strike to grim defeat.

61. Ward and Rogers, *Labor Revolt in Alabama*, 104–8. On the Pullman strike, see Salvatore, *Eugene V. Debs*, 114–46; Watts, *Order Against Chaos*, 37–85.

62. Birmingham *Age-Herald*, July 18, 1894 (Pratt City speech); Ward and Rogers, *Labor Revolt in Alabama*, 108–13.

63. From the earliest days of the strike, electoral politics was in the air. On April 23 an estimated 700–800 miners (the majority black) attended a pro-Kolb rally at Lake View Park in Birmingham (Vallens to Governor Jones, April 23, 1894, Jones Papers, "Pinkerton, 1893–94" File); see also Birmingham *Age-Herald*, February 9, 1894; Birmingham *Labor Advocate*, March 1, 1894; Ward and Rogers, *Labor Revolt in Alabama*, 118–21; Goodwyn, *Populist Moment*, 225–26. On the 1894 Alabama gubernatorial election generally, see Hackney, *Populism to Progressivism in Alabama*, 57–62; Rogers, Ward, Atkins, and Flynt, *Alabama*, 312–15.

64. Birmingham *Age-Herald*, June 28, 1894.

65. Vallens to Governor Jones, June 20, 23, 1894 (company offers, union leaders scramble); Vallens to Governor Jones, June 26, 1894 (results mixed); Birmingham *Age-Herald*, June 28, 1894 (Corona); JMP to Governor Jones, July 26, 1894 (White Elephant; see also July 28, 29, 1894); JMP to Governor Jones, August 9, 1894 ("If Kolb is elected"). See also

EW to Governor Jones, July 31, August 6, 1894 ("take up arms"). All letters in Jones Papers, "Pinkerton, 1893–94" File.

66. *Bessemerite*, June 10, 1896 (results); Birmingham *Age-Herald*, July 29, 1894 (Carlisle); Birmingham *Age-Herald*, July 22, 1894 (ministers; see also June 21, 28, July 13, 18, 20, 29, August 3, 1894); Birmingham *Age-Herald*, August 1, 4, 1894 (black Kolbites).

67. In Pratt City, Kolb received 58 percent of the vote; in Bessemer, 63 percent; Woodward, 65 percent; Dolomite, 82 percent; Warrior, 59 percent; Coalburg, 66 percent; Brookside, 84 percent; Johns, 56 percent. See Birmingham *Age-Herald*, August 12, 1894, and *Bessemerite*, June 10, 1896 (results); Rogers, *Alabama*, 312–14 (Kolbites' racial record). The official margin of victory was 27,583. The Kolbites claimed fraud on an even larger scale than in 1892, as the most ardent advocates of white supremacy managed once again to find overwhelming support among African Americans.

68. Birmingham *Labor Advocate*, September 8, 15, 1894; Ward and Rogers, *Labor Revolt in Alabama*, 131–36.

69. Birmingham *Labor Advocate*, August 18 ("agony over"), 25 ("Everything bright"), 1894; *UMWJ*, October 25, 1894, and Birmingham *Labor Advocate*, December 1, 1894, February 16, 23, March 16, May 4, June 8, 1895 (sporadic employment); Birmingham *Labor Advocate*, December 1, 1894, February 9, March 16, April 27, May 4, July 6, 13, 20, 27, 1895 (excess labor).

70. Average daily earnings around the district hovered between $1.00 and $1.50, but often fell below a dollar (Birmingham *Labor Advocate*, February 16, 23, June 29, 1895).

71. *UMWJ*, November 22, 1894, October 25, 1895; Birmingham *Labor Advocate*, December 1, 15, 1894, February 16, July 13, December 21, 1895. In his grim review of Alabama's labor conditions during a speech in Mobile in 1895, AFL president Samuel Gompers singled out the practices of the "pluck-me" establishments at Pratt City (Kaufman, Albert, and Palladino, *National Labor Movement Takes Shape*, 26–27).

72. Birmingham *Labor Advocate*, December 1, 1894, June 1, July 27, December 21, 1895 (dead work). *UMWJ*, November 22, 1894 (checkweighman); Birmingham *Labor Advocate*, December 1, 15, 1894, April 6, June 1, July 6, 27, August 3, December 21, 1895; Birmingham *Labor Advocate*, December 1, 1894, April 13, July 6, 1895 (dockage).

73. Birmingham *Labor Advocate*, March 2, 1895.

74. Ibid., October 20, 27 (quote), November 24, December 29, 1894; *UMWJ*, November 29, 1894.

75. Birmingham *Labor Advocate*, February 16, 1895 (Adger), December 1, 1894 (Corona).

76. Ibid., November 24, December 1, 1894 (quotes).

77. Ibid., November 24 ("stupor"), December 15 (Adger), 1894, February 2, 1895 (Mailly).

78. Ibid., March 23, 30, 1895; *UMWJ*, March 21, 1895.

79. Birmingham *Labor Advocate*, January 12, February 16, April 20, June 22, 29, July 6, November 2, 9, December 21, 1895.

80. Ibid., June 1, 22, 29, July 6, 13, August 3, 24, September 7, October 15, 1895; *UMWJ*, July 4, 18, August 8, 1895; Birmingham *Age-Herald*, June 28, 29, 1895; New Orleans *Daily Picayune*, September 2, 1895; *Wall Street Journal*, June 17, 1895; *National Labor Tribune*, August 1, 1895; Birmingham *State-Herald*, January 11, 1896.

Between April 18 and August 30, 1895, the selling price of TCI pig iron rose steadily from $7.50 to $11.50 per ton ("Statement Showing Comparative Prices of Pig Iron," in James

Bowron Scrapbook [1877–1895], BPLA). For all their criticisms of TCI's labor practices and conviction that the union spirit must not dissipate, UMW leaders praised the company for fulfilling its contractual obligations and pronounced the raises "the crucial point in the history of our contract between the Tennessee Coal, Iron, and Railroad Co. and the miners" (Birmingham *Labor Advocate*, June 29, July 6, 1895).

81. Birmingham *Labor Advocate*, July 20, 1895; Birmingham *News*, July 13, 15, 1895.

82. There were other flashpoints as well. During July, strikes broke out at Palos and Little Warrior in northern Jefferson County over the failure of the local operators to match the wage increases implemented by the major operators. By the end of the summer both strikes had been won. (On Palos: Birmingham *Age-Herald*, July 17, 1895; Birmingham *Labor Advocate*, July 20, August 3, 10, 24, 1895; *UMWJ*, July 25, August 8, 1895. On Little Warrior: Birmingham *Labor Advocate*, July 13, 20, 27, August 3, 24, 31, September 14, 21, 1895; Birmingham *News*, July 13, August 12, 1895.)

In the meantime, a longer and larger battle over the same issue had developed at Warrior, where over 500 miners struck against the Pearson, Watts, Moss, and Coaldale mines. The miners' bitterness was compounded by the introduction of "blacklegs" and deputies to protect them. After eight weeks the strike finally ground to a compromise agreement, as the miners received a portion of the raise they had requested, along with semimonthly pay (with the exception of the Pearson miners, who remained out). See Birmingham *Labor Advocate*, July 20, 27, August 3, 10, September 14, 21, October 8, 15, 29, November 2, 9, 1895; *UMWJ*, July 25, August 8, October 3, 17, November 14, 1895; Birmingham *Age-Herald*, September 11, 17, 19, 24, October 13, 15, 1895; Birmingham *State-Herald*, October 29, November 2, 7, 1895).

83. Birmingham *Age-Herald*, April 28, 1895; Birmingham *Labor Advocate*, March 23, April 20, 27, May 4, 11, 25, June 8, 15, July 6, 20, August 3, 10, 31, September 21, October 8, 1895; Birmingham *News*, July 6, 11, 1895; *UMWJ*, July 25, August 22, 1895.

84. The *Labor Advocate* contrasted the company with the more "benign" TCI. While the latter had signed a contract with its men following the 1894 strike, the Sloss company had declined to do so. While TCI now allowed checkweighmen, the Sloss company continued to deny its men this right. As a consequence, the *Labor Advocate* reported, "the men are having their weights stolen right and left" (July 20, 1894). While TCI recognized and dealt with miners' committees, the Sloss Company "has taken a fiendish delight in discharging every committee that the men would send to them" (August 10, 1895). Tensions were especially aggravated by the company's ongoing recruitment of outside labor, largely black, into a district already beset with material distress and overcrowding. Begun at the start of the year, the arrivals of train cars full of black labor at Sloss properties by July had become what the Birmingham *News* described as "a thing of daily occurrence." For reports of arrivals of black laborers at Sloss properties, see Birmingham *News*, July 6, 11, 19, 20, 23, 1895. For a scathing denunciation of this practice, see Birmingham *Labor Advocate*, August 3, 1895.

85. Birmingham *Labor Advocate*, August 3, 10 (quote), 1895; *UMWJ*, August 8, 22, 1895; Birmingham *Age-Herald*, August 6, 7, 10, 13, 1895; Birmingham *News*, August 3, 5, 6, 7, 9, 10, 12, 1895.

86. Birmingham *Labor Advocate*, September 7, 21 (quote), October 15, November 9, 1895.

87. "Report of Superintendent of Coal Mines at Brookside, Brazil and Cardiff for Fiscal Year Ending January 31, 1896," Sloss-Sheffield Steel and Iron Company, 1895–1902, 4.1.105.1, Sloss-Sheffield Records. For an extended discussion of the racial-ethnic assumptions and preferences of the coal employers, see chapter 1. There is no evidence that the plan to create a separate organization for black miners ever materialized.

88. On the January wage reduction, see Birmingham *State-Herald*, January 10, 1896; *UMWJ*, February 13, 1896. On the July increase, see *UMWJ*, June 11, 1896; *Bessemerite*, July 8, 1896.

89. *UMWJ*, March 26, 1896 (Mailly); *UMWJ*, April 9, 1896 ("Rapparee"). As a youth in Liverpool, Mailly had been awakened to issues of social justice through the works of Charles Dickens and Victor Hugo. He arrived at Adger, at the age of 22, in October 1893, just as unionism was returning to life in the Birmingham coal fields. It was during the miners' strike of the following year that the young union activist read Robert Blatchford's *Merrie England* and became a socialist (*Socialist Spirit* [Chicago], September 1901, 18–19).

90. The most extensive regular coverage of Populist appeal in the mining district over the summer and fall of 1896 is found in *Bessemerite*. On the 1896 gubernatorial election statewide, see Hackney, *Populism to Progressivism in Alabama*, 89–90; Rogers, Ward, Atkins, and Flynt, *Alabama*, 315–18.

91. Birmingham *Labor Advocate*, June 22, July 6, 1895 (first episode); Birmingham *Labor Advocate*, July 20, 1895 (second episode).

92. Quotes describing the riot are drawn from Birmingham *News*, July 31, August 1, 2, 3, 12, 22, 1895; *National Labor Tribune*, August 1, 1895; and Birmingham *State-Herald*, January 30, 1896.

93. *Alabama Sentinel*, January 3, 1891.

94. Birmingham *Labor Advocate*, December 1, 1894, June 1, 22, August 31, 1895, October 30, November 13, 1897; *UMWJ*, July 25, 1895.

95. Birmingham *Labor Advocate*, October 8, 29, 1895 (Warrior; see also discussion and sources in note 82); ibid., August 3, 1895 (Brookside).

96. Ibid., August 3, 10 (quote), 1895; *UMWJ*, August 8, 22, 1895; Birmingham *Age-Herald*, August 6, 7, 10, 13, 1895; Birmingham *News*, August 3, 5, 6, 7, 9, 10, 12, 1895.

97. Birmingham *Labor Advocate*, September 22, 1894, June 1 ("fraternal"), 29 ("white laborers"), 1895.

98. Ibid., December 1, 1895.

99. *UMWJ*, March 28, 1895 (Bergen); Birmingham *Labor Advocate*, July 27, 1895 ("Stuck Fast").

100. Birmingham *Labor Advocate*, June 29 ("Let us educate"), July 27 ("Stuck Fast"), 1895; see also October 8, 15, 1895.

101. Ibid., June 1 (material equality and social equality), 29 ("There need be no question"), August 31 ("Ree Verba"), 1895.

102. Ibid., April 27, June 8, 1895 (crimes); ibid., December 1, 1894 (dialect); *UMWJ*, October 25, 1894 (Bergen); Birmingham *Labor Advocate*, July 20, 1895 (Cardiff).

103. Birmingham *Labor Advocate*, August 31 (mine foreman), September 21 (Mailly), 1895; ibid., May 4, 1895 (new "colored man").

104. Ibid., September 21 (Corona), June 8 (Adger), August 3 (Spartacus; emphasis added), February 2 (Mailly), 1895.

105. *UMWJ*, November 12, 1896 ("the best work"); Birmingham *Labor Advocate*, November 11, 1895 ("Cardiff miner").

106. *UMWJ*, June 4, 1896.

CHAPTER FIVE

1. Bessemer *Weekly*, August 19 (McAdory quotes), September 2, 1905; Birmingham *Labor Advocate*, August 26, 1905.

2. *Mineral Belt Gazette*, March 4, 1905.

3. Bessemer *Journal*, July 22, 1897; *UMWJ*, July 15, 22, 1897; Birmingham *Labor Advocate*, July 3, 24, 31, 1897; Bessemer *Herald*, July 7, 1897; Birmingham *State-Herald*, July 3, 4, 14, 21, 1897; Tennessee Coal, Iron, and Railroad Company, *1898 Annual Report*, 3; *Proceedings of the Joint Convention*, 264–66.

4. Gowaskie, "From Conflict to Cooperation," 676–77; Taft, *A.F.L. in the Time of Gompers*, 138–40; Fox, *United We Stand*, 50–52.

5. Birmingham *Labor Advocate*, September 25, October 2, 16, 30, 1897; Birmingham *Age-Herald*, September 25, 1897; *UMWJ*, November 4, 1897; Bessemer *Journal*, October 28, 1897; UMWA, *Minutes of the Ninth Annual Convention, 1898*, 26.

6. On rising prosperity in the coal and iron trade and consequent rise of employment at the mines, see Birmingham *Age-Herald*, August 31, September 2, 1897; Birmingham *News*, January 11, 15, 1898; and the regular mine-by-mine survey of days worked per week, published in the Birmingham *Labor Advocate* for the fall and winter of 1897–98. On war: Birmingham *News*, March 7, 15, May 7, June 1, 1898; Birmingham *Age-Herald*, March 11, 1898. On building of locals: UMWA, *Minutes of the Ninth Annual Convention, 1898*, 20; Birmingham *Labor Advocate*, November 6, 13, December 25, 1897, February 5, 19, March 19, April 9, 1898; *UMWJ*, November 4, 1897, January 6, 1898; Birmingham *Age-Herald*, November 19, 1897, February 6, 1898. On miners' convention: Birmingham *Labor Advocate*, April 16, May 14, 21, 1898; Birmingham *News*, May 4, 16, 17, 1898; Birmingham *Age-Herald*, May 17, 18, 1898.

7. On organizing drive: Birmingham *Labor Advocate*, May 14, June 11, 18, 1898; Birmingham *Age-Herald*, June 2, 1898; Birmingham *News*, June 10, 20, 21, 1898. On advance: Birmingham *News*, June 29, 1898.

8. On the mounting receptiveness of American bituminous coal operators to the UMW as a means of stabilizing labor relations (as well as competition among the operators themselves): Gowaskie, "From Conflict to Competition"; Graebner, "Great Expectations," 53–55. On the trend toward union recognition in American industry: Brody, *Workers in Industrial America*, 24; Perlman and Taft, *Labor Movements, 1896–1932*, 13–19. On the rise of UMWA membership: UMWA, *Minutes of the Sixteenth Annual Convention, 1905*, 8.

9. On 1898 organizing drive: Birmingham *News*, September 2, 1898. For membership figures during 1898–1904: *UMWJ*, September 22, December 22, 1898, December 28, 1899, December 18, 1902; Birmingham *Labor Advocate*, May 26, June 23, 1900, April 26, 1902, December 12, 1903, June 14, 1907; UMWA, *Minutes of the Tenth Annual Convention, 1899*, 27; UMWA, *Minutes of the Twelfth Annual Convention, 1901*, 61; UMWA, *Minutes of the Thirteenth Annual Convention, 1902*, 65; UMWA, *Minutes of the Fourteenth Annual Convention, 1903*, 62, 448–50; UMWA, *Minutes of the Fifteenth Annual Convention, 1904*, 98–

99. On Local 664: UMWA, *Minutes of the Twelfth Annual Convention, 1901*, 20; Birmingham *Labor Advocate*, March 8, 1900, February 3, 1902.

10. On organization throughout the Birmingham trades: Head, "Development of the Labor Movement in Alabama," 174–80; Birmingham *News*, March 14, 1899; Birmingham *Labor Advocate* generally for 1898–1903; McKiven, *Iron and Steel*, 97–99; Worthman, "Black Workers and Labor Unions in Birmingham," 383, 395. On the Alabama State Federation of Labor: Head, "Development of the Labor Movement in Alabama," 184–87; Birmingham *Labor Advocate*, September 8, 1900, May 17, 1902, May 9, 1903. On the Union Labor League: Birmingham *Age-Herald*, November 20, 1897; Bessemer *Journal-Herald*, November 25, 1897. On the Socialist Party: *UMWJ*, October 20, 1898; Birmingham *Labor Advocate*, May 6, 1899, March 8, 29, 1902, March 7, December 19, 1903; Bessemer *Workman*, March 25, 1904. On visits by Debs: Birmingham *News*, May 25, 1899, September 16, 1904; Birmingham *Labor Advocate*, January 27, 1900.

Not all elements of the labor movement, it should be noted, operated to the UMW's benefit. The Knights of Labor found a second wind in pockets of the mineral district at the turn of the century, establishing footholds and emerging as a rival to the miners' union in coal towns in Jefferson County (particularly Warrior), Walker County (particularly Galloway and Littleton), and Tuscaloosa County (particularly Brookwood), and at the Ishkooda iron mines. The Knights enjoyed particularly strong support in these places among black miners, although the substance of their appeal, or of the differences between the two organizations, is difficult to discern from the record. By the early 1900s the Knights' presence in the Birmingham district had virtually disappeared. On Jefferson County: Birmingham *Labor Advocate*, February 11, October 14, 1899. On Walker County: Birmingham *News*, July 5, 19, 21, 24, 25, August 10, 1899; Birmingham *Labor Advocate*, October 28, 1899. On Tuscaloosa County: Birmingham *News*, July 3, 1900. On Knights of Labor–led strikes at the ore mines, see Birmingham *News*, April 1, 3, 4, 10, 14, June 15, 17, 21, July 13, 17, 18, 22, August 2, 17, 26, September 7, 18, 1899; Worthman, "Black Workers and Labor Unions," 397–98; Head, "Development of the Labor Movement," 171–72.

11. Between July 1898 and August 1899 the wage for run-of-mine coal rose in a series of 2½-cent increments from 37½ cents to 55 cents per ton. From July 1900 through June 1903 the Pratt scale ranged from a minimum of 45 cents to a maximum of 55 cents, depending on the price of pig iron. Shifts of wages paid on the Pratt scale from 1896 to 1908 are provided in UMWA, *Proceedings of the Nineteenth Annual Convention, 1908*, 518–19. On day laborers: Birmingham *News*, August 14, 1899.

12. *UMWJ*, July 18, 1901.

13. A firing deemed questionable might be challenged by the union. "We have . . . perhaps the best union camp in the district, carried on in strictly union lines," a miner wrote from Piper, "and when a man is discharged wrongfully a committee calls on the superintendent to rectify the matter. If this effort fails we then stop work" (*Mineral Belt Gazette*, May 7, 1904). See also *Proceedings of the Joint Convention*, 249.

14. "We are allowed to trade anywhere we please," wrote a miner from Adger in 1902 (*UMWJ*, March 13, 1902).

15. In 1899 the men at Blocton No. 3 mine forced the resignation of a mine boss who sold rooms to the highest bidder. One "Enos" recounted the episode lyrically in the Birmingham *Labor Advocate* (October 28, 1899):

There was a boss at No. 3
Who with his men couldn't agree,
And so the Blocton miners struck
And said this boss must light a shuck!
This smart boss was on the move
To try his innocence to prove,
But all his efforts came to naught
When men turned up whose room had bought;
They to the truth did firmly stick,
And this poor boss turned quite sick.
Then to the office straight he went
And in his resignation sent.
It was accepted and there will be
No more rooms sold at No. 3
A better chance we now will stand,
For our new boss is a very fine man.

For other examples of corrupt managerial practices underground, see Birmingham *Labor Advocate*, March 9, 1901; *UMWJ*, July 18, 1901.

16. On three-days-away rule: *Proceedings of the Joint Convention*, 247–48, 250. On funerals: *Proceedings of the Joint Convention*, 423–24; Birmingham *News*, June 23, 1899.

17. *UMWJ*, January 2, July 18, 1901; *Proceedings of the Joint Convention*, 257–63, 698, 705–6; Birmingham *Labor Advocate*, April 1, 1899; Birmingham *News*, March 22, 1899.

18. Birmingham *Labor Advocate*, September 30, 1905.

19. *Proceedings of the Joint Convention*, 931.

20. Ibid., 926–30.

21. Birmingham *News*, August 17, 22, 1899 (day laborers); Birmingham *News*, August 19, 1898, *UMWJ*, September 1, 1898 (subcontracting); Birmingham *News*, July 13, 14, 19, 21, August 30, 1899 (contested firing); Birmingham *News*, October 19, 20, 1898 (schools and doctors). On TCI strike: Birmingham *Daily Ledger*, September 23, 29–30, October 1–16, 1902; *UMWJ*, October 16, 1902; Birmingham *Labor Advocate*, October 4, 11, 1902; *Proceedings of the Joint Convention*, 647–48. On anthracite strike: Gowaskie, "John Mitchell and the Anthracite Mine Workers"; Cornell, *Anthracite Coal Strike of 1902*; Fox, *United We Stand*, 89–101.

22. *Mineral Belt Gazette*, July 2, 1905 (Brookwood); ibid., May 14, 1904 (Republic); Birmingham *News*, August 25, 1898 (Cardiff); Birmingham *News*, August 12, 1898 (Pratt City); Birmingham *News*, August 5, 1899, and *Proceedings of the Joint Convention*, 267 (hospital).

23. By 1900, African Americans made up approximately half of the state's 11,751 coal miners. See chapter 1, note 34.

On the spread around the South of disfranchisement and segregation codes at the turn of the century, see C. V. Woodward, *Origins of the New South*, 321–95; Kousser, *Shaping of Southern Politics*. On Alabama, see Hackney, *Populism to Progressivism in Alabama*; Mc-Millan, *Constitutional Development in Alabama*, 217–359; Rogers, Ward, Atkins, and Flynt, *Alabama*, 343–54.

24. Birmingham *News*, June 8, 1899. The ordinance did not go unenforced. In 1904, City Judge N. Feagin fined a young white man $25 and a black man $1 for playing dominoes together at the Davis Saloon in Birmingham (Birmingham *News*, July 7, 1904).

25. *UMWJ*, February 10, 1898. "If I get on a railway train," Davis noted elsewhere, "I must ride in a separate car, the same on the street car. If I want a drink I must go in to a separate bar, and I can only look at a hotel or restaurant" (*UMWJ*, November 25, 1897).

26. Worthman, "Black Workers and Labor Unions."

27. William Fairley testified before the 1903 arbitration board that District 20's membership was equally divided between black and white (*Proceedings of the Joint Convention*, 239).

28. The district vice presidency, the *Labor Advocate* occasionally noted, has been "conceded to the colored membership" (November 2, December 14, 1901, June 7, 1907). District 20 officers and board members were elected by delegates at conventions held in June of each year and sometimes in December as well. The racial breakdown of these officials can be found in coverage of these conventions in the *UMWJ* and the Birmingham *Labor Advocate* during June (and sometimes December) of each year. Important boards were often formally divided between black seats and white seats. On the District's most significant deliberative body, the executive board, blacks made up one of six members at the end of 1898 (Birmingham *News*, December 13, 1898); two of five in February 1899 (Birmingham *News*, February 4, 1899); two of seven in June 1900 (Birmingham *News*, June 19, 1900); and three of eight in March 1901 (Birmingham *Labor Advocate*, March 9, 1901), December 1903 (Birmingham *Labor Advocate*, December 26, 1903), and June 1904 (Birmingham *News*, June 10, 14, 1904). These same sources document similar mixes on other district committees.

29. At District 20's June 1898 convention, 20 out of 54 delegates were black (Birmingham *News*, June 20, 1898); at its December 1898 convention, the figure was 10 out of 24 (Birmingham *News*, December 12, 1898); in February 1899, 6 out of 33 (Birmingham *News*, February 4, 1899); in June 1899, 28 out of 64 (Birmingham *News*, June 19, 1899); in June 1900, 29 out of 77 (Birmingham *News*, June 20, 1900).

Black District 20 delegates to the UMW's national conventions included C. W. Cain and James Swinney (UMWA, *Minutes of the Eleventh Annual Convention, 1900*, 13; *Minutes of the Twelfth Annual Convention, 1901*, 21), Benjamin Greer and Silas Brooks (*Minutes of the Thirteenth Annual Convention, 1902*, 29; *Minutes of the Fourteenth Annual Convention, 1903*, 75; *Minutes of the Fifteenth Annual Convention, 1904*, 18).

30. *UMWJ*, May 3, 1900 (West); Birmingham *Age-Herald*, August 15, 1904 (Greer). On Washington's opposition to black involvement in labor unions, see Meier and Rudwick, "Attitudes of Negro Leaders toward the American Labor Movement," 39–41; Harlan, *Booker T. Washington: Wizard of Tuskegee*, 90–91.

31. *UMWJ*, July 18, 1901.

32. *UMWJ*, July 19, 1900, July 18, 1901; Birmingham *Labor Advocate*, June 30, 1900; Birmingham *News*, June 22, 25, 1900; Worthman, "Black Workers and Labor Unions," 391; Birmingham *News*, July 3, 1901; Worthman, "Black Workers and Labor Unions," 401. Black miners from Alabama also asserted their positions at national conventions, as when A. B. Thompson of Brookside submitted a resolution calling on the UMW to appoint a black national organizer to work among the "men of color [who] constitute a large per cent" of the miners in several states (UMWA, *Minutes of the Fifteenth Annual Convention, 1904*, 144–45).

The resolution was referred to the National Executive Board, whose subsequent action is not reflected in the records.

33. Birmingham *Labor Advocate*, June 23, 1900; Birmingham *News*, June 27, 1900.

34. Birmingham *Labor Advocate*, March 5, 1898 (Odd Fellows); May 24, 1902 (Bennett). On miners' saloons in the 1894 strike, see chapter 4.

35. Birmingham *Age-Herald*, June 25, 1904 (operators refuse to meet); Birmingham *Labor Advocate*, April 13, 1901 (Newcastle; also *UMWJ*, January 3, 1901).

36. *UMWJ*, August 20, 27 (quote), October 1, 1903; Worthman, "Black Workers and Labor Unions," 387. Socialist activist Mary White Ovington reported hearing of a similar (possibly the same) episode during a 1907 stop in Birmingham. There she visited with a man whom she identified only as the state secretary of the Socialist Party of Alabama, who told of a time he had met with a black coal miner at a railroad station. Local whites hated "me meeting a 'nigger' and practicing social equality. . . . They got hold of us, threw us together, made us kiss one another. They . . . ." The man broke off his story. Ovington, *Walls Came Tumbling Down*, 85–86.

37. Birmingham *Labor Advocate*, April 12, 1902 (Altoona); ibid., December 14, 1901 (Warner Mines); *UMWJ*, September 5, 1901 ("Joshua"). According to "Joshua," the approximately 500 miners at Warner—"about half colored and half white"—had until recently been meeting together, "but at present they hold separate meetings."

References to mixed and racially distinctive locals are scattered throughout the articles and labor directories of the Birmingham *Labor Advocate* for this period. The exceptions to this biracial arrangement were quite significant: among the several mixed locals was Pratt City Local No. 664, which at upwards of 1,500 boasted the largest membership in the United States. The racial structure of the Pratt local's leadership mirrored that of District 20 as a whole; in late 1898, for example, it included a white president, a black vice president, a white secretary-treasurer, and a black recording secretary (Birmingham *Labor Advocate*, November 26, 1898).

38. On banquets: Birmingham *Labor Advocate*, December 14, 1901, January 4, 1902; Bessemer *Workman*, June 20, 1902. On parks: Birmingham *Labor Advocate*, September 12, 1903. On convention delegations: Birmingham *Age-Herald*, June 11, 1904.

39. Birmingham *News*, June 18, 1904; Birmingham *Age-Herald*, June 19, 1904. Black convention delegates also voiced broad opposition to proposals for popular election of District 20 officers in 1902. In this case, though, the *Labor Advocate* suspected that they were motivated by a concern that such a reform would diminish black representation on the various committees (December 20, 1902).

40. Birmingham *Labor Advocate*, April 8, 1899 (Brookwood); ibid., December 30, 1899 (Blue Creek). See also *Mineral Belt Gazette*, June 25, 1904; Birmingham *Labor Advocate*, August 15, 1903; *UMWJ*, September 5, 1901.

41. Birmingham *Labor Advocate*, June 10, 1899 (criminals); ibid., December 24, 1898 (minstrel shows); *UMWJ*, July 28, 1898 ("colored dupes"); *UMWJ*, October 13, 1898, January 1, 1903, Birmingham *Labor Advocate*, March 25, 1899, May 4, 1901, May 17, 1902 (degrading lingo).

42. *UMWJ*, September 22, 1898; Birmingham *Labor Advocate*, October 14, 1899, August 11, 1900, February 22, March 15, 1902.

43. On *Labor Advocate* support for new constitution: March 23, October 12, 19 (quote),

October 26, November 9, 1901, November 7, 1903, April 30, 1904, January 13, February 3, 1906, January 18, 1908. The United Labor League of Alabama, with which District 20 was closely aligned, took the same position (Birmingham *Age-Herald*, January 15, 1905). On the protracted, ultimately successful campaign to reform the state constitution: Hackney, *Populism to Progressivism in Alabama*, 147–229; McMillan, *Constitutional Development in Alabama*, 233–359. On scattered support for black voting: Birmingham *Labor Advocate*, June 30, 1900, October 19, 1901. For syndicated pieces: Birmingham *Labor Advocate*, April 20, 1903, November 4, 1905.

44. Birmingham *Labor Advocate*, April 20 (Brooks quote), 27, October 12, 1901, September 12, 1903, September 4, 25, 1908; *UMWJ*, December 12, 1901.

45. Birmingham *Labor Advocate*, March 23, 1901.

46. Ibid., October 14, 1905.

47. *Proceedings of the Joint Convention*, 254–56, 263–64.

48. This cautious, often carefully calibrated expression of labor interracialism emerged at the national level of the UMW as well. How measured was the organization's commitment to racial equality can be glimpsed in the fate of an impassioned resolution submitted by several black delegates to the 1904 national convention. "We, the colored delegates," wrote J. S. Bell, W. J. Campbell, and L. A. Dowell, "watching the growing evils which tend to destroy the cardinal principles of trade unionism in America, to-wit: the disfranchising and discriminating against the union and non-union colored Americans, to the end of denying them the rights of American citizens by the enactment of laws to eliminate from them the rights guaranteed by the Constitution of the United States; and, believing as we do, that this is but a blow at the trade union movement of this country . . . and [that] the property qualification will come forward to drive the laboring white man from the ballot; we hereby petition this body, in the name of civilization, and in behalf of organized labor in America, to place this convention on record by condemning all laws and usages that have a tendency to perpetuate this un-American principle." After a "long discussion," the eloquent resolution was brushed aside by the convention in favor of a bloodless substitute: "Resolved, That the United Mine Workers of America favor adult suffrage, without regard to race, color or previous condition of servitude." UMWA, *Minutes of the Fifteenth Annual Convention, 1904*, 210–11.

On the UMW's racial record nationally, see Gutman, "The Negro and the United Mine Workers"; Hill, "Myth-Making as Labor History"; R. Lewis, *Black Coal Miners in America*; Spero and Harris, *The Black Worker*, 352–82 (1968 ed.).

49. Chaired by Delaware federal judge George B. Gray, who had conducted the landmark anthracite arbitration of the previous year, the board delivered a mixed judgment. It granted the miners a 2½-cent increase and semimonthly paydays while granting the operators stricter penalties for absenteeism. Both sides agreed that the miners had gotten the better of it. On the collapse of the 1903 negotiations: Birmingham *Labor Advocate*, July 27, 1903. On the formation and proceedings of the arbitration board: *Proceedings of the Joint Convention*; Birmingham *Labor Advocate*, August 1, 8, 15, 22, 29, 1903; *UMWJ*, August 13, 27, September 3, 1903. For the terms of settlement: *Proceedings of the Joint Convention*, 937–42. For UMW reaction: *UMWJ*, September 3, 1903. For operator reactions: Birmingham *Labor Advocate*, August 29, 1903.

50. On Walker County: Birmingham *Labor Advocate*, September 12, October 17, 1903; *UMWJ*, October 17, 1903. On Blue Creek: *UMWJ*, September 3, December 31, 1903;

Birmingham *Labor Advocate*, September 26, October 31, November 21, 1903, February 6, 13, March 5, 1904.

51. On the Pratt company conflict: Birmingham *News*, May 20, 24, June 3, 10, 13, 1904; Birmingham *Labor Advocate*, June 11, 1904; *Mineral Belt Gazette*, June 18, 1904. On the Little Warrior strike: Birmingham *Labor Advocate*, May 14, 28, 1904; Birmingham *News*, May 20, 24, 28, June 13, 1904; *Mineral Belt Gazette*, June 18, 1904. On the Blocton strike: *Mineral Belt Gazette*, May 7, 14, 21, 1904; Birmingham *Labor Advocate*, May 21, 28, June 4, 1904; Birmingham *Age-Herald*, May 14, 18, 20, 1904; Birmingham *News*, May 17, 23, 24, 26, 27, June 18, 1904; Bessemer *Workman*, May 20, 1904; *Mineral Belt Gazette*, May 14, 1904 (funeral).

52. On economic downturn: W. D. Lewis, *Sloss Furnaces*, 286; G. H. Moore, "Business Cycles, Panics, and Depressions," 152. On wage reductions: Birmingham *Labor Advocate*, October 10, 1903. On open-shop drive: Montgomery, *Fall of the House of Labor*, 269–75; Watts, *Order against Chaos*, 143–70; Perlman and Taft, *Labor Movements, 1896–1932*, 129–37; Ramirez, *When Workers Fight*, 93–95.

53. Birmingham *Labor Advocate*, October 3, November 7 (quote), 21, December 5, 1903, February 13, 1904; McKiven, *Iron and Steel*, 97–104.

54. Tennessee Coal, Iron, and Railroad Company, *1905 Annual Report*, 7–10 (Bacon); Birmingham *Labor Advocate*, July 2, 1904; Birmingham *News*, June 13, 20, 25, July 1, 2, 25, August 17 (Maben), 1904; Birmingham *Age-Herald*, June 29, July 2, 1904. On operators' disgruntlement with union recognition, see also W. A. Davis to Charles H. Marshall, July 29, 1904, in Alabama Mineral Land Company Collection, Letter Book 1903–04, BPLA.

55. On stockpiles: Birmingham *News*, June 30, July 2, 7, 1904; Birmingham *Labor Advocate*, July 9, 1904. On convicts: Birmingham *News*, July 2, 7, 12, 15, 19, 1904; Birmingham *Age-Herald*, July 12, 30, 1904. On strikebreakers: Birmingham *News*, July 16, 18, 1904.

56. Birmingham *News*, August 8, 9, 12, 13, 16, 23, 30, 1904, Birmingham *Age-Herald*, August 13, 17, 1904 (imported strikebreakers); Birmingham *News*, August 11, 1904, Birmingham *Age-Herald*, August 11, 13, 17, 1904 (production, evictions); Birmingham *Age-Herald*, August 27, 30, 1904, Birmingham *News*, August 30, 1904 (trespassing); Birmingham *News*, August 10, 1904 (deputy sheriffs); Birmingham *News*, August 11, 15, 22, 23, 27, 29, 30, 1904, Birmingham *Age-Herald*, August 11, 13, 17, 1904, *Mineral Belt Gazette*, September 3, 10, 17, 1904 (courts, tent colonies).

57. On leaving the district: Birmingham *Age-Herald*, July 30, 1904. On strikers' activities in the mining communities: Birmingham *News*, July 2, 4, 6, 9, 14, August 8, 13, 31, 1904; Birmingham *Age-Herald*, July 27, September 1, 1904; *Mineral Belt Gazette*, September 10, 1904. On material support: Birmingham *News*, July 2, August 27, 1904; Birmingham *Age-Herald*, July 12, 30, August 4, 6, 12, 17, 1904. On rallies: Birmingham *News*, August 5, 11, 15, 22, 23, 1904; Birmingham *Age-Herald*, August 15, 23, 28, 1904; *Mineral Belt Gazette*, September 3, 17, 1904.

58. Birmingham *Age-Herald*, August 26, 27, 30, September 1, 1904; Birmingham *News*, August 25, 26, 27, 30, 31, 1904. On Graves Mines: Birmingham *News*, August 26, 1904; Birmingham *Age-Herald*, August 25, 1904. On Brookside: Birmingham *Age-Herald*, August 29, 1904. On Adamsville: Birmingham *Age-Herald*, August 29–31, 1904.

59. Birmingham *News*, September 5, 1904 (Jones); Birmingham *News*, September 7,

1904 (state injunctions); Birmingham *News*, September 7, 8, 21, October 5, 1904, *Mineral Belt Gazette*, September 24, 1904 (deputies); Birmingham *Age-Herald*, September 14, 1904 ("yellow-dog" contracts); Birmingham *Age-Herald*, September 25, 1904 (Pratt); UMWA, *Minutes of the Sixteenth Annual Convention, 1905*, 8 (membership); *Mineral Belt Gazette*, October 22, December 31, 1904; Birmingham *Age-Herald*, December 9, 1904 (provisions).

60. On gatherings: Birmingham *Labor Advocate*, June 24, July 22, 1905; *Mineral Belt Gazette*, July 22, 1905, January 6, March 31, 1906. On resources and calls by leaders: Bessemer *Weekly*, June 17, 1905; Bessemer *Journal*, January 20, 1905; Birmingham *Age-Herald*, January 17, 1905; Birmingham *Labor Advocate*, January 20, 1906; UMWA, *Minutes of the Seventeenth Annual Convention, 1906*, 58, 204.

61. On strikebreakers: Birmingham *Labor Advocate*, January 6, 1905; *UMWJ*, February 23, March 2, June 17, 1905, June 28, 1906; Birmingham *Hot Shots*, April 22, 1905; UMWA, *Minutes of the Sixteenth Annual Convention, 1905*, 17. On arrests: Birmingham *Labor Advocate*, May 6, 1905. On violence: Birmingham *Labor Advocate*, January 7, April 15, 1905; *UMWJ*, December 1, 8, 1904, April 6, July 6, 1905; Birmingham *Age-Herald*, December 19, 1904; Birmingham *Hot Shots*, March 25, April 15, 1905. Perhaps the most noted episode of violence attributed to the strikers was the assassination of the widely hated TCI superintendent at Blocton, W. S. Lang (*Mineral Belt Gazette*, November 19, 1905). On calling off of strike: Birmingham *Labor Advocate*, August 25, 1906; UMWA, *Proceedings of the Eighteenth Annual Convention, 1907*, 52, 322.

62. The furnace operators worked to utilize antiunionism in the black community. In the local black paper *Hot Shots*, for example, there appeared an advertisement, signed by the operators, asking, with some creative license, "Why Remain Idle When You Can Make $4 and $5 Per Day?" (quoted in Birmingham *News*, August 17, 1904; the *Hot Shots* issue in which the advertisement appeared is no longer extant). On the introduction of strikebreakers: Birmingham *News*, August 9, 11, 15, 16, 18, 1904; Birmingham *Age-Herald*, August 8, 10, 13, 16, 1904. On evictions: Birmingham *Age-Herald*, August 13, 1904. On ministers: Birmingham *News*, August 15, 1904.

63. Birmingham *Age-Herald*, February 8, 1905; Bessemer *Workman*, September 16, 1904 ("Bessemerite"); Birmingham *Age-Herald*, August 13, 1904 (Blue Creek); *UMWJ*, July 27, 1905 (*Labor Advocate* quote).

64. Birmingham *Age-Herald*, August 12, 1904. At another meeting Greer expanded on these themes, maintaining that blacks could attain education and stability through the union. Social betterment through the labor movement, he argued, would elevate not only black miners but African Americans in general: "You must be men and when you convince [whites] of your integrity and determination you will serve to lift our race to a higher place among the people of the United States" (Birmingham *Age-Herald*, August 15, 1904).

65. On blacklist: Birmingham *Labor Advocate*, June 14, 1907. On newcomers: Bessemer *Weekly*, July 22, 1905. On panic: Brody, *Steelworkers in America*, 151; G. H. Moore, "Business Cycles, Panics, and Depressions," 153; Birmingham *Labor Advocate*, November 29, 1907. On U.S. Steel's antiunionism: Brody, *Steelworkers in America*. On the acquisition of TCI by U.S. Steel, see chapter 1 above.

66. For membership figures: Birmingham *Labor Advocate*, June 14, 1907; UMWA, *Proceedings of the Eighteenth Annual Convention, 1907*, 31–32, 103–4; UMWA, *Proceedings of the Nineteenth Annual Convention, 1908*, 402. On union contracts: *UMWJ*, June 22, 1905,

June 28, August 2, 1906, June 20, 1907; Birmingham *Labor Advocate*, June 24, July 1, 1905, June 23, 30, 1906, June 21, July 12, 19, 1907.

67. Birmingham *Labor Advocate*, January 10, 24, 1908; Birmingham *News*, January 17, June 13, 1908.

68. On launching of districtwide strike: *UMWJ*, July 16, 23, 1908; Birmingham *News*, June 26, July 2, 6, 1908. On strike activity: Birmingham *News*, July 7, 8, 11, 13, 1908; Blocton *Enterprise*, July 9, 1908; Straw, " 'This Is Not a Strike,' " 69–76. On second week of strike: Birmingham *Labor Advocate*, July 17, 1908; Birmingham *News*, July 14, 1908. The story of the 1908 coal strike, briefly outlined here, has been told by a number of historians, most extensively in Straw, " 'This Is Not a Strike' " and "The Collapse of Biracial Unionism"; and Elmore, "The Birmingham Coal Strike of 1908."

69. On strikebreakers and evictions: Birmingham *News*, July 22, 23, 27, 29, August 3, 1908; Blocton *Enterprise*, August 20, 1908; New Orleans *Times-Democrat*, July 24, 1908. On tent colonies: Birmingham *News*, July 25, 1908. On deputy sheriffs and posted statements: E. L. Higdon to Governor Braxton Comer, July 17, 1908, in Comer Papers, Correspondence, RC2:G66, 1908 Box C–H, File G, ADAH; J. O. Long to Governor Comer, July 19, 1908, in Comer Papers, Correspondence, RC2:G67, 1908 Box I–R, File K–L, ADAH.

70. On gatling guns: Republic Iron and Steel Company telegram to Adjutant General Bibb Graves, July 15, 1908, in Comer Papers, Correspondence, RC2: G67, 1908 Box I–R, File I–J, ADAH. (When Alabama Adjutant General Graves declined its request, the Tennessee company spent $600 for a Colt rapid-fire gun of its own, which it mounted on a mine car that it used to bring its strikebreakers to work. See Lucien C. Brown, Chief Deputy [of Jefferson County] to Adjutant General Graves, July 15, 1908, in Graves Papers, SG15234, Correspondence, Box 1905–08, File July 15–31, 1908, ADAH.) On strikers surrounding the mines: Birmingham *News*, July 15, 17, 29, 1908. On women: Birmingham *News*, August 1, 1908. On attacks on trains: Birmingham *Labor Advocate*, August 7, 14, 1908; Birmingham *News*, July 14, 1908; *UMWJ*, August 20, 27, 1908. On denunciations of violence and the governor's proclamation: *UMWJ*, August 20, 1908; Straw, "Soldiers and Miners in a Strike Zone."

71. Birmingham *News*, July 18 ("The train was"), July 25, 27 (Fairley), 29, 1908; *UMWJ*, August 6 ("It seems"), 20 ("tramp of soldiers"), 1908; Birmingham *Labor Advocate*, August 28, 1908, and Straw, "Soldiers and Miners in a Strike Zone," 297–300 (embracing troops); New Orleans *Daily Picayune*, July 18, 1908 (rumors).

72. On union membership growth: Birmingham *Labor Advocate*, August 7, 1908; UMWA, *Proceedings of the Twentieth Annual Convention, 1909*, 82; Birmingham *News*, August 15, 1908. Sixty-nine new locals were created in 1908, the vast majority during the strike (UMWA, *Proceedings of the Twentieth Annual Convention, 1909*, 214–15). On arrival of White and national union support: Birmingham *Labor Advocate*, August 14, 1908; Birmingham *Truth*, August 15, 1908; Birmingham *News*, August 12, 1908; UMWA, *Proceedings of the Twentieth Annual Convention, 1909*, 80, 163, 867. On ACOA: Birmingham *News*, July 25, August 8, 1908; *UMWJ*, August 20, 1908.

73. For black middle-class voices: Birmingham *Blade*, August 1, 1908; *Hot Shots*, July 11, 15, 23, 30, August 26, 1908. On miners of each race: Birmingham *News*, July 25, 1908; UMWA, *Proceedings of the Twentieth Annual Convention, 1909*, 865. On Adamsville: Birmingham *News*, July 30, 1908. On Republic: Birmingham *News*, July 20–24, 27, 1908.

74. Birmingham *News*, August 19, 1908 (ACOA); Birmingham *Age-Herald*, August 24, 1908 ("Social Equality Horror"); Birmingham *Age-Herald*, August 25, 29, 1908 (Comer); Birmingham *News*, August 31, 1908 (tent colonies); Birmingham *Age-Herald*, August 8, 25, 1908 (Evans; see also August 22, 26). Black speakers, Evans suggested cryptically, dared to allude in the presence of white women "to those delicate matters of social status which can only be discussed properly with fair women in the private home and by husband and father" (Birmingham *Age-Herald*, August 8, 1908). Evans was quietly paid $500 by the ACOA "for services rendered as newspaper correspondent" during the strike. The ACOA also paid *Age-Herald* editor Edward W. Barrett $3,000 "for the assistance rendered by that paper" (ACOA minute book no. 1, September 24, 1908, 25, ACOA Collection). Comer also likened the union to the carpetbaggers of the Reconstruction era, a comment Fairley promptly denounced as "the most vicious and inflammatory [utterance] that I have ever read coming from the lips of a governor" (Birmingham *News*, August 4, 5, 1908).

75. The situation, Evans added, smacked not only of Reconstruction but of the recent riot against blacks in Springfield, Illinois, an event "which came from criminal assault on a white woman by a negro" (Birmingham *Age-Herald*, August 7, 1908). Comer quote: Birmingham *News*, August 25, 31, 1908; see also August 4, 1908.

76. Evans quote: Birmingham *Age-Herald*, August 22, 1908. See also letter by Walter Moore and Guy Johnson, Birmingham *Age-Herald*, August 24, 1908; Birmingham *News*, August 31, 1908. Beyond brief references such as this, little is known about the activities of the women's auxiliaries that arose during the strike. Dalrymple quote: Birmingham *Age-Herald*, August 30, 1908.

77. Birmingham *Labor Advocate*, September 4, 1908 (McDonald), September 25, 1908 (White), August 28, 1908 (editorial).

78. Birmingham *Labor Advocate*, September 25, 1908 (White); Long, *Where the Sun Never Shines*, 43; Gutman, "The Negro and the United Mine Workers of America," 58. On the active role of women in major coal strikes around the United States during the late nineteenth and early twentieth centuries, see Long, *Where the Sun Never Shines*, 140–65, 217–304; Montgomery, *Fall of the House of Labor*, 338.

79. Birmingham *News*, July 21, 22, 25, 27, August 13, 1908; UMWA, *Proceedings of the Twentieth Annual Convention, 1909*, 866.

80. On "racial problem": UMWA, *Proceedings of the Twentieth Annual Convention, 1909*, 873. On meetings prohibited: ibid., 86. On tent colonies: Birmingham *News*, August 27, 1908; Birmingham *Labor Advocate*, August 28, 1908; UMWA, *Proceedings of the Twentieth Annual Convention, 1909*, 865–66. On Republic: Gilson Gardner, "Charged Governor Whipped Miners—Destroyed Homes" (typescript), in Wieck Collection, Box 1, Folder L-32 to L-33, Wayne State University Archives of Labor and Urban Affairs, Detroit, Mich. On hygiene: Birmingham *News*, August 28, 1908; Birmingham *Labor Advocate*, August 28, 1908. On "niggers idle": UMWA, *Proceedings of the Twentieth Annual Convention, 1909*, 865.

81. *UMWJ*, September 10, 1908 ("it is hard"); Birmingham *News*, August 28, 1908 (prayer); UMWA, *Proceedings of the Twentieth Annual Convention, 1909*, 865 (ultimatum), 867, 870 (warning). On calling off of strike: *UMWJ*, September 3, 1908; Birmingham *Labor Advocate*, September 4, 1908; Birmingham *News*, August 31, 1908; UMWA, *Proceedings of the Twentieth Annual Convention, 1909*, 86.

82. Birmingham *Labor Advocate*, September 4, 1908, and *UMWJ*, October 8, 1908 (work

rules); Birmingham *Labor Advocate*, October 30, 1908 (dockage); Birmingham *Labor Advocate*, October 23, 1908 (store); Birmingham *Labor Advocate*, October 2, 1908 (local movement); UMWA, *Proceedings of the Twentieth Annual Convention, 1909*, 504 (checkweighman); Blocton *Enterprise*, October 29, 1908, and UMWA, *Proceedings of the Twentieth Annual Convention, 1909*, 863–64 (discharged); *UMWJ*, March 11, 1909 ("goodbye"). On the large-scale departure of miners from the district following the strike, see ACOA minutes book no. 1, entry for September 24, 1908, 19, 21–22, ACOA Collection, BPLA.

83. "The loss of that strike practically demoralized the district," an internal document of the national UMW noted the following summer. "Industrial conditions have not been favorable to inaugurate an organizing campaign in that state this year" ("Official to the United Mine Workers of America" [typescript], Indianapolis, August 5, 1909, 5, in Wieck Collection, Box 2, Folder "United Mine Workers of America Circulars, 1897–1930," Wayne State University Archives of Labor and Urban Affairs, Detroit, Mich.). For membership figures: UMWA, *Proceedings of the Twentieth Annual Convention, 1909*, 70; *UMWJ*, January 21, 1909.

84. *UMWJ*, February 4, 1909. Beyond conveying the poignant aftermath of the 1908 defeat, Leach's letter substantially revises our knowledge of the origins of the civil rights anthem "We Shall Overcome." Historians and folk singers have long traced the evolution of the old southern spiritual "I Will Overcome" into a song of collective protest (first "We Will Overcome," then "We Shall Overcome") to 1940s labor struggles among textile workers of Piedmont Carolina and black tobacco workers of Charleston, South Carolina. (Activists in the latter campaign brought it to the Highlander Folk School in Tennessee, which in turn passed it on, via organizers of the Student Nonviolent Coordinating Committee, to the civil rights movement of the 1960s.) For this generally accepted history of the song, see C. R. Wilson, "We Shall Overcome," 230–31. Leach's letter, however, indicates that "I Will Overcome" had been transformed into a movement standard (with the significant shift in pronoun from "I" to "We"), at least in the Birmingham coal fields, as far back as the early twentieth century. His passing reference to "that good old song" is in some ways tantalizing. While he hints that the singing of the song at union gatherings was customary, this study has uncovered no other mention of it (although it might be added that the available sources cast all too little light on the use of song in the Alabama miners' organizations, not to mention church services, generally for this period). Nor even is it self-evident that the miners described in Leach's letter were all or predominantly black; to the contrary, the absence of any reference in his long letter to color or issues of race suggests—though it does not prove—just the opposite, for the substantial presence of African Americans was usually noted in such correspondence, whether written by blacks or whites.

85. "When they saw that the United Mine Workers had completely tied up the industrial situation in that country," Vice President White told the national convention, "they went to the old closet and brought out the ghastly specter of racial hatred and held it before the people of Alabama" (UMWA, *Proceedings of the Twentieth Annual Convention, 1909*, 872).

86. See C. V. Woodward, *Origins of the New South*, 323, 326, 352; C. V. Woodward, *Tom Watson: Agrarian Rebel*; Palmer, *"Man Over Money,"* 197; McMath, *American Populism*, 174, 197–98.

87. UMWA, *Proceedings of the Twentieth Annual Convention, 1909*, 864.

88. C. V. Woodward, *Origins of the New South*, 392–95; Hackney, *Populism to Progressivism in Alabama*, 122–323; Grantham, *Southern Progressivism*, 46–50.

89. For an expanded discussion of the argument presented here, see Letwin, "Interracial Unionism, Gender, and 'Social Equality' in the Alabama Coalfields."

90. Hall, *Revolt Against Chivalry*, 145; Painter, " 'Social Equality,' Miscegenation, Labor, and Power," 49. See also McMillen, *Dark Journey*, 14; Brundage, *Lynching in the New South*, 4–5, 58–72.

The linkage made between "social equality" and a wide range of black or interracial activities is a theme that has arisen through much of nineteenth- and twentieth-century American history. George M. Fredrickson develops the point when he describes how the "social equality" charge could provide segregationists with a facile bridge between black political empowerment and the ruination of white womanhood: "The granting of political rights, it was argued, had led to dreams of 'social equality' and had encouraged blacks to expropriate white women by force. Thus the Negro's overpowering desire for white women was often described as the central fact legitimating the whole program of legalized segregation and disfranchisement" (Fredrickson, *Black Image in the White Mind*, 282). Martha Hodes traces this connection to the turbulent era of Reconstruction, noting "the way in which extreme white anxiety over sexual liaisons between white women and black men was linked to fears of black men's political and economic independence" (Hodes, "The Sexualization of Reconstruction Politics," 407). See also Litwack, *Been in the Storm So Long*, 265; Rabinowitz, *Race Relations in the Urban South*, 186–87; Gilmore, "Gender and Jim Crow," 164–65, 188–95. Gunnar Myrdal would describe the same mentality a half century later: "Every single measure [of segregation and discrimination] is defended as necessary to block 'social equality' which in its turn is held necessary to prevent 'intermarriage' " (Myrdal, *American Dilemma*, 587).

91. Ayers, *Promise of the New South*, 140; Green, "The Brotherhood of Timber Workers, 1910–1913"; Arnesen, *Waterfront Workers of New Orleans*.

CHAPTER SIX

1. Fies to DeBardeleben, December 1, 1916 (quote; emphasis in original), and Taylor to DeBardeleben, May 19, 1919, December 15, 1920, DeBardeleben Papers, Box 72, File "Ta-Te (Misc)." For background on Taylor, see Harlan, *Booker T. Washington: Making of a Black Leader*, 229–30; Harlan, *Booker T. Washington: Wizard of Tuskegee*, 37, 54; Harlan, *Booker T. Washington Papers*, 3: 55, 351–52, 4: 21–23; Fies, "Man with a Light on His Cap," 70.

2. Robert W. Taylor to Fies, October 30, 1916, DeBardeleben Papers, Box 72, File "Ta-Te (Misc)."

3. Fies to DeBardeleben, December 1, 1916, DeBardeleben Papers, Box 72, File "Ta-Te (Misc)."

4. Henry DeBardeleben to Peoples Ice Company, December 21, 1916, DeBardeleben Papers, Box 4, File "Pe."

5. At Piper, for example, the Little Cahaba Coal Company miners walked off in 1913 for the right to select their own checkweighman; after a short time, the company granted their demand. See *UMWJ*, January 16, 1913. On economic trends: W. D. Lewis, *Sloss Furnaces*, 330. For an example of union contracts, see *UMWJ*, August 8, 1912.

6. On existence and enforcement of "yellow-dog" contracts: *UMWJ*, October 27, December 22, 1910, March 9, April 13, May 11, 1911; Birmingham *Labor Advocate*, April 21, 1911. On

spies: *UMWJ*, July 10, 1913. For membership figures: *UMWJ*, January 20, 1910, January 26, 1911, January 18, 1912, February 20, 1913; UMWA, *Proceedings of the Twenty-second Annual Convention* (Indianapolis: Cheltenham Press, 1911), 164. Statham statement: *UMWJ*, March 6, 1913. For committee report: "Report of Committee on Conditions in the Alabama Coal Fields," by James S. Moran, Paul J. Pauson, and E. T. Fitzgibbons, to President John P. White, Philip Taft Papers, BPLA.

7. For panoramic depictions of the miners' conditions: *UMWJ*, November 25, 1909, August 4, October 27, December 8, 1910, March 9, 23, April 13, 1911, July 31, 1913; Birmingham *Labor Advocate*, November 10, 1911; *American Federationist*, October 1910, 924. For evidence that Alabama Fuel and Iron Company miners were being "held and worked as Peons," Fairley presented the U.S. Department of Labor with several samples of passes scrawled out to miners seeking company permission to visit Birmingham. See Fairley to William B. Wilson, Secretary of Labor, June 2, 1913, and John P. White to William B. Wilson, June 5, 1913, Chief Clerk's Files, Case 20/35, General Records of the Department of Labor, RG 174, NA.

8. Ward and Rogers, *Convicts, Coal, and the Banner Mine Tragedy*; W. D. Lewis, *Sloss Furnaces*, 322–24.

9. *UMWJ*, July 17, 1916 (Townley). See also ibid., October 27, 1910, March 9, 1911, March 6, April 24, 1913; Birmingham *Labor Advocate*, April 18, 1913.

10. Both the 1910 and the 1920 censuses list African Americans as comprising 54 percent of Alabama's coal mine labor force (11,189 of 20,778 in 1910, 14,097 of 26,204 in 1920; see chapter 1). While no state-level statistics exist for the years between 1910 and 1920, there are bases for conjecture that the proportion of blacks rose higher than 54 percent during the intervening period. First, impressionistic descriptions from the 1910s often describe the coal fields as approximately two-thirds, and occasionally three-quarters, black. Second, as discussed later in this chapter, the Great Migration northward starting in 1916 drew large numbers of African Americans from the mineral district between 1916 and 1920.

11. ACOA minute book no. 1, November 10, 1908, 57, ACOA Collection.

12. Rikard, "Experiment in Welfare Capitalism," 54–59, 133–37; W. D. Lewis, *Sloss Furnaces*, 316–17.

13. On the welfare programs developed under the auspices of U.S. Steel in the early twentieth century, see Brody, *Steelworkers in America*, 88–92, 110–11, 154–58, 165–69; Rikard, "Experiment in Welfare Capitalism," 30–34; Tarbell, *Life of Elbert H. Gary*, 152–77; Brandes, *American Welfare Capitalism*, 30, 32, 84, 86. For background on Crawford, see Rikard, "Experiment in Welfare Capitalism," 34–35.

14. Rikard, "Experiment in Welfare Capitalism," 91–268; Hart, *Social Problems of Alabama*, 81–85; W. D. Lewis, *Sloss Furnaces*, 317–22.

15. Interestingly, McCormack was referring to the widespread abuses of the "county fee system" of law enforcement (see chapter 1), which, he observed, yielded the "indiscriminate shooting and arresting of people by the officers of the law for trivial offenses" (ACOA minute book no. 1, September 7, 1910, 89; June 6, 1911, 99–100, ACOA Collection).

16. Launched in 1911, these activities fell under the auspices of the Alabama Safety Association, established by the ACOA in 1914. See ACOA minute book no. 1, entries for 1911–1916, ACOA Collection; *Annual Report of Coal Mines: State of Alabama (1912)* (Birmingham: Alabama Mineral Map Co., 1912), 12–14; *Annual Report of Coal Mines: State of*

Alabama (1914) (Birmingham: Alabama Mineral Map Co, 1914), 7; *Annual Report of Coal Mines: State of Alabama (1915)* (Birmingham: Alabama Mineral Map Co., 1915), 7.

17. ACOA minute book no. 1, 1913 annual report, 168 (McCormack), 163 and 188 (Davidson), ACOA Collection. Several explanations for the divergent approaches of the Tennessee and the Sloss companies suggest themselves. Although Sloss was the region's second largest operator, it could scarcely begin to approach the enormous capital resources available to Alabama's subsidiary of U.S. Steel. Having chosen to bypass steel production and cast its future instead on pig iron, the Sloss company did not feel so intensely the urgency of acquiring a stable, motivated labor force. Finally, as mentioned above, Sloss remained distinctly more grounded in the southern style of managing labor (particularly black labor) and thus less responsive to the Progressive Era welfare philosophy spearheaded in the Birmingham district by TCI. On Sloss's lack of interest in corporate welfare, see W. D. Lewis, *Sloss Furnaces*, 298-99, 322.

18. *Annual Report of Coal Mines: State of Alabama (1912)*, 14; Rikard, "An Experiment in Welfare Capitalism" (in this vivid 1983 study, Rikard drew on a rich cache of TCI records which unfortunately have not otherwise been open to scholarly inspection); ACOA minute book no. 1, 1916 Secretary-Treasurer's Report, 237, ACOA Collection (Davidson).

19. For Armes's critique, see *UMWJ*, July 31, 1913. A year earlier the ACOA had discontinued a contract (of a nature unspecified in its minutes) with Armes (ACOA minute book no. 1, June 4, 1912, 135, ACOA Collection). Whether this action reflected, or triggered, disenchantment between the two parties is not known; either way, it may have figured in the striking shift in tone from Armes's 1910 *Story of Coal and Iron* and the sharp attack on the operators cited here. For other harsh judgments from the labor press, see Birmingham *Labor Advocate*, June 21, 1912; July 25, 1913; June 12, 1914; May 13, 1916.

20. Montgomery, *Fall of the House of Labor*, 332 (quote), 370-410; Brody, *Workers in Industrial America*, 39-45; McCartin, " 'An American Feeling,' " 67-86. On the surge in demand for Alabama pig iron following the outbreak of war in Europe, see W. D. Lewis, *Sloss Furnaces*, 338.

21. Coverage of labor organizing among Birmingham workers during these years abounds in the Birmingham *News*, the Birmingham *Age-Herald*, the Birmingham *Labor Advocate*, and the *American Federationist*. See also McKiven, *Iron and Steel*, 105-10, 125-31; McCartin, "Labor's 'Great War,' " 138-91; McCartin, " 'Americanism,' Federal Intervention, and the Limits of Interracial Unionism"; Spero and Harris, *The Black Worker*, 246-49, 358-66.

22. On the Great Migration of the World War I era, see Scott, *Negro Migration during the War*; U.S. Department of Labor, *Negro Migration in 1916-1917*; Scott, "Letters of Negro Migrants," and "Additional Letters of Negro Migrants"; Henri, *Black Migration*; Gottlieb, *Making Their Own Way*, 12-62; Grossman, *Land of Hope*; Trotter, *The Great Migration*; Jacqueline Jones, *Labor of Love, Labor of Sorrow*, 152-95.

23. Some historians, however, have noted that the passage of most African Americans from the rural South to the urban North was not always direct and immediate; many, they hold, made their way into the realm of urban and wage labor only by degrees, expanding their range of mobility from local, to regional, and finally long-distance. See Gottlieb, *Making Their Own Way*, 12-38; Grossman, *Land of Hope*, 13-37; Jacqueline Jones, *The Dispossessed*, esp. 21-24, 75-78, 104-8.

24. "Negro laborers are leaving the district by the hundreds," one coal operator remarked

in early fall (Birmingham *Age-Herald*, September 26, 1916). On black departure from Birmingham before 1916, see U.S. Immigration Commission, *Immigrants in Industries: Bituminous Coal Mining*, 7: 217. Newspaper reports of black departure from the district were extensive. According to one estimate, 75 blacks were leaving Birmingham each day for the coal fields of Kentucky and West Virginia (Birmingham *Age-Herald*, September 22, 1916; see also November 27, 1916). At times, this steady outflux was punctuated by extraordinarily large departures. In July 1916, for instance, a special train took 610 African Americans north to Pennsylvania (U.S. Department of Labor, *Negro Migration*, 53). The exodus persisted into 1917. During the first three weeks of June, for example, 857 blacks were reported to have left Birmingham for the North (U.S. Department of Labor, *Negro Migration*, 57).

25. Birmingham *Age-Herald*, November 27, 1916, March 21, 1917; U.S. Department of Labor, *Negro Migration*, 64.

26. U.S. Department of Labor, *Negro Migration*, 63. The prevalence of coal miners among migrants from the Birmingham district is both indicated and explained by the main destinations. Although many migrants went to large cities such as Chicago, Detroit, East St. Louis, Cincinnati, Cleveland, Pittsburgh, and Philadelphia (U.S. Department of Labor, *Negro Migration*, 55), large numbers headed for coal districts in central or upper Appalachia. For references to other coal fields as a major end point for Birmingham migrants, see Birmingham *Age-Herald*, September 22, 23, October 21, 1916; U.S. Department of Labor, *Negro Migration*, 63.

27. The Bessemer Coal, Iron, and Land Company, for example, had to shut down for 134 days that year for lack of railroad cars (B. D. Williams to H. L. Badham, August 10, 1917, in Bessemer Coal, Iron, and Railroad Company papers, private collection of Pat Cather, Birmingham, Ala.). For further evidence on the mining companies' need to shut down operations or cut back hours on short notice, see U.S. Department of Labor, *Negro Migration*, 65–66; Birmingham *Age-Herald*, September 13, October 27, 1916; DeBardeleben to Fies, August 25, 1916, DeBardeleben Papers, Box 87, File "M. H. Fies, Vice President."

28. "I am beginning to suffer greater apprehension now with reference to the effect of the car shortage on our labor," DeBardeleben vice president Fies reflected that fall, adding that the condition of miners who could work only three to five hours per day was "really becoming pitiable. . . . I am of the opinion that we are going to irreparable [*sic*] lose if this situation does not improve" (Fies to DeBardeleben, November 30, 1916, in DeBardeleben Papers, Box 87, File "M. H. Fies"; see also DeBardeleben to Fies, August 25, 1916, DeBardeleben Papers, Box 87, File "M. H. Fies, Vice President"). Where mines were running only three days a week, the *Age-Herald* added in fall 1916, "the workers are having great difficulties in making ends meet" (October 8, 1916).

29. U.S. Department of Labor, *Negro Migration*, 65.

30. Birmingham *Age-Herald*, September 26, November 20, 1916, January 23, 1917 (quote). Data on the causes, mechanics, and effects of black migration to, through, and out of the Birmingham district can be found in U.S. Department of Labor, *Negro Migration*, 53–65; Scott, *Negro Migration*, 3, 19–21; and regular articles in the Birmingham *Age-Herald*, Birmingham *News*, and Birmingham *Reporter*.

31. J. F. McCurdy to David Moore, December 4, 1916, and DeBardeleben to Fies, December 18, 1916, DeBardeleben Papers, Box 87, File "M. H. Fies."

32. Birmingham *Labor Advocate*, January 20, 27, 1917; McKiven, "Class, Race, and

Community," 181-85. Coverage of these various drives abounds in the Birmingham *Labor Advocate*, Birmingham *Age-Herald*, Birmingham *News*, and the *American Federationist*.

33. *UMWJ*, August 3 (Margaret), July 27 (Townley), June 1 (Acton), 1916. See also ibid., May 4, 1916; Birmingham *Labor Advocate*, August 19, 1916.

34. For general background on this outlook, see Meier, *Negro Thought in America*, esp. 100-118; Harlan, *Booker T. Washington: Making of a Black Leader* and *Booker T. Washington: Wizard of Tuskegee*; Fredrickson, *Black Liberation*, 94-136; Gaines, *Uplifting the Race*.

35. On Adams: Birmingham *Reporter*, January 9, February 13, 1915, August 19, 1916. On Rameau: Birmingham *Age-Herald*, March 29, 1917. On Goodgame: Birmingham *Reporter*, January 16, 1915. On Eason: Birmingham *Reporter*, September 30, 1916, July 13, 1918. On Mason: Harlan, *Booker T. Washington: Wizard of Tuskegee*, 230. On Republican connections: Birmingham *Reporter*, October 21, November 4, 1916. For regular mention of the black social and economic institutions of early-twentieth-century Birmingham, see such black newspapers as the Birmingham *Wide-Awake*, *Hot Shots*, Birmingham *Reporter*, and *Workmen's Chronicle*.

36. On the Sixteenth Street Baptist Church: Birmingham *Reporter*, June 3, September 23, November 4, 1916. During the first wave of the black exodus of 1916, the church, under the leadership of its new, activist pastor, Dr. A. C. Williams, intensified and consolidated these efforts by launching a Social Service Association, which broke down into a kindergarten department, a religious services department, a physical culture department, a community extension department, and an industrial department. On the Sixth Avenue Baptist Church: Birmingham *Reporter*, October 6, 1917, June 8, 1918. W. H. Mixon, state presiding elder of the African Methodist Episcopal Church, and James B. Carter, pastor of the Payne Chapel A.M.E. Church, were also major forces in the city's black middle class (Birmingham *Reporter*, November 4, 1916, January 6, 1917, June 8, 1918). On the integral and expansive function of the church in urban black communities of the early-twentieth-century South, see Lincoln and Mamiya, *Black Church*; Dittmer, *Black Georgia*, 50-55; Rabinowitz, *Race Relations in the Urban South*, 198-225; Meier, *Negro Thought in America*, 130-34.

37. *Baptist Leader*, January 23, February 6, March 20, April 3, 17, 24, May 8, July 10, 17, August 7, November 27, December 18, 1909, January 30, February 19, 26, March 12, 1910.

38. *Baptist Leader*, May 8, July 10 (quote), 17, August 14, 21, September 11, 1909, March 12, 1910. Reverend Goodgame, influential pastor of the Sixth Avenue Baptist Church, was chairman of the Institute's Board of Ministers, and meetings were held at the Sixteenth Street Baptist Church (*Baptist Leader*, November 20, 1909). For more on the Ministerial and Economical Institute, see *Baptist Leader*, July 10, 17, August 14, September 9, November 6, 1909, February 26, 1910.

39. See, for example, Birmingham *Reporter*, March 18, 1916; ACOA minute book no. 1, January 31, 1917, 254, ACOA Collection; and two letters in the DeBardeleben Papers: A. C. Williams to DeBardeleben, July 24 and October 6, 1917, Box 59, File "W"; and Pierce S. Moten to DeBardeleben, Box 2, File "Mo."

40. Harlan, *Booker T. Washington: Wizard of Tuskegee*, esp. 238-65; Meier, *Negro Thought in America*, esp. 85-99.

41. On newspaper circulation: Treasurer to Fies (re *Workmen's Chronicle*), September 11, 1917, DeBardeleben Papers, Box 87, File "M. H. Fies"; Treasurer to Oscar Adams (re

Birmingham *Reporter*), July 13, 1917, DeBardeleben Papers, Box 76, File "Aa–Ad Misc." Mason quote: Birmingham *Reporter*, August 4, 1917. "Don't bite" quote: Birmingham *Reporter*, August 18, 1917. Williams quote: Birmingham *Reporter*, July 21, 1917; see also July 28, 1917. On Booker T. Washington's attitude toward unions: Meier and Rudwick, "Attitudes of Negro Leaders toward the American Labor Movement," 39- 41.

42. *UMWJ*, June 1, 1916 (Acton); Birmingham *Labor Advocate*, July 1, 1916 (Republic); *UMWJ*, May 4, 1916 (Flat Creek). Ostensibly concerned with discouraging absenteeism among the black miners, "shack rousters" became the focal point of resentment in the black communities on account of abuses that smacked of slavery days. The Acton miner wrote of "the white brute in human form . . . who rides . . . with his billy and revolver hanging to his saddle, going house to house beating up negro men and using their women to suit his fancy, and if any negro opens his mouth in protest . . . they are either beaten almost to death or shot down like a dog."

43. *UMWJ*, May 4, 1916 (white miner); Birmingham *Labor Advocate*, June 17, 1916 ("Colored Miners"). The miner from Acton (see note 42) argued the same imperative.

44. Birmingham *Labor Advocate*, June 17, 1916.

45. Ibid., June 24 (Wood; Statham on "Rameau problem"), July 8 (Statham on driving likes of Rameau from district), 1916.

46. *UMWJ*, June 1, 1916 ("We, the negro miners"); Birmingham *Labor Advocate*, July 16, 1916 ("There is no use").

47. Letters in DeBardeleben Papers: Fies to DeBardeleben, May 31, 1917, Box 87, File "M. H. Fies" ("extremely acute," one in five men); DeBardeleben to Phoenix-Girard C. & W. Co., July 21, 1917, Box 4, File "Phoenix-Girard Coal and Wood Co" ("precarious"); Fies to DeBardeleben, October 18, 1917, Box 87, File "M. H. Fies" (six of thirty-six men); DeBardeleben to S. P. Kennedy, December 10, 1917, Box 59, File "U.S. Fuel Administration" (barely 50 percent); W. S. Blauvelt to J. P. White, Labor Advisor to U.S. Fuel Administration, June 11, 1918, File "H. L. Kerwin," Box 1343, General Correspondence, Administrative Division, Bureau of Labor, General Records of the U.S. Fuel Administration, RG 67.

48. R. Galloway to DeBardeleben, April 25, 1917, Box 2, File "F'" (Galloway quote); Fies to DeBardeleben, July 31, 1918, Box 87, File "M. H. Fies"; DeBardeleben to Fies, April 8, 1918, Box 87, File "M. H. Fies"; all in DeBardeleben Papers.

49. H. C. Selheimer to "All Coal Operators of Alabama," September 19, 1918, Box 59, File "U.S. Fuel Commission Administration"; Fies to Edgar L. Adler, September 27, 1917, Box 87, File "M. H. Fies"; both in DeBardeleben Papers. ("Demoralization" was a term commonly used by late-nineteenth- and early-twentieth-century southern employers to denote scarce or irregular labor supply.) This phenomenon was by no means unique to the Birmingham district; during the war boom especially, competition through increased rates, greater frequency, and new forms of pay, coupled with direct labor raiding, sprang up from the textile mills of the Piedmont to telephone companies around the country and in similar fashion fueled turnover. For these industries, see Hall et al., *Like a Family*, 110–11, 184; Greenwald, *Women, War, and Work*, 208–9.

50. Empire Coal Company to Southern Railway & Light Co., August 27, 1917, Box 30, File "Sn–Ste" (see also DeBardeleben to A. C. Danner, November 8, 1917, Box 2, File "Mobile Coal Co."); DeBardeleben to Kennedy, December 10, 1917, Box 59, File "U.S. Fuel Adminis-

tration"; DeBardeleben to A. C. Williams, November 26, 1917, Box 59, File "Milton H. Fies"; all in DeBardeleben Papers. The black professional elite, for its part, widely shared this assessment of black responses to wage increases. Tuskegee University Professor Monroe Work, for example, argued during the war that black labor tended, "because of increased pay, to work fewer days in the week or month." This tendency, he added, "is more largely true of unskilled than of skilled labor. In general, unskilled labor is more or less ignorant. Its wants are few and easily satisfied" (Monroe N. Work, "Southern Labor as Affected by the War and Migration," manuscript for paper to be delivered to the Southern Sociological Congress, Birmingham, April 15, 1918, Monroe N. Work Papers, Box 4, Tuskegee University Archives).

51. All sources in DeBardeleben Papers: Fies to DeBardeleben, May 31, 1917, Box 87, File "M. H. Fies"; Fies, draft of speech to ACOA, n.d., Box 87, File "M. H. Fies," 6–7; DeBardeleben to Danner, November 8, 1917, Box 2, File "Mobile Coal Co."; DeBardeleben to R. Galloway, April 26, 1917, Box 2, File [no name].

52. McCartin, "Abortive Reconstruction"; McCartin, "Labor's 'Great War,'" 138–91; Kennedy, *Over Here*, 258–79; Dubofsky, "Abortive Reform," 197–220; Dubofsky, *State and Labor*, 61–74; O'Conner, *National War Labor Board*. On the suppression of labor radicalism, see Dubofsky, *State and Labor*, 67–69.

53. Birmingham *Age-Herald*, June 4, 12, 17–19, 1917; Birmingham *Labor Advocate*, June 9, 16, 1917; *UMWJ*, June 14, 21, 1917.

54. Birmingham *News*, August 9, 1917 (DeBardeleben); *UMWJ*, June 28, 1917 (Moore); Birmingham *Age-Herald*, July 26, 1917.

55. Birmingham *Reporter*, August 18, 1917; Birmingham *Age-Herald*, July 29, 1917 (Mixon).

56. Birmingham *Reporter*, August 4 (Kennamer), July 28 ("should be given"), 1917; P. Colfax Rameau to President Wilson, September 12, 1917, Casefile 33/618, Box 28, Dispute Case Files, General Records of the Federal Mediation and Conciliation Service, RG 280.

In one instance described by the Birmingham *Reporter* (July 28, 1917) during the organizing drive of 1917, a black miner at the predominantly black mining camp of Praco was elected president of the local and promptly asked by whites from the District 20 leadership to step aside in favor of the white runner-up. "On what grounds must I do that?" the black miner reportedly demanded. "Well, you know we white men are not going to let you preside over us nigger." "Well I know I am not going to resign." "You are an impudent d— - black s—— of a b——."

57. Some opponents of the UMW wove the two themes together. One Mrs. G. H. Mathis toured the mineral district delivering speeches to the effect that "considerable German influence" was "getting a hold on the negroes" in the coal fields. Such propaganda, she warned in a letter to a Department of Labor official, "appears to be using the miners Union as a channel" (Mathis to Lewis Post, September 3, 1917, Casefile 33/618, RG 280).

58. Birmingham *Labor Advocate*, September 29, 1917 ("The non-union miners"); *UMWJ*, October 25, 1917 ("feudal principalities"); Birmingham *Labor Advocate*, June 30, 1917 (Statham). See also Birmingham *Labor Advocate*, June 23, August 4, 11, November 24, 1917; *UMWJ*, June 21, October 25, 1917, April 18, 1918. Organized labor as a whole around the Birmingham district invoked the rhetoric of wartime patriotism on behalf of its cause. See McCartin, "Labor's 'Great War,'" 146–47.

59. McCartin makes this point in "Abortive Reconstruction," 11–12.

60. Birmingham *Labor Advocate*, July 7, 1917.

61. Ibid., June 23–July 21, 1917; *UMWJ*, June 28–July 26, 1917; Birmingham *Age-Herald*, June–July 1917 issues. For a list of newly organized District 20 locals during June and July of 1917, see UMWA, *Proceedings of the Twenty-sixth Consecutive and Third Biennial Convention* (Indianapolis: Bookwalter-Ball Printing Co., 1918), 1: 261-63, 265-66. According to District 20's books, Kennamer reported, the union had organized "upwards of 23,000 miners and mine laborers" as of July 31, 1917 (J. R. Kennamer to W. R. Fairley, June 8, 1918, Casefile 33/618, RG 280; see also W. R. Fairley, "Preliminary Report of Commissioner of Conciliation, October 9, 1917," Casefile 33/618, RG 280).

62. J. R. Kennamer to W. B. Wilson, Secretary of Labor, August 8, 1917, Casefile 33/618, RG 280; Birmingham *Labor Advocate*, July 28, August 18, 1917.

63. "Reasons for Organizing," anonymous typescript, Casefile 33/618, RG 280; Birmingham *Labor Advocate*, July 28, 1917 (attendance figures); Birmingham *Age-Herald*, July 31, 1917 ("very few old faces").

64. Birmingham *Age-Herald*, July 31 (black and white representation), August 1 (independent positions), August 2 (Kennamer), 1917.

65. Birmingham *Labor Advocate*, July 28, August 18, 1917; J. R. Kennamer, William Harrison, and J. L. Clemo, "To the Alabama Coal Operators, Collectively and Individually," July 23, 1917, Casefile 33/618, RG 280. For operators' rebuff to requests from the U.S. Committee on Coal Production that they meet with UMW officials, see the following, all in Casefile 33/618, RG 280: F. S. Peabody, Chairman, U.S. Committee on Coal Production, to Erskine Ramsey, August 6, 9, 1917; Ramsey to Peabody, August 7, 1917. For operators' warnings about a predominantly black UMW, Ramsey to Peabody, June 14, 1917; see also Ramsey to Peabody, August 3, 1917, and G. B. McCormack to W. B. Wilson, August 18 and September 8, 1917, all in Casefile 33/618, RG 280. For U.S. Attorney Bell's report, see Robert N. Bells to Attorney General, August 6, 1917, Class #16–17, 1917 and 1920–21 Strikes, Alabama Coal, General Records of the U.S. Department of Justice, RG 60, NA.

66. Birmingham *Age-Herald*, August 11, 15, August 17–September 9, 1917; Birmingham *Labor Advocate*, September 8, 15, 1917. Documents all in Casefile 33/618, RG 280: W. B. Wilson, "Memorandum to the Alabama Coal Operators' Association"; Wilson to Alabama Coal Operators' Association, August 27, 1917; J. R. Kennamer and J. L. Clemo to President Wilson, August 29, 1917; Kennamer to W. B. Wilson, August 20, 1917; Kennamer, Clemo, William Harrison, Thomas King, and John L. Lewis to W. B. Wilson, August 25, 1917; W. R. Fairley to W. B. Wilson, August 29, 31, 1917; William Diamond, Kennamer, and Clemo, "To the Officers and Delegates of the Special Convention of District 20 U.M.W. of A.," September 7, 1917.

67. Birmingham *Age-Herald*, December 20–22, 1917; *UMWJ*, December 27, 1917; Birmingham *Labor Advocate*, December 29, 1917; ACOA minute book no. 1, resolution regarding Fuel Administrator's ruling, December 1917, 280–88, ACOA Collection. Documents in Casefile 33/618, RG 280: W. R. Fairley to Hugh L. Kerwin, October 27 and December 29, 1917; J. R. Kennamer and J. L. Clemo to Dr. H. A. Garfield, Coal Administrator, October 8, 1917; W. R. Fairley, "Preliminary Report of Commissioner of Conciliation," October 9, 1917; W. R. Fairley to W. B. Wilson, December 15, 1917; Rembrandt Peale and John L. Lewis to "Mine Workers of Alabama" (text of Garfield Agreement), December 14, 1917. On the origins and performance of the U.S. Fuel Administration, see Johnson, *Politics of Soft Coal*,

17–94; Cuff, "Harry Garfield, the Fuel Administration, and the Search for a Cooperative Order"; Kennedy, *Over Here*, 123–25.

68. Birmingham *Labor Advocate*, February 18, March 2, 9, April 13, 1918; Birmingham *Age-Herald*, January 14, 15, February 2, 13, 15–20, 23, 24, 26, 1918; *UMWJ*, May 23, 1918. Documents in Casefile 33/618, RG 280: W. R. Fairley to Hugh L. Kerwin, January 3, February 15, 16, 18, 1918; W. R. Fairley to W. B. Wilson, June 11, 1918. In United Mine Workers of America file, Box 17, General Correspondence, Executive Officer, Entry 1, RG 67: Rembrandt Peale to J. R. Kennamer, February 20, 1918; Harry Garfield to J. R. Kennamer, February 26, 1918. In Alabama, Oct., Nov., Dec., 1917 file, Box 745, General Correspondence, Records of the Bureau of State Organizations, 1917–1919, Administrative Division, RG 67: J. R. Kennamer to John P. White, February 12, 23, 1918.

69. Birmingham *Age-Herald*, May 2–4, 17; Birmingham *Labor Advocate*, May 4, 11, 18, 1918; *UMWJ*, April 25, May 9, 23, 1918. Documents in Casefile 33/618, RG 280: "Important Decisions of the Umpire, Judge H. C. Selheimer, on the Garfield Agreement"; "Recommendations of Fuel Administration as Adopted By Miners in Convention Held in Birmingham, Alabama, December 21, 1917, and Supplement Wage Agreement Adopted May 3rd, 1918."

70. UMWA, *Proceedings of the Twenty-seventh Consecutive and Fourth Biennial Convention* (Indianapolis: Bookwalter-Ball Printing Co., 1919), 1: 63, 350 (McGuire). In United Mine Workers of America file, Box 17, General Correspondence, Executive Officer, Entry 1, RG 67: J. R. Kennamer and William Harrison to Dr. H. A. Garfield, August 26 ("unrest and discontent") and September 7 ("below living wage"), 1918; Birmingham *Labor Advocate*, September 21, 1918 (Brown).

71. For propaganda activities of black leaders, see regular coverage during 1917 and 1918 in Birmingham *Reporter*, *Workmen's Chronicle*, and Birmingham *Age-Herald*. For examples of requests for and acknowledgments of material support from the operators for propaganda efforts by leading black orators and editors, see the following, all in DeBardeleben Papers: Mixon to Fies, May 11, 1918, and Mixon to DeBardeleben, July 3, 1918, Box 2, File "Mi"; Williams to DeBardeleben, October 26, 1917, Box 59, File "W"; Rameau to DeBardeleben, April 27, 1918, Box 52, File "Ra–Re."

72. UMWA, *Proceedings of the Twenty-seventh Consecutive and Fourth Biennial Convention* (Indianapolis: Bookwalter-Ball Printing Co., 1919), 1: 125 (appeals effective); Birmingham *Labor Advocate*, June 15, 1918 (union leaders); see also Birmingham *Age-Herald*, June 14, 1918.

73. Unless otherwise indicated, this narrative of the 1918 steelworkers' strike draws upon McKiven, "Class, Race, and Community," 185–99; McCartin, "Labor's 'Great War,'" 158–78; Spero and Harris, *The Black Worker*, 247–49; Birmingham *Labor Advocate*, February 23, March 2, June 15, 1918. In addition to TCI, the struck operators included the United States Cast Iron Pipe Company at Bessemer, Birmingham Machine and Foundry, Bessemer Machine and Foundry, and Woodward Iron Company.

74. On the traditional Jim Crow orientation of Birmingham's iron and steel unions, see McKiven, *Iron and Steel*. McKiven argues that the initiative by the Metal Trades Council to organize black workers into a different union (Mine Mill) also reflected "just another way to secure whites' monopoly of the best jobs" ("Class, Race, and Community," 194).

75. Raymond Swing, Examiner, report on "The Birmingham Case," July 1918, Entry 4, Casefile 2, General Records of the National War Labor Board, RG 2, NA.

76. McCartin, "Abortive Reconstruction," 16.

77. *UMWJ*, October 15, 1919; see also May 15, 1919.

78. McCartin, "Abortive Reconstruction," 12–16.

79. On the 1919 strike nationally: Brody, *Workers in Industrial America*, 44–45; Dubofsky, *State and Labor*, 76–79; McCartin, "Labor's 'Great War,'" 401–49; Montgomery, *Fall of the House of Labor*, 399–410; Taft, *The A.F.L. in the Time of Gompers*, 406–11. On the strike in Alabama: Birmingham *Labor Advocate*, October 18–December 6, 20, 1919 (Statham quote from November 1, 1919), January 24, 1920. Documents in File 13/161, Bituminous Coal Strike, Oct. 1919, Box 3, General Subject Files, 1913–1921, Office of the Secretary, RG 174, NA: J. R. Kennamer to W. B. Wilson, October 23, 1919; W. B. Wilson to J. R. Kennamer, October 31, 1919. On Kennamer's protests: Birmingham *Labor Advocate*, April 3, 1920;. *UMWJ*, April 1, 15, 1920; Taft, *Organizing Dixie*, 52–53.

80. On economic downturn: W. D. Lewis, *Sloss Furnaces*, 364; Soule, *Prosperity Decade*, 96–106. On post-Garfield negotiating: *UMWJ*, May 1, June 1, 1920; Birmingham *Labor Advocate*, April 3, 10, 24, May 22, June 12, 1920.

81. *UMWJ*, July 1, 1920; Birmingham *Labor Advocate*, June 12–26, July 3, 24, 31, August 7, 14, 28, 1920; see also regular coverage in Birmingham *Age-Herald*.

82. The story of the 1920–21 strike has been recounted by several historians more extensively than it is here. Except where otherwise indicated, this narrative draws upon Straw, "The United Mine Workers of America and the 1920 Coal Strike in Alabama"; Taft, *Organizing Dixie*, 53–58; R. Lewis, *Black Coal Miners in America*, 59–70.

83. UMWA, *Proceedings of the Twenty-eighth Consecutive and Fifth Biennial Convention* (Indianapolis: Bookwalter-Ball-Greathouse Printing Co., 1921), 105.

84. Birmingham *Labor Advocate*, June 19, 1920; *UMWJ*, July 1, 1920.

85. P. Colfax Rameau to Governor Thomas E. Kilby, September 6, October 12, 14, 1920, in File "Coal Strike—Sept.–Nov. 1920—C," Governor Thomas E. Kilby Papers, ADAH.

86. For significant scholarship on southern race and labor in the CIO era, see introduction, note 2.

Selected Bibliography

MANUSCRIPTS

Birmingham, Ala.
   Birmingham Public Library Archives
      Alabama Coal Operators' Association Collection
      Alabama Mineral Land Company Collection
      DeBardeleben Coal Company Papers
      Alfred M. Shook Papers
      Sloss-Sheffield Steel and Iron Company Records
      Philip Taft Papers
   Private collection of Pat Cather
Detroit, Mich.
   Wayne State University Archives of Labor and Urban Affairs
      Lester J. Cappon Collection
      Edward A. Wieck Collection
Miscellaneous
   Terence V. Powderly Papers, Knights of Labor microfilm edition (manuscript collection
      held at Catholic University, Washington, D.C.)
Montgomery, Ala.
   Alabama Department of Archives and History
      Governor Braxton B. Comer Papers
      Adjutant General Bibb Graves Papers
      Governor Thomas G. Jones Papers
      Governor Thomas E. Kilby Papers
      Governor Thomas Jefferson Seay Papers
Tuscaloosa, Ala.
   William Stanley Hoole Special Collections Library, University of Alabama, Tuscaloosa,
      Archives
      Nicholas B. Stack Papers
Tuskegee, Ala.
   Tuskegee University Archives
      Monroe N. Work Papers
Washington, D.C.
   National Archives
      U.S. Department of Justice, Record Group 60
      U.S. Department of Labor, Record Group 174
      U.S. Federal Mediation and Conciliation Service, Record Group 280
      U.S. Fuel Administration, Record Group 67
      U.S. National War Labor Board, Record Group 2

Great Britain

Board of Trade. *Report of an Enquiry by the Board of Trade into Working Class Rents, Housing, and Retail Prices.* London: His Majesty's Stationary Office, 1911.

State of Alabama

Alabama Commissioner of Industrial Resources. *Alabama: A Few Remarks upon Her Resources and the Advantages She Possesses as Inducements to Immigration.* Montgomery: John G. Stokes and Company, 1869.

——. *Report of the Commissioner of Industrial Resources to the Governor.* Montgomery: W. W. Screws, 1871.

——. *Report of the Commissioner of Industrial Resources of Alabama to the Governor.* Montgomery: W. W. Screws, 1874.

Alabama Department of Agriculture and Industries. *The Alabama Opportunity.* N.p.: n.p., 1906.

Annual Report of Coal Mines: State of Alabama. Birmingham: various publishers, 1912–21.

Biennial Report of the Inspector of Mines. Birmingham: various publishers, 1893–1902.

Biennial Report of the Inspectors of Convicts to the Governor. Montgomery: various publishers, 1884–1906. (Reports for 1894–1906 are titled *Biennial Report of the Board of Inspectors of Convicts to the Governor.*)

Quadrennial Report of the Board of Inspectors of Convicts for the State of Alabama, to the Governor. Montgomery: Brown Printing Co., 1906–18.

Report of Inspector of Alabama Coal Mines. Birmingham: various publishers, 1906, 1908, 1909.

United States

Bureau of the Census. *Mines and Quarries: 1902.* Washington, D.C.: Government Printing Office, 1905.

——. *Ninth Census of the United States, 1870.* Vol. 3, *The Statistics of Wealth and Industry.* Washington, D.C.: Government Printing Office, 1872.

——. *Tenth Census of the United States, 1880. Compendium, Part 2* (1883). Vol. 2, *Report on the Manufactures, Part 2* (1883). Vol. 15, *Report on the Mining Industries (Exclusive of Precious Metals)* (1886). Vol. 21, *Report on the Defective, Dependent, and Delinquent Classes of the Population* (1888). Washington, D.C.: Government Printing Office.

——. *Eleventh Census of the United States, 1890.* Vol. 1, *Population, Part 1* (1892). Vol. 5, *Report on Crime, Pauperism, and Benevolence* (1895). Vol. 14, *Report on the Mineral Industries* (1892). Washington, D.C.: Government Printing Office.

——. *Twelfth Census of the United States, 1900. Abstract of the Twelfth Census* (1902). Vol. 1, *Population, Part 1* (1901). Vol. 13, *Occupations* (1904). Washington, D.C.: Government Printing Office.

——. *Thirteenth Census of the United States, 1910. Statistics for Alabama* (1913). Vol. 4, *Occupational Statistics* (1914). Washington, D.C.: Government Printing Office.

——. *Fourteenth Census of the United States, 1920. State Compendium, Alabama.* Washington, D.C.: Government Printing Office, 1924.

Commissioner of Labor. *Third Annual Report of the Commissioner of Labor, 1887: Strikes and Lockouts.* Washington, D.C.: Government Printing Office, 1888.

——. *Tenth Annual Report of the Commissioner of Labor, 1894.* Vol. 1, *Strikes and Lock-outs.* Washington, D.C.: Government Printing Office, 1896.

Department of Labor. *Negro Migration in 1916–1917.* Washington, D.C.: Government Printing Office, 1919.

Department of Treasury. *Statistical Abstract of the United States for 1889.* Washington, D.C.: Government Printing Office, 1890.

——. *Statistical Abstract of the United States for 1900.* Washington, D.C.: Government Printing Office, 1901.

——. *Statistical Abstract of the United States for 1905.* Washington, D.C.: Government Printing Office, 1905.

Immigration Commission. *Immigrants in Industries: Part 1, Bituminous Coal Mining.* Vols. 6 and 7 of *Reports of the United States Immigration Commission.* Washington, D.C.: Government Printing Office, 1911.

——. *Immigrants in Industries: Part 2, Iron and Steel.* Vols. 8 and 9 of *Reports of the United States Immigration Commission.* Washington, D.C.: Government Printing Office, 1911.

Industrial Commission. *Agriculture and Agricultural Labor.* Vol. 10 of *Reports of the Industrial Commission.* Washington, D.C.: Government Printing Office, 1901.

Senate. *Report of the Committee of the Senate upon the Relations between Labor and Capital.* Vol. 4. Washington, D.C.: Government Printing Office, 1885.

NEWSPAPERS

American Federationist
American Manufacturer
Alabama Sentinel
Alabama True Issue
Baptist Leader (Birmingham)
Bessemer
Bessemer *Herald*
Bessemer *Journal*
Bessemer *Journal-Herald*
Bessemer *Weekly*
Bessemer *Workman*
Bessemerite
Birmingham *Advance*
Birmingham *Age*
Birmingham *Age-Herald*
Birmingham *Blade*
Birmingham *Chronicle*
Birmingham *Daily Ledger*
Birmingham *Daily News*
Birmingham *Evening Chronicle*
Birmingham *Iron Age*
Birmingham *Labor Advocate*

Birmingham *News*
Birmingham *Observer*
Birmingham *Reporter*
Birmingham *Semi-Weekly Review*
Birmingham *State-Herald*
Birmingham *Sunday Chronicle*
Birmingham *Truth*
Birmingham *Weekly Iron Age*
Birmingham *Wide-Awake*
Blocton *Enterprise*
Chattanooga *Republican*
Chattanooga *Tradesman*
Cullman Advance and Guide
Engineering and Mining Journal
Ensley *Industrial Record*
Free Lance (Birmingham)
Hot Shots (Birmingham)
Huntsville *Gazette*
Jefferson Enterprise
Jefferson Independent
John Swinton's Paper
Journal of United Labor
Labor Union (Birmingham)
Mineral Belt Gazette (Birmingham)
Montgomery *Advertiser*
National Labor Tribune
Negro American (Birmingham)
New Orleans *Daily Picayune*
Pratt Mines *Advertiser*
Red Mountain Magnet (Birmingham)
Socialist Spirit
Union Labor Review (Birmingham)
United Mine Workers Journal
Wall Street Journal
Warrior *Advance*
Warrior *Advance and Guide*
Warrior *Index*
West End Banner (Birmingham)
Workmen's Chronicle (Birmingham)

BOOKS

Armes, Ethel. *The Story of Coal and Iron in Alabama*. Birmingham: Birmingham Chamber
 of Commerce, 1910.

Arnesen, Eric. *Waterfront Workers of New Orleans: Race, Class, and Politics, 1863–1923.* New York: Oxford University Press, 1991.

Aronowitz, Stanley. *False Promises: The Shaping of American Working Class Consciousness.* Durham, N.C.: Duke University Press, 1992.

Ayers, Edward L. *The Promise of the New South: Life after Reconstruction.* New York: Oxford University Press, 1992.

——. *Vengeance and Justice: Crime and Punishment in the Nineteenth-Century American South.* New York: Oxford University Press, 1984.

Bennett, David H. *The Party of Fear: From Nativist Movements to the New Right in American History.* 1988. Reprint, New York: Vintage, 1990.

Billings, Dwight B., Jr. *Planters and the Making of the New South: Politics and Development in North Carolina, 1865–1900.* Chapel Hill: University of North Carolina Press, 1979.

Blassingame, John W. *The Slave Community: Plantation Life in the Antebellum South.* New York: Oxford University Press, 1972.

Bond, Horace Mann. *Negro Education in Alabama: A Study in Cotton and Steel.* 1939. Reprint, New York: Atheneum, 1969.

Brandes, Stuart D. *American Welfare Capitalism, 1880–1940.* Chicago: University of Chicago Press, 1976.

Brody, David. *Steelworkers in America: The Nonunion Era.* 1960. Reprint, New York: Harper and Row, 1969.

——. *Workers in Industrial America: Essays on the Twentieth Century Struggle.* 2d ed. New York: Oxford University Press, 1993.

Brophy, John. *A Miner's Life.* Madison: University of Wisconsin Press, 1964.

Brundage, W. Fitzhugh. *Lynching in the New South: Georgia and Virginia, 1880–1930.* Urbana: University of Illinois Press, 1993.

Burke, J. W. *The Coal Fields of Alabama.* Mobile: n.p., ca. 1885.

Caldwell, Henry M. *History of the Elyton Land Company and Birmingham, Ala.* 1892. Reprint, Birmingham: Southern University Press, 1972.

Cayton, Horace R., and George S Mitchell. *Black Workers and the New Unions.* Chapel Hill: University of North Carolina Press, 1939.

Cell, John W. *The Highest Stage of White Supremacy: The Origins of Segregation in South Africa and the American South.* Cambridge: Cambridge University Press, 1982.

Chandler, Alfred D. *The Visible Hand: The Managerial Revolution in American Business.* Cambridge: Harvard University Press, 1977.

Chapman, H. H. *The Iron and Steel Industries of the South.* University: University of Alabama Press, 1953.

Clark, Victor S. *History of Manufacturers in the United States, 1860–1914.* 3 vols. Washington, D.C.: Carnegie Institute of Washington, 1928.

Cohen, William. *At Freedom's Edge: Black Mobility and the Southern White Quest for Racial Control, 1861–1915.* Baton Rouge: Louisiana State University Press, 1991.

Coleman, John S. *Josiah Morris (1818–1891): Montgomery Banker Whose Faith Built Birmingham.* New York: Newcomen Society of North America, 1948.

Commons, John R., David J. Saposs, Helen Sumner, E. B. Mittelman, H. E. Hoagland,

John B. Andrews, and Selig Perlman. *History of Labour in the United States.* Vol. 2. New York: Macmillan, 1936.

Corbin, David Alan. *Life, Work, and Rebellion in the Coal Fields: The Southern West Virginia Miners, 1880–1922.* Urbana: University of Illinois Press, 1981.

Cornell, Robert J. *The Anthracite Coal Strike of 1902.* Washington, D.C.: Catholic University Press of America, 1957.

Crane, Mary Powell. *The Life and Times of James R. Powell and Early History of Alabama and Birmingham.* Brooklyn: Braunworth, 1930.

Dittmer, John. *Black Georgia in the Progressive Era, 1900–1920.* Urbana: University of Illinois Press, 1980.

Dix, Keith. *Work Relations in the Coal Industry: The Hand-Loading Era, 1880–1930.* Morgantown: Institute for Labor Studies, West Virginia University, 1977.

Dombhart, John Martin. *History of Walker County: Its Towns and Its People.* Thornton, Ark.: Cayce, 1937.

Doster, James F. *Railroads in Alabama Politics, 1875–1914.* Tuscaloosa: University of Alabama Press, 1957.

Dubofsky, Melvyn. *The State and Labor in Modern America.* Chapel Hill: University of North Carolina Press, 1994.

Dubose, John W. *Jefferson County and Birmingham, Alabama: Historical and Biographical.* Birmingham: Teeple & Smith, 1887.

———. *The Mineral Wealth of Alabama and Birmingham Illustrated.* Birmingham: N. T. Green, 1886.

Ellison, Rhoda Coleman. *Bibb County, Alabama: The First Hundred Years, 1818–1918.* Tuscaloosa: University of Alabama Press, 1984.

Evans, Chris. *History of the United Mine Workers of America: 1860 to 1890.* Indianapolis: n.p., 1900.

———. *History of the United Mine Workers of America from the Year 1890 to 1900.* Indianapolis: n.p., n.d.

Fels, Rendig. *American Business Cycles, 1865–1897.* Chapel Hill: University of North Carolina Press, 1959.

Fink, Gary M. *Espionage, Labor Conflict, and New South Industrial Relations: The Fulton Bag and Cotton Mills Strike of 1914–1915.* Ithaca: ILR Press, 1993.

Fink, Leon. *Workingmen's Democracy: The Knights of Labor and American Politics.* Urbana: University of Illinois Press, 1983.

Fishback, Price V. *Soft Coal, Hard Choices: The Economic Welfare of Bituminous Coal Miners, 1890–1930.* New York: Oxford University Press, 1992.

Flynt, Wayne. *Dixie's Forgotten People: The South's Poor Whites.* Bloomington: Indiana University Press, 1979.

———. *Mine, Mill, and Microchip: A Chronicle of Alabama Enterprise.* Northridge, Calif.: Windsor Publications, 1987.

———. *Poor But Proud: Alabama's Poor Whites.* Tuscaloosa: University of Alabama Press, 1989.

Foner, Eric. *Reconstruction: America's Unfinished Revolution, 1863–1877.* New York: Harper & Row, 1988.

Foner, Philip S. *The Great Labor Uprising of 1877.* New York: Monad, 1977.

———. *History of the Labor Movement in the United States: From Colonial Times to the Founding of the American Federation of Labor.* 1947. Reprint, New York: International Publishers, 1982.

———. *Organized Labor and the Black Worker, 1619–1973.* New York: International Publishers, 1974.

Foner, Philip S., and Ronald L. Lewis, eds. *The Black Worker: A Documentary History from Colonial Times to the Present.* Vol. 4, *The Era of the American Federation of Labor and the Railroad Brotherhoods.* Philadelphia: Temple University Press, 1979.

Fox, Maier B. *United We Stand: The United Mine Workers of America, 1890–1990.* Washington, D.C.: United Mine Workers of America, 1990.

Fredrickson, George. *The Black Image in the White Mind: The Debate on Afro-American Character and Destiny, 1817–1914.* New York: Oxford University Press, 1971.

———. *Black Liberation: A Comparative History of Black Ideologies in the United States and South Africa.* New York: Oxford University Press, 1995.

Gaines, Kevin K. *Uplifting the Race: Black Leadership, Politics, and Culture in the Twentieth Century.* Chapel Hill: University of North Carolina Press, 1996.

Garlock, Jonathan. *Guide to the Local Assemblies of the Knights of Labor.* Westport, Conn.: Greenwood, 1982.

Gaston, Paul M. *The New South Creed: A Study in Southern Mythmaking.* New York: Knopf, 1970.

Genovese, Eugene D. *The Political Economy of Slavery: Studies in the Economy and Society of the Slave South.* New York: Vintage, 1967.

———. *Roll, Jordan, Roll: The World the Slaves Made.* New York: Vintage, 1974.

Going, Allen Johnston. *Bourbon Democracy in Alabama, 1874–1890.* University: University of Alabama Press, 1951.

Goodrich, Carter. *The Miner's Freedom.* Boston: Marshall Jones, 1925.

Goodwyn, Lawrence. *The Populist Moment: A Short History of the Agrarian Revolt in America.* New York: Oxford University Press, 1978.

Gottlieb, Peter. *Making Their Own Way: Southern Blacks' Migration to Pittsburgh, 1916–30.* Urbana: University of Illinois Press, 1987.

Grantham, Dewey W. *Southern Progressivism: The Reconciliation of Progress and Tradition.* Knoxville: University of Tennessee Press, 1983.

Greenberg, Stanley B. *Race and State in Capitalist Development: Comparative Perspectives.* New Haven: Yale University Press, 1980.

Greenwald, Maurine Weiner. *Women, War, and Work: The Impact of World War I on Women Workers in the United States.* Westport, Conn.: Greenwood, 1980.

Grossman, James R. *Land of Hope: Chicago, Black Southerners, and the Great Migration.* Chicago: University of Chicago Press, 1989.

Gutman, Herbert G. *Work, Culture, and Society in Industrializing America.* New York: Vintage, 1977.

Hackney, Sheldon. *Populism to Progressivism in Alabama.* Princeton: Princeton University Press, 1969.

Hahn, Steven. *The Roots of Southern Populism: Yeoman Farmers and the Transformation of the Georgia Upcountry, 1850–1890.* New York: Oxford University Press, 1983.

Hair, William Ivy. *Bourbonism and Agrarian Protest: Louisiana Politics, 1877–1900*. Baton Rouge: Louisiana State University Press, 1969.

Hall, Jacquelyn Dowd. *Revolt Against Chivalry: Jessie Daniel Ames and the Women's Campaign Against Lynching*. New York: Columbia University Press, 1979.

Hall, Jacquelyn Dowd, James Leloudis, Robert Korstad, Mary Murphy, Lu Ann Jones, and Christopher B. Daly. *Like a Family: The Making of a Southern Cotton Mill World*. Chapel Hill: University of North Carolina Press, 1987.

Hamilton, Virginia Van der Veer. *Alabama: A History*. 1977. Reprint, New York: W. W. Norton, 1984.

Harlan, Louis R. *Booker T. Washington: The Making of a Black Leader, 1856–1901*. Oxford: Oxford University Press, 1972.

——. *Booker T. Washington: The Wizard of Tuskegee, 1901–1915*. New York: Oxford University Press, 1983.

——, ed. *The Booker T. Washington Papers*. 14 vols. Urbana: University of Illinois Press, 1972–89.

Harris, Carl V. *Political Power in Birmingham, 1871–1921*. Knoxville: University of Tennessee Press, 1977.

Harris, William H. *The Harder We Run: Black Workers since the Civil War*. New York: Oxford University Press, 1982.

Hart, Hastings H. *Social Problems of Alabama*. Montgomery, Ala.: n.p., 1918.

Henri, Florette. *Black Migration: Movement North, 1900–1920*. Garden City, N.J.: Anchor, 1976.

Higham, John. *Strangers in the Land: Patterns of American Nativism, 1860–1925*. 1955. Reprint, New Brunswick: Rutgers University Press, 1988.

Historical and Statistical Review and Mailing and Shipping Guide of North Alabama (Illustrated). New York and Birmingham: Southern Commercial Publishing Co., 1888.

Honey, Michael K. *Southern Labor and Black Civil Rights: Organizing Memphis Workers*. Urbana: University of Illinois Press, 1993.

Hyman, Michael R. *The Anti-Redeemers: Hill-Country Political Dissenters in the Lower South From Redemption to Populism*. Baton Rouge: Louisiana State University Press, 1990.

Jackson, Carlton. *The Dreadful Month*. Bowling Green, Ohio: Bowling Green State University Popular Press, 1982.

Johnson, James P. *The Politics of Soft Coal: The Bituminous Industry from World War I through the New Deal*. Urbana: University of Illinois Press, 1979.

Jones, Jacqueline. *The Dispossessed: America's Underclasses from the Civil War to the Present*. New York: Basic, 1992.

——. *Labor of Love, Labor of Sorrow: Black Women, Work, and the Family from Slavery to the Present*. New York: Basic, 1985.

Jones, James Pickett. *Yankee Blitzkrieg: Wilson's Raid through Alabama and Georgia*. Athens: University of Georgia Press, 1976.

Jones, Walter B. *History and Work of Geological Surveys and Industrial Development in Alabama*. University: Geological Survey of Alabama, 1935.

Kaufman, Stuart B., Peter J. Albert, and Grace Palladino, eds. *A National Labor Movement Takes Shape, 1895–1898*. Vol. 4 of *The Samuel Gompers Papers*. Urbana: University of Illinois Press, 1991.

Kealey, Gregory S., and Bryan D. Palmer. *Dreaming of What Might Be: The Knights of Labor in Ontario, 1880–1900*. 1982. Reprint, Toronto: New Hogtown Press, 1987.

Kelley, Robin D. G. *Hammer and Hoe: Alabama Communists during the Great Depression*. Chapel Hill: University of North Carolina Press, 1990.

——. *Race Rebels: Culture, Politics, and the Black Working Class*. New York: Free Press, 1994.

Kennedy, David M. *Over Here: The First World War and American Society*. Oxford: Oxford University Press, 1980.

King, Edward. *The Great South: A Record of Journey*. Vol. 1. 1875. Reprint, New York: Burt Franklin, 1969.

Kirwan, Albert D. *Revolt of the Rednecks: Mississippi Politics, 1876–1925*. 1951. Reprint, Gloucester, Mass.: Peter Smith, 1964.

Klein, Maury. *History of the Louisville & Nashville Railroad*. New York: Macmillan, 1972.

Knights of Labor. *Proceedings of the General Assemblies*. Various places and publishers, 1881–1888.

Kousser, J. Morgan. *The Shaping of Southern Politics: Suffrage Restriction and the Establishment of the One-Party South, 1880–1910*. New Haven: Yale University Press, 1974.

Krause, Paul. *The Battle for Homestead, 1880–1892: Politics, Culture, and Steel*. Pittsburgh: University of Pittsburgh Press, 1992.

Lamoreaux, Naomi R. *The Great Merger Movement in American Business, 1895–1904*. New York: Cambridge University Press, 1985.

Laurie, Bruce. *Artisans into Workers: Labor in Nineteenth-Century America*. New York: Noonday, 1989.

Lewis, Ronald. *Black Coal Miners in America: Race, Class, and Community Conflict, 1780–1980*. Lexington: University Press of Kentucky, 1987.

——. *Coal, Iron, and Slaves: Industrial Slavery in Maryland and Virginia, 1715–1865*. Westport, Conn.: Greenwood, 1979.

Lewis, W. David. *Sloss Furnaces and the Rise of the Birmingham District: An Industrial Epic*. Tuscaloosa: University of Alabama Press, 1994.

Lincoln, C. Eric, and Lawrence H. Mamiya. *The Black Church in African American Experience*. Durham, N.C.: Duke University Press, 1990.

Litwack, Leon F. *Been in the Storm So Long: The Aftermath of Slavery*. New York: Vintage, 1979.

Long, Priscilla. *Where the Sun Never Shines: A History of America's Bloody Coal Industry*. New York: Paragon, 1989.

Lyell, Charles. *A Second Visit to the United States of North America*. New York: Harper & Brothers, 1850.

McCalley, Henry. *On the Warrior Coal Field*. Part of *Geological Survey of Alabama*, edited by Eugene Allen Smith. Montgomery: Barrett & Co., 1886.

McKee, Thomas Hudson. *The National Conventions and Platforms of All Political Parties, 1789–1905*. New York: Burt Franklin, 1906.

McKiven, Henry M., Jr., *Iron and Steel: Class, Race, and Community in Birmingham, Alabama, 1875–1920*. Chapel Hill: University of North Carolina Press, 1995.

McLaurin, Melton A. *The Knights of Labor in the South*. Westport, Conn.: Greenwood, 1978.

McMath, Robert C., Jr. *American Populism: A Social History, 1877–1898*. New York: Hill and Wang, 1993.

McMillan, Malcolm C. *Constitutional Development in Alabama, 1798–1901: A Study in Politics, the Negro, and Sectionalism*. Chapel Hill: University of North Carolina Press, 1955.

———. *Yesterday's Birmingham*. Miami: E. A. Seeman Publishing, 1975.

McMillen, Neil. *Dark Journey: Black Mississippians in the Age of Jim Crow*. Urbana: University of Illinois Press, 1989.

Meier, August. *Negro Thought in America, 1880–1915*. Ann Arbor: University of Michigan Press, 1963.

Mills, J. Thornton, III. *Politics and Power in a Slave Society: Alabama, 1800–1860*. Baton Rouge: Louisiana State University Press, 1978.

Milner, John T. *Alabama: As It Was, As It Is, As It Will Be*. Montgomery: Barrett & Brown, 1876.

Montgomery, David. *Beyond Equality: Labor and the Radical Republicans, 1862–1872*. 1967. Reprint, Urbana: University of Illinois Press, 1981.

———. *The Fall of the House of Labor: The Workplace, the State, and American Labor Activism, 1865–1925*. Cambridge: Cambridge University Press, 1987.

———. *Workers' Control in America: Studies in the History of Work, Technology, and Labor Struggles*. Cambridge: Cambridge University Press, 1979.

Moore, Albert Burton. *History of Alabama*. 1934. Reprint, Tuscaloosa: Alabama Book Store, 1951.

Moore, Barrington, Jr. *Social Origins of Dictatorship and Democracy: Lord and Peasant in the Making of the Modern World*. Boston: Beacon, 1966.

Morgan, Edmund S. *American Slavery, American Freedom: The Ordeal of Colonial Virginia*. New York: Norton, 1975.

Moss, Florence Hawkins Wood. *Building Birmingham and Jefferson County*. Birmingham: Birmingham Printing Co., 1947.

Myrdal, Gunnar. *An American Dilemma: The Negro Problem and Modern Democracy*. New York: Harper & Brothers, 1944.

Newby, I. A. *Plain Folk in the New South: Social Change and Cultural Persistence, 1880–1915*. Baton Rouge: Louisiana State University Press, 1989.

Norrell, Robert J. *James Bowron: The Autobiography of a New South Industrialist*. Chapel Hill: University of North Carolina Press, 1991.

———. *A Promising Field: Engineering in Alabama, 1837–1987*. Tuscaloosa: University of Alabama Press, 1990.

Northrup, Herbert R. *Organized Labor and the Negro*. New York: Harper & Brothers, 1944.

O'Conner, Valerie Jean. *The National War Labor Board: Stability, Social Justice, and the Voluntary State in World War I*. Chapel Hill: University of North Carolina Press, 1983.

Oestreicher, Richard Jules. *Solidarity and Fragmentation: Working People and Class Consciousness in Detroit, 1875–1900*. Urbana: University of Illinois Press, 1986.

Ovington, Mary White. *The Walls Came Tumbling Down*. New York: Harcourt, Brace & World, 1947. Reprint, New York: Schocken Books, 1970.

Palmer, Bruce. *"Man Over Money": The Southern Populist Critique of American Capitalism*. Chapel Hill: University of North Carolina Press, 1980.

Perkins, Crawford A. *The Industrial History of Ensley, Alabama: Past, Present, and Future.* N.p.: n.p., ca. 1907.

Perlman, Selig, and Philip Taft. *Labor Movements, 1896–1932.* Vol. 4 of *History of Labor in the United States.* New York: Macmillan, 1935.

Perman, Michael. *The Road to Redemption: Southern Politics, 1869–1879.* Chapel Hill: University of North Carolina Press, 1984.

Phillips, William Battle. *Iron Making in Alabama.* 3d ed. University, Alabama: Brown Printing Co., 1912.

Proceedings of the Joint Convention of the Alabama Coal Operators Association and the United Mine Workers of America, 1903, and Arbitration Proceedings. Birmingham: Roberts & Son, 1903.

Rabinowitz, Howard N. *Race Relations in the Urban South: 1865–1890.* New York: Oxford University Press, 1978.

Rachleff, Peter J. *Black Labor in the South: Richmond, Virginia, 1865–1890.* Philadelphia: Temple University Press, 1984.

Ramirez, Bruno. *When Workers Fight: The Politics of Industrial Relations in the Progressive Era, 1898–1916.* Westport, Conn.: Greenwood, 1978.

Ricker, Ralph R. *The Greenback-Labor Movement in Pennsylvania.* Bellefonte, Pa.: Pennsylvania Heritage, Inc., 1966.

Riley, Rev. Benjamin F. *Alabama As It Is: The Immigrant's and Capitalist's Guide Book to Alabama.* Atlanta: Constitution Publishing Co., 1888.

———. *Makers and Romance of Alabama History.* N.p.: n.p., ca. 1915.

Rogers, William Warren. *The One-Gallused Rebellion: Agrarianism in Alabama, 1865–1896.* Baton Rouge: Louisiana State University Press, 1970.

Rogers, William Warren, with Robert David Ward, Leah Rawls Atkins, and Wayne Flynt. *Alabama: The History of a Deep South State.* Tuscaloosa: University of Alabama Press, 1994.

Roy, Andrew. *A History of the Coal Miners of the United States.* 3d ed. Columbus, Ohio: J. L. Trauger Printing Co., 1907.

Salutos, Theodore. *Farmer Movements in the South, 1865–1933.* Berkeley: University of California Press, 1960.

Salvatore, Nick. *Eugene V. Debs: Citizen and Socialist.* Urbana: University of Illinois Press, 1982.

Scott, Emmett J. *Negro Migration during the War.* 1920. Reprint, New York: Arno and New York Times, 1969.

Sharkey, Robert P. *Money, Class, and Party: An Economic Study of Civil War and Reconstruction.* Baltimore: Johns Hopkins Press, 1959.

Shifflett, Crandall. *Coal Towns: Life, Work, and Culture in Company Towns of Southern Appalachia, 1880–1960.* Knoxville: University of Tennessee Press, 1992.

Sobel, Robert. *Panic on Wall Street: A History of America's Financial Disasters.* New York: Macmillan, 1968.

Soule, George. *The Prosperity Decade: From War to Depression, 1917–1929.* New York: Rinehart & Co., 1947.

Spero, Sterling D., and Abram L. Harris. *The Black Worker: The Negro and the Labor Movement.* New York: Columbia University Press, 1931. Reprint, New York: Atheneum, 1968.

Stampp, Kenneth M. *The Peculiar Institution: Slavery in the Ante-Bellum South*. New York: Vintage, 1956.

Starobin, Robert S. *Industrial Slavery in the Old South*. New York: Oxford University Press, 1970.

Stover, John F. *The Railroads of the South, 1865–1900*. Chapel Hill: University of North Carolina Press, 1955.

Suffern, Arthur E. *The Coal Miners' Struggle for Industrial Status*. New York: Macmillan, 1926.

Taft, Philip. *The A.F.L. in the Time of Gompers*. New York: Harper & Brothers, 1957.

———. *Organizing Dixie: Alabama Workers in the Industrial Era*. Westport, Conn.: Greenwood, 1981.

Tarbell, Ida M. *The Life of Elbert H. Gary: The Story of Steel*. New York: D. Appleton and Co., 1925.

Tennessee Coal, Iron, and Railroad Company. *Annual Reports*. N.p: n.p., 1890, 1893, 1898, 1905.

Thomas, Emory. *The Confederate Nation, 1861–1865*. New York: Harper & Row, 1979.

Trotter, Joe William, Jr. *Coal, Class, and Color: Blacks in Southern West Virginia, 1915–32*. Urbana: University of Illinois Press, 1990.

———, ed. *The Great Migration in Historical Perspective: New Dimensions of Race, Class, and Gender*. Bloomington: Indiana University Press, 1991.

Unger, Irwin. *The Greenback Era: A Social and Political History of American Finance, 1865–1879*. Princeton: Princeton University Press, 1964.

United Mine Workers of America (UMWA). *Minutes of the Annual Conventions of the United Mine Workers of America*. Indianapolis: various publishers, 1898–1906.

———. *Proceedings of the Annual Conventions of the United Mine Workers of America*. Indianapolis: various publishers, 1907–1921.

Voss, Kim. *The Making of American Exceptionalism: The Knights of Labor and Class Formation in the Nineteenth Century*. Ithaca: Cornell University Press, 1993.

Walker, Clarence. *Deromanticizing Black History: Critical Essays and Reappraisals*. Knoxville: University of Tennessee Press, 1991.

Walkowitz, Daniel J. *Worker City, Company Town: Iron and Cotton-Worker Protest in Troy and Cohoes, New York, 1855–84*. Urbana: University of Illinois Press, 1981.

Wallace, Anthony F. C. *St. Clair: A Nineteenth-Century Coal Town's Experience with a Disaster-Prone Industry*. New York: Alfred A. Knopf, 1987.

Ward, Robert David, and William Warren Rogers. *Convicts, Coal, and the Banner Mine Tragedy*. Tuscaloosa: University of Alabama Press, 1987.

———. *Labor Revolt in Alabama: The Great Strike of 1894*. University: University of Alabama Press, 1965.

Warren, Kenneth. *The American Steel Industry, 1850–1970: A Geographical Interpretation*. Pittsburgh: University of Pittsburgh Press, 1973.

Watts, Sarah Lyons. *Order against Chaos: Business Culture and Labor Ideology in America, 1880–1915*. Westport, Conn.: Greenwood, 1991.

Wesley, Charles H. *Negro Labor in the United States, 1850–1925*. New York: Vanguard, 1927.

Wharton, Vernon Lane. *The Negro in Mississippi, 1865-1890*. New York: Harper & Row, 1947.

White, Marjorie Longenecker. *The Birmingham District: An Industrial History and Guide*. Birmingham: Birmingham Historical Society, 1981.

Wiener, Jonathan M. *Social Origins of the New South: Alabama, 1860-1885*. Baton Rouge: Louisiana State University Press, 1978.

Williamson, Joel. *The Crucible of Race: Black-White Relations in the American South since Emancipation*. New York: Oxford University Press, 1984.

———, ed. *The Origins of Segregation*. Boston: Heath, 1968.

Woodward, C. Vann. *Origins of the New South, 1877-1913*. Baton Rouge: Louisiana State University Press, 1951.

———. *The Strange Career of Jim Crow*. London: Oxford University Press, 1955.

———. *Tom Watson: Agrarian Rebel*. New York: Rinehart, 1938.

Woodward, Joseph H., II. *Alabama's Blast Furnaces*. Woodward, Ala.: Woodward Iron Company, 1940.

Wright, Gavin. *Old South, New South: Revolutions in the Southern Economy since the Civil War*. New York, Basic, 1986.

ARTICLES

Abramowitz, Jack. "The South: Arena for Greenback Reformers." *Social Education* 17 (March 1953): 108-10.

Alexander, Major Winthrop. "Ten Years of Riot Duty." *Journal of the Military Service Institution of the United States* 19, no. 82 (July 1896).

Amsden, Jon, and Stephen Brier. "Coal Miners on Strike: The Transformation of Strike Demands on the Formation of a National Union." *Journal of Interdisciplinary History* 7, no. 4 (Spring 1977): 583-616.

Arnesen, Eric. "Following the Color Line of Labor: Black Workers and the Labor Movement Before 1930," *Radical History Review* 55 (Winter 1993): 53-87.

———. " 'It Aint Like They Do in New Orleans': Race Relations, Labor Markets, and Waterfront Labor Movements in the American South, 1880-1923." In *Racism and the Labour Market: Historical Studies*, edited by Marcel Van Der Linden and Jan Lucassen, 57-100. Bern: Peter Land AG, 1995.

———. " 'Like Banquo's Ghost, It Will Not Down': The Race Question and the American Railroad Brotherhoods, 1880-1920." *American Historical Review* 99, no. 5 (December 1994): 1601-34.

———. " 'What's on the Black Worker's Mind?' African-American Workers and the Union Tradition." *Gulf Coast Historical Review* 10, no. 1 (Fall 1994): 5-18.

Banta, Brady. "Henderson Steel & Manufacturing Company." In *Iron and Steel in the Nineteenth Century*, edited by Paul F. Paskoff, 154-55. New York: Facts on File, 1989.

Barjenbruck, Judith. "The Greenback Political Movement: An Arkansas View." *Arkansas Historical Quarterly* 36, no. 2 (Summer 1977): 107-22.

Bernstein, Irving. "Herbert G. Gutman as Labor Historian." *International Journal of Politics, Culture, and Society* 2, no. 3 (Spring 1989): 396-99.

Brainerd, Alfred F. "Colored Mine Labor." *Transactions of the American Institute of Mining Engineers* 14 (1885–86): 78–80.

Brier, Stephen. "In Defense of Gutman: The Union's Case." *International Journal of Politics, Culture, and Society* 2, no. 3 (Spring 1989): 382–95.

———. "Interracial Organizing in the West Virginia Coal Industry: The Participation of Black Mine Workers in the Knights of Labor and the United Mine Workers, 1880–1894." In *Essays in Southern Labor History*, edited by Gary M. Fink and Merl E. Reed, 18–41. Westport, Conn.: Greenwood, 1977.

Cappon, Lester J. "Trend of the Southern Iron Industry under the Plantation System." *Journal of Economic and Business History* 2 (February 1930): 352–81.

Cobb, James C. "Beyond Planters and Industrialists: A New Perspective on the New South." *Journal of Southern History* 54, no. 1 (February 1988): 45–68.

Cox, LaWanda. "From Emancipation to Segregation: National Policy and Southern Blacks." In *Interpreting Southern History: Historiographical Essays in Honor of Sanford W. Higginbotham*, edited by John B. Boles and Evelyn Thomas Nolen, 199–253. Baton Rouge: Louisiana State University Press, 1987.

Cuff, Robert. "Harry Garfield, the Fuel Administration, and the Search for a Cooperative Order during World War I." *American Quarterly* 30, no. 1 (Spring 1978): 39–53.

Daniel, Pete. "The Tennessee Convict War." *Tennessee Historical Quarterly* 34 (Fall 1975): 273–92.

Dean, Lewis S. "Michael Tuomey and the Pursuit of a Geological Survey of Alabama, 1847–1857." *Alabama Review* 44, no. 2 (April 1991): 101–11.

"Discussion: Race, Ethnicity and Organized Labor." *New Politics* 1, no. 3 (Summer 1987): 22–71.

Dubofsky, Melvyn. "Abortive Reform: The Wilson Administration and Organized Labor, 1913–1920." In *Work, Community, and Power: The Experience of Labor in Europe and America, 1900–1925*, edited by James E. Cronin and Carmen Sirianni, 197–220. Philadelphia: Temple University Press, 1983.

Fields, Barbara J. "Ideology and Race in American History." In *Region, Race, and Reconstruction: Essays in Honor of C. Vann Woodward*, edited by J. Morgan Kousser and James M. McPherson, 143–77. New York: Oxford University Press, 1982.

Fies, Milton H. "The Man with a Light on His Cap: A Brief Chronicle of Coal Mining in Walker County, 1912–1960." In *Annals of Northwest Alabama*, vol. 3, compiled by Carl Elliott. Tuscaloosa: privately published, 1965.

Fitch, John A. "Birmingham District: Labor Conservation." *Survey* 27 (January 6, 1912): 1449–1556.

French, John D. "'Reaping the Whirlwind': The Origins of the Allegheny County Greenback Labor Party in 1877." *Western Pennsylvania Historical Magazine* 64, no. 2 (April 1981): 97–119.

Fried, Albert. "The Gutman School; At What Intellectual Price." *International Journal of Politics, Culture, and Society* 2, no. 3 (Spring 1989): 400–403.

Fuller, Justin F. "Boom Towns and Blast Furnaces: Town Promotion in Alabama, 1885–1893." *Alabama Review* 29, no. 1 (January 1976): 37–48.

———. "From Iron to Steel: Alabama's Industrial Evolution." *Alabama Review* 17, no. 2 (April 1964): 137–48.

——. "Henry F. DeBardeleben, Industrialist of the New South." *Alabama Review* 39, no. 1 (January 1986): 3–18.

Gould, Jeffrey. "The Strike of 1887: Louisiana Sugar War." *Southern Exposure* 12, no. 6 (November–December 1984): 45–55.

Gowaskie, Joseph M. "From Conflict to Cooperation: John Mitchell and Bituminous Coal Operators, 1898–1908." *The Historian* 38, no. 4 (August 1976): 669–88.

——. "John Mitchell and the Anthracite Mine Workers: Leadership and Rank-and-File Militancy." *Labor History* 27, no. 1 (Winter 1986): 54–83.

Graebner, William. "Great Expectations: The Search for Order in Bituminous Coal, 1890–1917." *Business History Review* 48, no. 1 (Spring 1974): 49–72.

Green, James R. "The Brotherhood of Timber Workers, 1910–1913: A Radical Response to Industrial Capitalism in the Southern U.S.A." *Past and Present* 60 (August 1973): 161–200.

Gutman, Herbert G. "Black Coal Miners and the Greenback Labor Party in Redeemer Alabama, 1878–79: The Letters of Warren D. Kelley, Willis Johnson Thomas, 'Dawson,' and Others." *Labor History* 10, no. 3 (Summer 1969): 506–35.

——. "The Negro and the United Mine Workers of America: The Career and Letters of Richard L. Davis and Something of Their Meaning: 1890–1900." In *The Negro and the American Labor Movement*, edited by Julius Jacobson, 49–127. Garden City: Anchor, 1968.

Hahn, Kenneth. "The Knights of Labor and the Southern Black Worker." *Labor History* 18, no. 1 (Winter 1977): 49–70.

Halpern, Rick. "Interracial Unionism in the Southwest: Fort Worth's Packinghouse Workers, 1937–1954." In *Organized Labor in the Twentieth-Century South*, edited by Robert H. Zieger, 158–82. Knoxville: University of Tennessee Press, 1991.

——. "Organized Labor, Black Workers, and the Twentieth-Century South: The Emerging Revision." In *Race and Class in the American South since 1890*, edited by Melvyn Stokes and Rick Halpern, 43–76. Oxford: Berg, 1994.

Herr, Kincaid A. "The Louisville & Nashville Railroad, 1850–1940, 1941–1959." Originally published April 1943. Reprint, *L. & N. Magazine*, October 1959, 27–39.

Hill, Herbert. "Myth-Making as Labor History: Herbert Gutman and the United Mine Workers of America." *International Journal of Politics, Culture, and Society* 2 (Winter 1988): 132–200.

——. "Race, Ethnicity and Organized Labor: Opposition to Affirmative Action." *New Politics* 1, no. 2 (Winter 1987): 31–82.

Hine, William B. "Black Organized Labor in Reconstruction Charleston." *Labor History* 25, no. 4 (Fall 1984): 504–17.

Hodes, Martha. "The Sexualization of Reconstruction Politics: White Women and Black Men in the South after the Civil War." *Journal of the History of Sexuality* 3, no. 3 (1993): 402–17.

Jones, John H. "Coal-Mining Industry of Alabama in 1889." *Census Bulletin No. 27*, published by U.S. Bureau of the Census, January 30, 1891, 3–5. Washington, D.C.: Government Printing Office, 1891.

Karson, Marc, and Ronald Radosh, "The American Federation of Labor and the Negro Worker, 1894–1949." In *The Negro and the American Labor Movement*, edited by Julius Jacobson, 155–87. Garden City, N.J.: Anchor, 1968.

Keith, Jean E. "Sand Mountains and Sawgrass Marshes." *Alabama Review* 7, no. 3 (April 1954): 99–101.

Kelley, Robin D. G. " 'We Are Not What We Seem': Re-thinking Black Working-Class Opposition in the Jim Crow South." *Journal of American History* 80, no. 3 (June 1993): 75–112.

Kessler, Sidney. "The Organization of the Negro in the Knights of Labor." *Journal of Negro History* 37, no. 3 (July 1952): 248–76.

Klein, Maury. "The Strategy of Southern Railroads." *American Historical Review* 73, no. 4 (April 1968): 1058–59.

Kleppner, Paul. "The Greenback and Prohibition Parties." In *The Gilded Age of Politics*, vol. 2 of *History of U.S. Political Parties, 1860–1910*, edited by Arthur M. Schlesinger Jr., 1549–1614. New York: Chelsea House, 1973.

Knowles, Morris. "Water and Waste: The Sanitary Problems of a Modern Industrial District." *Survey* 27 (January 6, 1912): 1485–92.

Korstad, Robert, and Nelson Lichtenstein. "Opportunities Found and Lost: Labor, Radicals, and the Early Civil Rights Movement." *Journal of American History* 75, no. 3 (December 1988): 786–811.

Kulik, Gary. "Black Workers and Technological Change in the Birmingham Iron Industry, 1881–1931." In *Southern Workers and Their Unions, 1880–1975*, edited by Merl E. Reed, Leslie S. Hough, and Gary M. Fink, 23–42. Westport, Conn.: Greenwood, 1981.

Laslett, John H. M. "Samuel Gompers and the Rise of American Business Unionism." In *Labor Leaders in America*, edited by Melvyn Dubofsky and Warren Van Tine, 62–88. Urbana: University of Illinois Press, 1987.

Letwin, Daniel L. "Interracial Unionism, Gender, and 'Social Equality' in the Alabama Coalfields, 1878–1908." *Journal of Southern History* 61, no. 3 (August 1995): 519–54.

Lewis, Ronald L. "From Peasant to Proletarian: The Migration of Southern Blacks to Central Appalachian Coalfields." *Journal of Southern History* 55, no. 1 (February 1989): 77–102.

Lichtenstein, Alex. "Racial Conflict and Racial Solidarity in the Alabama Coal Strike of 1894: New Evidence for the Gutman-Hill Debate." *Labor History* 36, no. 1 (Winter 1995): 63–76.

———. " 'Through the Rugged Gates of the Penitentiary': Convict Labor and Southern Coal, 1870–1900." In *Race and Class in the American South since 1890*, edited by Melvyn Stokes and Rick Halpern, 3–42. Oxford: Berg, 1994.

Luckie, J. B. "The Cholera." In *Early Days in Birmingham*, by Pioneers Club (Birmingham), 20–24. Birmingham: n.p., 1968.

Lyell, Charles. "Coal Fields of Tuscaloosa, Alabama." *American Journal of Science and Arts*, 2d ser., 1 (May 1846): 371–76.

McCartin, Joseph A. " 'An American Feeling': Workers, Managers, and the Struggle over Industrial Democracy in the World War I Era." In *Industrial Democracy in America: The Ambiguous Promise*, edited by Nelson Lichtenstein and Howell John Harris, 67–86. Cambridge: Cambridge University Press, 1993.

McKenzie, Robert H. "The Economic Impact of Federal Operations in Alabama during the Civil War." *Alabama Historical Quarterly* 38, no. 1 (Spring 1976): 59–65.

——. "Horace Ware: Alabama Iron Pioneer." *Alabama Review* 26, no. 3 (July 1973): 157–64.

——. "Reconstruction of the Alabama Iron Industry, 1865–1880." *Alabama Review* 25, no. 3 (July 1972): 178–91.

Meier, August, and Elliott Rudwick. "Attitudes of Negro Leaders toward the American Labor Movement from the Civil War to World War I." In *The Negro and the American Labor Movement*, edited by Julius Jacobson, 27–48. Garden City, N.J.: Anchor, 1968.

Miller, Randall M. "Daniel Pratt's Industrial Urbanism: The Cotton Mill Town in Antebellum Alabama." *Alabama Historical Quarterly* 34, no. 1 (Spring 1972): 5–35.

Montgomery, David. "William H. Sylvis and the Search for Working-Class Citizenship." In *Labor Leaders in America*, edited by Melvyn Dubofsky and Warren Van Tine, 13–28. Urbana: University of Illinois Press, 1987.

Moore, A. B. "Railroad Building in Alabama during the Reconstruction Period." *Journal of Southern History* 1, no. 4 (November 1935): 422–41.

Moore, Geoffrey H. "Business Cycles, Panics, and Depressions." In *Encyclopedia of American History*, vol. 1, edited by Glenn Porter, 151–56. New York: Charles Scribner's Sons, 1980.

Nelson, Bruce. "Class and Race in the Crescent City: The ILWU, from San Francisco to New Orleans." In *The CIO's Left-Led Unions*, edited by Steve Rosswurm, 19–45. New Brunswick: Rutgers University Press, 1992.

——. "Organized Labor and the Struggle for Black Equality in Mobile during World War II." *Journal of American History* 80, no. 3 (December 1993): 952–88.

Norrell, Robert J. "Caste in Steel: Jim Crow Careers in Birmingham, Alabama." *Journal of American History* 73, no. 3 (December 1986): 669–94.

Oestreicher, Richard. "Terence Powderly, the Knights of Labor, and Artisanal Republicanism." In *Labor Leaders in America*, edited by Melvyn Dubofsky and Warren Van Tine, 30–61. Urbana: University of Illinois Press, 1987.

Painter, Nell Irvin. "The New Labor History and the Historical Moment." *International Journal of Politics, Culture, and Society* 2, no. 3 (Spring 1989): 367–70.

——. " 'Social Equality,' Miscegenation, Labor, and Power." In *The Evolution of Southern Culture*, edited by Numan V. Bartley, 47–67. Athens: University of Georgia Press, 1988.

Rabinowitz, Howard N. "More than the Woodward Thesis: Assessing *The Strange Career of Jim Crow*." *Journal of American History* 75, no. 4 (December 1988): 842–56.

Reed, Merl. "The Augusta Textile Mills and the Strike of 1886." *Labor History* 14, no. 2 (Spring 1973): 228–46.

Roberts, Francis. "William Manning Lowe and the Greenback Party in Alabama." *Alabama Review*, 5, no. 2 (April 1952): 100–121.

Roediger, David. "History Making and Politics." *International Journal of Politics, Culture, and Society* 2, no. 3 (Spring 1989): 371–72.

——. " 'Labor in White Skin': Race and Working-Class History." In *Reshaping the U.S. Left: Popular Struggles in the 1980s*, edited by Mike Davis and Michael Sprinker, 287–308. London: Verso, 1988.

Scott, Emmett J. "Additional Letters of Negro Migrants of 1916–1918." *Journal of Negro History* 4, no. 4 (October 1919): 412–65.

———. "Letters of Negro Migrants of 1916–1918." *Journal of Negro History* 4, no. 3 (July 1919): 290–340.

Shulman, Steven. "Racism and the Making of the American Working Class." *International Journal of Politics, Culture, and Society* 2, no. 3 (Spring 1989): 361–66.

Stein, Judith. "Southern Workers in National Unions: Birmingham Steelworkers, 1936–1951." In *Organized Labor in the Twentieth-Century South*, edited by Robert H. Zieger, 183–222. Knoxville: University of Tennessee Press, 1991.

Stockham, Richard J. "Alabama Iron for the Confederacy: The Selma Works." *Alabama Review* 21, no. 3 (July 1968): 163–72.

Straw, Richard A. "The Collapse of Biracial Unionism: The Alabama Coal Strike of 1908." *Alabama Historical Quarterly* 37, no. 2 (Summer 1975): 92–114.

———. "Soldiers and Miners in a Strike Zone: Birmingham, 1908." *Alabama Review* 38, no. 4 (October 1985): 289–308.

———. "The United Mine Workers of America and the 1920 Coal Strike in Alabama." *Alabama Review* 28, no. 2 (April 1975): 104–28.

Surface, George T. "The Negro Mine Laborer: Central Appalachian Coal Field." *Annals of the American Academy of Political and Social Science* 33 (January–June 1909): 338–52.

Vandiver, Frank E. "The Shelby Iron Company in the Civil War: A Study of a Confederate Industry" (Part 1), *Alabama Review* 1, no. 1 (January 1948): 12–26; (Part 2), *Alabama Review* 1, no. 2 (April 1948): 111–27; (Part 3), *Alabama Review* 1, no. 3 (July 1948): 203–17.

Wiener, Jonathan. "Class Structure and Economic Development in the South, 1865–1955" *American Historical Review* 84, no. 5 (December 1979): 970–92.

Wilson, Charles Reagan. "We Shall Overcome." In *Encyclopedia of Southern Culture*, edited by Charles Reagan Wilson and William Ferris, 230–31. Chapel Hill: University of North Carolina Press, 1989.

Wilson, Francille Rusan. "Black Workers' Ambivalence toward Unions." *International Journal of Politics, Culture, and Society* 2, no. 3 (Spring 1989): 378–81.

Woodward, C. Vann. "*Strange Career* Critics: Long May They Persevere." *Journal of American History* 75, no. 4 (December 1988): 857–68.

Woodward, Joseph, II. "Alabama Iron Manufacturing, 1860–1865." *Alabama Review* 7, no. 3 (July 1954): 199–207.

Worthman, Paul B. "Black Workers and Labor Unions in Birmingham, Alabama, 1897–1904." *Labor History* 10, no. 3 (Summer 1969): 375–407.

———. "Working Class Mobility in Birmingham, Alabama, 1880–1914." In *Anonymous Americans: Explorations in Nineteenth-Century Social History*, edited by Tamara K. Hareven, 172–213. Englewood Cliffs, N.J.: Prentice-Hall, 1971.

Worthman, Paul B., and James R. Green. "Black Workers in the New South, 1865–1915." In *Key Issues in the Afro-American Experience*, vol. 1, edited by Nathan Huggins, Martin Kilson, and Daniel Fox, 47–69. New York: Harcourt Brace, 1971.

DISSERTATIONS, THESES, AND UNPUBLISHED PAPERS

Abernathy, John H., Jr., "The Knights of Labor in Alabama." Master's thesis in Commerce, University of Alabama, 1960.

Elmore, Nancy Ruth. "The Birmingham Coal Strike of 1908." Master's thesis, University of Alabama, 1966.

Fuller, Justin. "History of the Tennessee Coal, Iron and Railroad Company, 1852–1907." Ph.D. diss., University of North Carolina, 1966.

Gilmore, Glenda Elizabeth. "Gender and Jim Crow: Women and the Politics of White Supremacy in North Carolina, 1896–1920." Ph.D. diss., University of North Carolina, 1993.

Gutman, Herbert G. "Citizen-Miners and the Erosion of Traditional Rights: A Study of the Coming of Italians to the Western Pennsylvania Coal Mines, 1873–1878." Unpublished paper in author's possession, 1967.

Head, Holman. "The Development of the Labor Movement in Alabama Prior to 1900." Master's thesis, University of Alabama, 1955.

Lichtenstein, Alex. "The Political Economy of Convict Labor in the New South." Ph.D. diss., University of Pennsylvania, 1990.

McCartin, Joseph A. "Abortive Reconstruction: Federal War Labor Policies, Union Organization, and the Politics of Race, 1917–1920." Paper presented to Pennsylvania State University Labor History Workshop, University Park, Pa., March 1995.

——. " 'Americanism,' Federal Intervention, and the Limits of Interracial Unionism in Birmingham, Alabama, 1917–1919." Paper presented at the annual meeting of the Organization of American Historians, Anaheim, Calif., 1993.

——. "Labor's 'Great War': American Workers, Unions, and the State, 1916–1920." Ph.D. diss., State University of New York at Binghamton, 1990.

McKiven, Henry M. "Class, Race, and Community: Iron and Steel in Birmingham, Alabama, 1875–1920." Ph.D. diss., Vanderbilt University, 1990.

Rikard, Marlene Hunt. "An Experiment in Welfare Capitalism: The Health Care Services of the Tennessee Coal, Iron, and Railroad Company." Ph.D. diss., University of Alabama, 1983.

Shapiro, Karin A. "The Tennessee Coal Miners' Revolts of 1891–92: Industrialization, Politics, and Convict Labor in the Late Nineteenth-Century South." Ph.D. diss., Yale University, 1991.

Straw, Richard A. " 'This Is Not a Strike, It is Simply a Revolution': Birmingham Miners and the Struggle for Power." Ph.D. diss., University of Missouri-Columbia, 1980.

Index

Bessemer, 17, 69, 99, 110, 111, 165, 233 (n. 67)

Bessemer Coal, Iron, and Land Company, 250 (n. 27)

Bessemer Machine and Foundry, 255 (n. 73)

Bills, Robert, 178

Birmingham: founding of, 3, 9–10; as industrial centerpiece of New South, 3, 12, 16–17; diversity of labor force in, 3–4, 23; sporadic growth and development of, 4, 11–13, 15–20, 58, 197 (n. 7), 199 (n. 20), 200 (n. 25); mineral resources in area of, 10, 13; early coal and iron companies in, 15–16, 17, 18, 19; steelmaking in, 18–20, 200 (n. 25); black sections of, 28, 54; as hub of southern Knighthood, 69–71; District 20 of UMW founded in, 90; and 1894 railroad strike, 109; turn-of-the-century union resurgence in, 126, 128; District 20 of UMW reestablished in, 127; Jim Crow in, 130, 131–32, 153, 239 (n. 24); working-class ferment in during World War I, 164–66; and Great Migration, 165, 250 (n. 26); black middle class in, 166–67; metal trades workers in, 181–83; rebirth of Ku Klux Klan in, 182

Birmingham *Age*, 24, 52, 225 (n. 71)

Birmingham *Age-Herald*, 36–37, 53, 92, 109, 110, 177, 228 (n. 14); and white supremacy, 94, 148; pro-company, anti-union stance of, 102, 143, 144, 148, 175, 245 (n. 74); on labor shortages in World War I era, 165, 250 (n. 28)

Birmingham *Evening Chronicle*, 213 (n. 47)

Birmingham *Free Lance*, 49

Birmingham *Independent*, 218 (n. 22)

Birmingham *Iron Age*, 10, 15, 21, 64, 65, 73, 79, 218 (n. 19)

Birmingham *Labor Advocate*, 32, 36, 113, 134, 141, 176, 240 (n. 39); on convict lease system, 49; supports "Working-man's" Democrats, 97; and interracialism, 100, 103, 104, 107, 118–19, 120, 121, 132, 133, 135, 137–39, 177; criticizes use of black strikebreakers, 102, 103; and 1894 strike, 102, 103, 104, 107, 111–12; criticizes operators, 102, 141; and "social equality" issue, 120–21, 149; as voice of unionism in coal fields, 123, 160; and black disfranchisement, 136; on diversity of UMW members, 144; on outcome of 1908 strike, 151

Birmingham Machine and Foundry, 255 (n. 73)

Birmingham Mineral Railroad, 17

Birmingham *News*, 51, 117, 118, 131, 144, 147; on industrial potential of Birmingham, 19; on racial tension in mining communities, 37–38, 39; and white supremacy, 148

Birmingham *Observer*, 65

Birmingham *Reporter*, 166, 168, 169, 175, 176, 186, 253 (n. 56)

Birmingham Rolling Mill, 16

Birmingham Trades Council (BTC), 97, 119, 128, 131–32, 182

Bittner, Van A., 186, 187

Bituminous Coal Commission, 184

Black Belt, 57, 168: as source of labor for coal fields, 3, 26–27, 51, 92, 185

"Blacklegs." *See* Strikebreakers

Blacks: as union organizers/leaders, 1–2, 6, 58, 61–62, 64, 75–76, 118, 130–31, 134, 135, 152, 186, 239 (nn. 28, 29); varying relations with whites in working class, 2–3, 6, 35–40; number of relative to whites in coal fields, 3–4, 23, 26, 114, 116, 120, 130, 153, 159, 161, 176, 202 (nn. 34, 36), 248 (n. 10); separation from whites in larger society, 4, 5, 35, 38, 40, 81, 82, 125–26, 130, 134, 135; work together with whites in mines, 4, 23, 40, 130; union loyalty and work of extolled by labor press, 6, 61, 92–93, 100–101, 121, 133, 135, 136; demeaning portraits of in labor press, 6, 118, 120–21, 135–36; predominance of in union membership, 6, 118, 130, 178, 186; skill levels and work of in coal fields, 23–24, 26, 27–28, 77, 122–23,

of, 68, 70–71, 75; source of affects miners' attitudes toward operators, 74–75;
labor press criticism of, 100, 115; color
line works to benefit of, 119, 122; black
middle-class sympathy with, 158. *See
also* Operators

Carbon Hill, 69, 90, 98, 207 (nn. 3, 4), 222
(n. 47), 227 (n. 28)

Cardiff, 69, 111, 113, 114, 207 (n. 3), 209
(n. 25), 210 (n. 31), 227 (n. 28)

Carlisle, E. E., 110

Carnegie, Andrew, 18

Carter, James B., 251 (n. 36)

Catholic Church, 32

Cheatham, S. P., 127

Checkweighman: miners demand to
choose, 45, 113, 119, 177, 230 (n. 40), 247
(n. 5); operators refuse miners' choice
of, 112, 113, 151; miners win control of,
128, 179, 180, 184

Churches: in mining communities, 32, 35,
41, 163; as social core of black Birmingham, 167

Cincinnati *Gazette*, 16

Class: and race in workers' lives and consciousness, 3, 4, 5, 30, 54, 56, 81, 100,
126, 137, 145, 171–72, 192–93; and conflict between miners and operators, 4, 5,
30, 54, 58, 145, 159; miners united by
common experience of, 7, 34–35, 40, 54,
56, 60, 79, 122, 137, 145, 192; and convict
lease system, 49, 97; and effects of federal intrusion in South, 59–60, 175, 182,
188; and "social equality," 82

Clemo, J. L., 181

Closed shop, 141, 142, 153–54, 178

Coalburg, 41, 98, 110, 113, 207 (n. 3), 233
(n. 67); convict labor at, 29, 41, 50; company store at, 41, 210 (n. 31); Knights in,
69; strikes at, 84, 108, 142, 222 (n. 47),
227 (n. 28); UMW organizing efforts at,
99, 112

Coal Valley, 112

Coker, C. M., 127

Coketon, 206 (n. 2). *See also* Pratt City

Colonial economy: economy of South as,
12, 30, 57, 200–201 (n. 28)

Comer, Braxton B., 148, 150–51, 152, 153,
187, 245 (n. 74)

Commissary. *See* Company store

Company housing, 4, 111, 123; miners compelled to live in, 32; eviction of striking
miners from, 42, 84, 91, 102, 103, 108,
114, 142, 143, 146, 185; conditions in, 42,
161, 162, 163; miners' complaints about,
42, 161, 230 (n. 40); advantages of for
operators, 44, 51; blacks tend to occupy
more than whites, 114; as concern for
UMW, 149

Company store, 4, 24, 123, 163, 233 (n. 71);
GLP criticism of, 5, 59; Knights' criticism of, 5, 71, 72; miners compelled to
patronize, 32, 43, 112, 151, 161; miners'
complaints about, 43–44, 80, 86, 113,
161; advantages of for operators, 44, 51;
as strike grievance, 73; blacks tend to
patronize more than whites, 114; profits
from, 210 (n. 31)

Congress of Industrial Organizations, 189

Conley, John L., 86, 90

Convict lease system, 4, 89, 123; GLP
opposition to, 5, 58–59, 64; Knights'
opposition to, 5, 68, 71, 72, 78, 221
(n. 45); Redeemer Democrats' support
of, 28–29, 59, 64; arrangements of, 29;
provides cheap labor for operators,
29–30; miners' opposition to, 48–50,
58–59, 60, 73, 77, 86, 99, 161, 184, 224
(n. 59); general public's opposition to,
49–50, 53, 108, 161; UMW opposition
to, 90, 128, 136; Populist opposition to,
97, 109, 229 (n. 31); lessening use of, 161,
177. *See also* Convicts

Convicts: as source of labor in coal fields,
28, 29; predominance of blacks among
in coal fields, 29, 50, 77; conditions and
conduct of in coal fields, 50, 52–54, 213
(n. 47), 214 (n. 53); used to replace striking miners, 102, 109, 111, 142, 152. *See
also* Convict lease system

Fairfield, 163

Fairley, William R., 129, 131, 161, 179, 239 (n. 27); as District 20 president, 127, 132–33; attacks subcontracting system, 144; and suppression of 1908 strike, 147; and "social equality" charges, 149–50, 245 (n. 74)

Farley, Charley, 132, 133

Farmers: as source of labor in coal fields, 26, 27, 204 (n. 48); GLP seeks to align workers/miners with, 56–57, 59, 217 (n. 10); small, and planters, 57, 66

Farmers' Alliance, 72, 85, 96, 97–98, 99, 229 (n. 26), 230 (n. 33)

Farming: movement of workers between mining and, 26, 27, 51, 114, 205 (n. 51)

Fies, Milton H., 158, 172, 173, 174, 205 (n. 51), 250 (n. 28)

Fileno, Salvator, 74

Fink, Leon, 71

Flynn, Ed, 133

Ford, Allen, 36

Galloway, 237 (n. 10)

Galloway Coal Company, 172

Garfield, Harry, 179, 180, 184

"Garfield Agreement," 179–80, 183, 184

Gary, Elbert, 162

Gary, George B., 241 (n. 49)

Gender: and interracial unionism in mines, 7, 154–56, 186; and "social equality" issue, 148–49, 154–55, 247 (n. 90)

Gentry, A. H., 127

Georgia Pacific Railway, 207 (n. 3)

Glover, Walter, 108

Gompers, Samuel, 182, 226 (n. 1), 233 (n. 71)

Goodgame, John W., 167, 181, 251 (n. 38)

Goodrich, Carter, 45

Grange, 72, 96

Graves Mines, 142

Great Migration. See Blacks: and Great Migration

Greenbackism/Greenback-Labor Party

(GLP), 130, 218 (n. 17), 220 (n. 31); inter-racialism of, 1–2, 3, 5, 54, 55, 56, 60–63, 66–67, 76, 94, 99, 152, 154, 217 (n. 14), 218 (n. 19); and Democrats, 1–2, 5, 55, 57, 58, 62, 63, 64, 65, 66; and white supremacy, 2, 55, 56, 57, 62, 81; as fore-runner of UMW, 5–6, 152, 188; as fore-runner of Knights, 55–56, 67, 68, 71, 72; program of, 56, 58, 60; national origins of, 56–57; southern forms of, 57, 58, 60; and Republicans, 57, 58, 64, 65; origins of in Alabama, 57–58; political activities and electoral campaigns of, 63–66, 68, 96, 216 (n. 4), 219 (n. 25); decline of, 66, 67; as forerunner of Populists, 97

Greer, Benjamin L., 131, 134, 145–46, 239 (n. 29), 243 (n. 64)

Gutman, Herbert, 149

Hale, Ulysses S., 182

Hall, Jacquelyn Dowd, 154–55

Hallier, Joe, 134

Handy, W. F., 64

Hannigan, W. S., 100, 105, 108

Harden, John, 35–36

Harper, Jim, 64

Harris, Edward Ronald, 71, 73–74, 223 (n. 53)

Harris, Gus, 117

Harrison, W. L., 184

Helena, 67, 69, 98, 206 (n. 2), 222 (n. 47)

Henry Ellen, 69, 98, 227 (n. 28)

Higdon, E. L., 146, 149–50

Hill, George, 117

Hill, Herbert, 3

Holman, Chat, 108

Hooper, Richard, 133

Horse Creek. See Dora

Hospun, Henry, 220 (n. 31)

Huntsville Gazette, 75

Immigrants: as workers in coal fields, 23, 24–26, 73–75, 114–15, 144, 201–2 (n. 33), 203 (n. 40), 203–4 (n. 44); as strikebreak-ers, 143, 145

Operators: conflicting interests of miners and, 4, 7, 30, 40, 45–46, 54, 72–73, 140–42; and convict labor, 4, 28–30, 50, 213 (n. 47); GLP and, 5; Knights and, 5, 35; efforts to divide miners along racial lines, 6, 7, 56, 89, 115, 118, 121, 123, 131, 140, 150; labor press denunciations of, 6, 93, 94–95, 102, 107, 164; necessity of low production costs for, 18, 24, 29–30, 41, 126; efforts to keep wages low, 18, 24, 44–45, 115; and recognition of union, 19, 127, 134, 140–41, 153–54, 185; ideal labor force for, 24; and company store, 24, 43–44; antiunionism of, 24, 112–13, 115, 129, 134, 140, 141–42, 151, 159, 160, 164, 184; views on native white workers, 24, 114, 203 (n. 40); views on immigrant workers, 24–26, 114–15, 203 (n. 40), 203–4 (n. 44); views on black workers, 25, 27, 28, 105, 114–15, 139–40, 144, 159, 161–62, 174, 203 (n. 40), 203–4 (n. 44), 214 (n. 51); and miners' mobility, 25, 27, 205 (n. 51), 206 (n. 53), 214 (n. 51); and social and economic control of miners, 32, 40–44, 115, 205 (n. 51), 209 (n. 23); paternalism of, 41–42, 50, 67, 113, 159, 162–64; and company housing, 42; and recruitment of excess labor, 44, 51, 112, 123, 234 (n. 84); and "dead work," 45, 46, 112; and working conditions, 48; and racial diversity of workforce, 54, 114–15; view contest with labor as matter of who will run the mines, 74, 141–42; decline to recognize UMW, 91, 93, 115, 141–42, 153–54, 178, 179, 185; and use of black strikebreakers, 91–95, 102–7, 116, 123, 142, 143–44, 146, 150; and 1894 strike, 101–11, 116; divergent labor practices among, 113–15, 163, 249 (n. 17); and 1904–6 strike, 142–45; and 1908 strike, 145–51; and resurrection of ACOA, 147; accuse union of encouraging "social equality," 148, 149, 153, 186, 187; and effects of World War I on labor situation, 159, 165, 172–74; black middle-class

leaders as allies of, 158, 166–68, 175–76, 177, 181; and federal agencies regulating wartime labor relations, 178, 188; and "Garfield Agreement," 179–80, 183; and Bituminous Coal Commission, 184–85; and 1920–21 strike, 185–88. *See also* Alabama Coal Operators' Association; *and specific companies*
Orange, James H., 129
O'Reilly, Tom, 80, 224 (n. 65)
Orr, Jack, 132
Ovington, Mary White, 240 (n. 36)
Owens, Mrs. Lewis, 106
Oxmoor, 64, 65

Painter, Nell Irvin, 155
Palmer, A. Mitchell, 184
Palos, 234 (n. 82)
Parry, David M., 141
Paternalism. *See* Operators: paternalism of
Patton Junction, 207 (n. 3)
Patton Mines, 69, 84, 207 (n. 3)
Penny, Anna E., 215 (n. 57)
People's (Populist) Party. *See* Populist (People's) Party
Percy, Walker, 139–40
Pettiford, W. R., 167
Pinkertons: use of in labor conflicts, 97, 103, 104, 133, 231 (n. 48)
Pioneer Mining and Manufacturing Company, 17, 209 (n. 23)
Piper, 237 (n. 13)
Planters: and industrial development, 4, 13, 14, 20, 197 (n. 7), 198 (n. 9), 201 (n. 28); complain that mines drain labor supply, 26, 28, 92; and control of African Americans, 28–29; and Redeemer Democrats, 28–29, 57; support convict lease system, 50; and small farmers, 57, 66; Reconstruction and, 59; Populist challenges to, 96
"Pluck-me" stores. *See* Company stores
Populist (People's) Party, 90, 96, 97, 99, 115, 136, 152, 153. *See also* Jeffersonian Democrats

Pow, Adam, 139–40
Powderly, Terence V., 68, 71, 80, 85
Powderly, 221 (n. 38)
Pratt, Daniel, 15, 16, 198 (n. 11)
Pratt City (Pratt Mines), 32, 91, 98, 99, 110, 113, 206 (n. 2), 207 (n. 6), 233 (n. 67); convict labor at, 29, 50, 52–53; as "good" mining town, 35; company housing at, 42; GLP in, 65, 66; Knights in, 67, 69; strikes at, 79, 84, 85, 103, 104, 105, 106, 108, 109, 222 (n. 47), 227 (n. 28); UMW organizing efforts in, 91, 112; UMW local in, 128, 160, 240 (n. 37); racial composition of workforce at, 202 (n. 36); population of, 207 (n. 4)
Pratt Coal and Coke Company, 16
Pratt Coal and Iron Company, 16, 17, 25, 29, 49, 84, 140
Pratt Consolidated Coal Company, 178
"Pratt Massacre," 109
Prattville, 198 (n. 11)
Price, Thomas, 207 (n. 3)
Prohibition Reform Party, 217 (n. 16)

Rameau, P. Colfax, 167, 168, 169–71, 176, 181, 186, 187
Ramsey, Erskine, 178
Randall, J. H., 65
Reconstruction, 175, 182, 188, 245 (nn. 74, 75); Redeemer Democrats deliver South from, 2, 55, 60; and convict lease system, 28; end of in Alabama, 57; nature and legacy of, 59–60, 82, 148, 220 (n. 31), 247 (n. 90)
Redeemers: deliver South from Reconstruction, 2, 55, 60; and white supremacy, 2, 60, 62, 94; GLP and Knights challenges to, 5, 67, 76–77; come to power in 1874 elections, 15; and convict lease system, 28–29. See also Democrats
Red Men, 32, 129
Republic, 147, 151, 169
Republicans, 98, 216 (n. 4), 228 (n. 14); and industrial development, 14; Redeemers contrast themselves to, 28; and GLP, 57,

58, 64, 65; weakness of in Alabama after Reconstruction, 58, 63–64, 115; and defeat of Reconstruction, 60; and black voters, 66, 76, 80; and Knights of Labor, 76, 80; and Populists, 115; and Birmingham's black middle class, 167
Republic Iron and Steel Company, 19, 129, 141, 180
Richson, Granson, 76
Riley, Benjamin F., 10
Russell, W. E., 94
Russellville, 182

Sayreton, 142
Scott, E. S., 132
Seay, Thomas, 53
Seddon, Thomas, 114
Segregation/Jim Crow: labor interracialism and, 2, 3, 7, 32, 123, 126–27, 130, 137, 143–44, 147–48, 150, 153, 154–56, 188, 191–92; UMW challenge to, 6, 130, 131–32, 144–45, 153, 154–56, 188; and shaping of miners' world, 32, 40, 130; AFL and, 90, 226 (n. 1); in Birmingham, 130, 239 (n. 24); gender and, 154–56, 247 (n. 90). See also White supremacy
Selheimer, H. C., 180
"Shack rousters," 166, 169, 171, 185, 252 (n. 42)
Sharit, H. J., 64, 65, 66, 69, 219 (n. 25)
Shepherd, John, 37–38, 208–9 (n. 18)
Shook, Alfred M., 91, 92, 228 (n. 21)
Sloss, James W., 15–16, 27, 28
Sloss Furnace Company, 16
Sloss Iron and Steel Company, 18, 99, 202 (n. 36); formation of, 17; northern capital behind, 17; growth of, 19; use of convict labor by, 29, 161; 1888 strike against, 84–85; use of black strikebreakers by, 103; and 1894 strike, 103, 110, 111; recruitment of excess labor by, 112, 123, 234 (n. 84); 1895 strike against, 113, 114, 119, 120; and 1904–6 strike, 141, 142; antiunion stance of, 141; uninterested in welfare programs, 163, 249 (n. 17); com-

pany stores of, 210 (n. 31); 1890 strike against, 227 (n. 8); 1893 strike against, 230 (n. 35)

Smith, John G., 112, 113

"Social equality": Knights and, 81–83, 225 (n. 71); charges of as strategy to impede unionism and worker solidarity, 82, 123, 143, 148–49, 154–55, 188; black acceptance of strictures against, 82, 137; significance of threat of for southerners, 82–83, 154–55, 247 (n. 90); UMW denies promoting, 120, 123, 127, 130, 137, 149–50, 159, 178

Socialist Party, 128

Sorsby, J. F., 186

South & North Railroad, 9, 12, 13, 14, 15, 16

Stack, Nicholas, 35, 84, 85, 221 (n. 45), 225 (n. 73)

Statham, R. A., 160, 170–71, 177, 184

Stephens, Uriah S., 69

Stockton, 112

Strikebreakers, 84, 145, 149; abundance of helps defeat strikes, 6, 72, 111, 143, 187; miners' resentment of and actions against, 77–78, 79, 103, 106–7, 108, 142–43, 146, 185; operators' divide-and-rule strategy in using blacks as, 91–95, 102–7, 116, 118, 123, 142, 143–44, 146, 150, 185

Subcontracting system, 51, 139; miners' opposition to, 46, 113, 122, 129, 144, 177, 179, 184, 230 (n. 40); miners' acceptance and use of, 46, 122; nature of, 46, 211 (n. 37); discontinuation of, 46–47, 129; blacks rely on more than whites, 122–23; revival of, 161

Sumter, 98, 99, 104–5, 116, 118, 122, 207 (n. 3). See also Blue Creek

Swing, Raymond G., 182, 183

Swinney, James, 239 (n. 29)

Taft, William Howard, 175

Taylor, Robert Wesley, 157–59, 166, 167, 181

Tennessee Coal, Iron, and Railroad Company (TCI), 99, 127, 165, 200 (n. 23), 228 (n. 21), 230 (n. 35); entry into Birmingham district, 17; growth of, 17, 18, 19, 29, 102; during 1893–97 depression, 18–19; taken over by U.S. Steel, 19, 145, 162; and steel production in Birmingham, 19, 162; use of convict labor by, 29, 49, 161, 215 (n. 57); company housing of, 42, 162, 163; 1888 strike against, 84–85; declines to recognize UMW, 91; use of black strikebreakers by, 92, 103, 142; and 1894 strike, 102–3, 105, 108, 109, 110, 111, 121; antiunion stance of, 112, 113, 141–42, 143, 181; adopts labor philosophy more accommodating than most operators', 113, 115, 116–17, 234 (nn. 80, 84); paternalism of, 113, 162, 162–63, 164, 249 (n. 17); 1902 strike against, 129; 1903 strike against, 140–41; and 1904–6 strike, 141–42, 143; 1918 strike against, 180, 255 (n. 73); 1890 strike against, 227 (n. 8)

Thomas, Edward, 209 (n. 23)

Thomas, Willis J., 1–2, 58, 61–62, 63, 69, 217 (n. 10), 217–18 (n. 17), 220 (n. 31)

Thomas, 17, 209 (n. 23)

Thompson, A. B., 239 (n. 32)

Thompson, E. P., 193

Thompson, Richard, 41

Thrasher, Bart, 36

Tuomey, Michael, 13, 20

Tutwiler, Julia S., 52

Union Labor League, 128, 141, 241 (n. 43)

Union Labor Party, 71, 76

United Kingdom: miners from, 23, 24, 25–26, 201–2 (n. 33). See also Immigrants

United Mine Workers (UMW), 234 (n. 80); interracialism of, 2, 3, 5, 54, 90, 93, 94, 99–101, 103–7, 118–23, 127, 130–40, 143–45, 147–49, 152–56, 175–76, 177–78, 183, 186, 188, 223 (n. 54), 239 (nn. 28, 29, 32), 240 (n. 37), 241 (n. 48); and white supremacy, 2, 6, 93, 94, 101, 107, 118, 119, 120–21, 126, 134, 135, 136,

Whites: respect of for black unionists, 2, 62–63, 79, 81, 100–101, 106, 121–22, 133, 140, 152; varying relations with blacks in working class, 2–3, 6, 35–40; number of relative to blacks in coal fields, 3–4, 23, 114, 116, 120, 161, 202 (nn. 34, 36), 248 (n. 10); separation from blacks in larger society, 4, 5, 35, 38, 40, 81, 82, 125–26, 130, 134, 135; work together with blacks in mines, 4, 23, 40, 130; disproportionate number of in union leadership, 6, 135, 176, 191–92, 239 (nn. 28, 29); ethnic background of among miners, 23, 114; skill levels and work of in coal fields, 23–24, 25–26, 77, 122–23, 202 (n. 37), 202–3 (n. 38); operators' views of as workers, 24–25, 54, 114, 203 (n. 40), 203–4 (n. 44); movement of between farming and mining, 26, 51; and violence against blacks, 36–38, 126; as convict laborers, 50; and Democratic Party, 63, 66–67; supremacist attitudes of within union, 63, 79–81, 92, 94–95, 101, 103–4, 107, 120–21, 123, 134, 135–36, 139–40, 149–50, 152, 176, 192, 253 (n. 50); align with blacks against "outsiders," 73–75, 77–78, 79, 104–5, 106, 118–19, 121–22, 144, 147, 152, 187, 224 (n. 59); as strikebreakers, 77–78, 93, 94–95, 105, 106, 107, 122, 143, 181–82; and black strikebreakers, 92, 93–94, 103, 106, 107, 108, 109, 228 (n. 14); displacement of in coal fields, 92, 102, 103, 107, 110, 116, 122; and company housing, 114; and company store, 114; and subcontracting system, 122–23, 139, 161; as metal trades workers, 181. *See also* Blacks; Greenbackism/Greenback-Labor Party: interracialism of; Knights of Labor: interracialism of; Segregation/ Jim Crow; "Social equality"; United Mine Workers: interracialism of; White supremacy

White supremacy: interracial unionism and, 2, 3, 7, 126–27, 134, 147–48, 188, 192; UMW and, 2, 6, 93, 94, 101, 107, 118, 119, 120–21, 126, 134, 135, 136, 139, 152; GLP and, 2, 55, 56, 57, 62; Knights and, 2, 56, 67, 80–81; labor press and, 94, 107, 120–21, 135–36, 137–39; and sexuality, 154–55. *See also* Segregation/Jim Crow

Williams, A. C., 169, 173, 181, 251 (n. 36)

Wilson, William B., 174–75, 178–79

Wilson, Woodrow, 176, 179, 184

Women. *See* Gender

Wood, J. B., 170, 171, 182

Woods (deputy sheriff), 117

Woodward, C. Vann, 2, 17, 57, 193

Woodward, 110, 111, 233 (n. 67)

Woodward Iron Company, 16, 38, 118, 202 (n. 36), 255 (n. 73)

Work, Monroe N., 253 (n. 50)

Working conditions: as concern for GLP, 5, 58; as concern for Knights, 5, 67, 71, 72, 76, 213 (n. 44); dangerous nature of in mines, 48, 51; miners' complaints about, 48, 60, 99, 160, 160; as strike grievance, 73; as concern for Populists, 97; as concern for UMW, 128, 136, 149, 226 (n. 2); deterioration of after 1908 strike defeat, 151; effect of World War I on, 159

Workingmen's Chronicle, 167, 168, 169–70, 175, 186

World War I: and increased demand for labor, 159, 172, 172, 174; effect of federal agencies on labor relations during, 160, 174–75, 178–80, 183, 188; produces unprecedented working-class ferment, 164–65; miners' patriotic reaction to, 181; strike wave following, 183–84; shortage of work in aftermath of, 184

World War II, 189

Worthman, Paul B., 130

"Yellow-dog" contracts, 143, 160